The U.S. Army/Marine Corps
Counterinsurgency Field Manual

The contents of this publication are solely those of the authors and con-
tributors, and not of the publisher, editor(s), or employees of Echo Point
Publishing & Media. Echo Point Publishing & Media and employees
disclaim all responsibility for any injury or adverse effects of any kind
to persons or property resulting from any ideas, methods, instructions,
or products referred to in this publication. We present this book solely
as an historical document and do not condone or endorse violence of
any kind. Quite to the contrary, Echo Point Publishing & Media urges
anyone considering acts of violence as a solution to seek professional
assistance immediately.

Published by Echo Point Books & Media
Brattleboro, Vermont
www.EchoPointBooks.com

FM 3-24 Counterinsurgency
ISBN: 978-1-62654-423-9 (paperback)
       978-1-62654-424-6 (casebound)
       978-1-62654-425-3 (spiralbound)

Cover design by Rachel Boothby Gualco
Editorial and proofreading assistance by Ian Straus,
Echo Point Books & Media

Printed and bound in the United States of America

FM 3-24

COUNTER-INSURGANCY FIELD MANUAL

ECHO POINT BOOKS & MEDIA, LLC

Foreword

This manual is designed to fill a doctrinal gap. It has been 20 years since the U.S. Army published a manual devoted to counterinsurgency operations, and 25 since the Marine Corps published its last such manual. With our Soldiers and Marines fighting insurgents in both Afghanistan and Iraq, it is thus essential that we give them a manual that provides principles and guidelines for counterinsurgency operations (COIN). Such guidance must be grounded in historical studies. However, it also must be informed by contemporary experiences.

This manual takes a general approach to COIN. The Army and the Marine Corps recognize that every insurgency is contextual and presents its own set of challenges. You cannot fight former Saddamists and Islamic extremists the same way you would have fought the Viet Cong, the Moros, or the Tupamaros; the application of principles and fundamentals to deal with each vary considerably. Nonetheless, all insurgencies, even today's highly adaptable strains, remain wars amongst the people, employ variations of standard themes, and adhere to elements of a recognizable revolutionary campaign plan. This manual therefore addresses the common characteristics of insurgencies. It strives to provide those carrying out a counterinsurgency campaign a solid foundation on which to build in seeking to understand and address specific insurgencies.

A counterinsurgency campaign is, as described in this manual, a mix of offensive, defensive, and stability operations, conducted along multiple lines of operation. It requires Soldiers and Marines to employ a mix of both familiar combat tasks and skills more often associated with nonmilitary agencies, with the balance between them varying depending on the local situation. This is not easy. Leaders at all levels must adjust their approach constantly, ensuring that their elements are ready each day to be greeted with a handshake or a hand grenade, to take on missions only infrequently practiced until recent years at our combat training centers, to be nation builders as well as warriors, to help re-establish institutions and local security forces, to assist in the rebuilding of infrastructure and basic services, and to facilitate the establishment of local governance and the rule of law. The list of such tasks is a long one and involves extensive coordination and cooperation with a myriad of intergovernmental, indigenous, and international agencies. Indeed, the responsibilities of leaders in a counterinsurgency campaign are daunting – and the discussions in this manual endeavor to alert them to the challenges of such campaigns and to suggest general approaches for grappling with those challenges.

Conducting a successful counterinsurgency campaign thus requires a flexible, adaptive force led by agile, well-informed, culturally astute leaders. It is our hope that this manual provides the necessary guidelines to succeed in such a campaign, in operations that, inevitably, are exceedingly difficult and complex. Our Soldiers and Marines deserve nothing less.

DAVID H. PETRAEUS
Lieutenant General, USA
Commander
U.S. Army Combined Arms Center

JAMES N. MATTIS
Lieutenant General, USMC
Commanding General
Marine Corps Combat Development Command

Field Manual
No. 3-24

Fleet Marine Force Manual
No. 3-24

Headquarters
Department of the Army
Washington, DC

Headquarters
Marine Corps Combat Development Command
Department of the Navy
Headquarters
United States Marine Corps
Washington, DC

16 June 2006 (Final Draft)

COUNTERINSURGENCY
(Final Draft—Not for Implementation)

Contents

		Page
	PREFACE	vi
Chapter 1	INSURGENCY AND COUNTERINSURGENCY	1-1
	Overview	1-1
	Aspects of Insurgency	1-1
	Aspects of Counterinsurgency	1-15
	Summary	1-24
Chapter 2	UNITY OF EFFORT: INTEGRATING CIVILIAN AND MILITARY ACTIVITIES	2-1
	Integration	2-1
	Key Counterinsurgency Participants and their Likely Roles	2-4
	Key Responsibilities in Counterinsurgency	2-8
	Civilian and Military Integration Mechanisms	2-9
	Tactical-level Interagency Considerations	2-13
	Summary	2-13
Chapter 3	INTELLIGENCE IN COUNTERINSURGENCY	3-1
	Section I – Intelligence Characteristics in Counterinsurgency	3-1

Section II – Predeployment Planning and Intelligence Preparation of the Battlefield 3-2
Define the Operational Environment...3-2
Describe the Effects of the Operational Environment 3-3
Evaluate the Threat...3-11
Determine Threat Courses of Action...3-12

Section III – Intelligence, Surveillance, and Reconnaissance Operations 3-14
The Intelligence–Operations Dynamic..3-15
Human Intelligence and Operational Reporting..3-16
Surveillance and Reconnaissance Considerations...3-17
Considerations for Other Intelligence Disciplines 3-18

Section IV – Counterintelligence and Counterreconnaissance.............3-19

Section V – Analysis 3-20
Current Operations..3-20
Network Analysis...3-21
Intelligence Reach...3-21
Continuity 3-21

Section VI – Intelligence Collaboration and Fusion 3-22
Intelligence Cell and Working Groups...3-22
Protecting Sources..3-23
Host-nation Integration..3-23

Section VII – Summary..3-23

Chapter 4   DESIGNING COUNTERINSURGENCY OPERATIONS 4-1
The Importance of Campaign Design 4-1
The Relationship Between Design and Planning...4-2
The Nature of Design 4-3
Elements of Design 4-3
Campaign Design for Counterinsurgency...4-4
Summary 4-9

Chapter 5   EXECUTING COUNTERINSURGENCY OPERATIONS 5-1
The Nature of Counterinsurgency Operations 5-1
Logical Lines of Operations in Counterinsurgency 5-2
Common Logical Lines of Operations in Counterinsurgency 5-7
Counterinsurgency Approaches...5-16
Assessment of Counterinsurgency Operations 5-24
Targeting 5-27
Learning and Adaptation 5-29
Summary 5-29

Chapter 6   DEVELOPING HOST-NATION SECURITY FORCES 6-1
Overview 6-1
Challenges, Resources, and End State 6-2
Framework for Development..6-7
Police in Counterinsurgency 6-18

Chapter 7	**LEADERSHIP AND ETHICS FOR COUNTERINSURGENCY**	**7-1**
	Leadership in Counterinsurgency	7-1
	Ethics	7-5
	Warfighting Versus Policing	7-5
	Detention and Interrogation	7-7
	The Learning Imperative	7-9
	Summary	7-9
Chapter 8	**BUILDING AND SUSTAINING CAPABILITY AND CAPACITY**	**8-1**
	Logistic Considerations in Counterinsurgency Operations	8-1
	Logistic Support to Logical Lines of Operations	8-4
	Theater Support Contractors	8-16
	Summary	8-19
Appendix A	**A GUIDE FOR ACTION: PLAN, PREPARE, EXECUTE, AND ASSESS**	**A-1**
	Plan	A-1
	Prepare	A-2
	Execute	A-4
	Ending the Tour	A-8
	Three "What Ifs"	A-9
	Conclusion	A-9
Appendix B	**INTELLIGENCE PREPARATION OF THE BATTLEFIELD**	**B-1**
	Section I – Intelligence Preparation of the Battlefield Steps	**B-1**
	Section II – Define the Operational Environment	**B-1**
	Section III – Describe the Effects of the Operational Environment	**B-1**
	Civil Considerations (ASCOPE)	B-2
	Terrain Analysis	B-10
	Weather Analysis	B-10
	Military Aspects of Terrain (OAKOC) and Civil Considerations	B-11
	Section IV – Evaluate the Threat	**B-11**
	Insurgency-related Threats	B-11
	Opportunities	B-12
	Objective and Motivation Identification	B-12
	Popular Support or Tolerance	B-12
	Support Activities—Capabilities, and Vulnerabilities	B-14
	Information and Media Activities—Capabilities, and Vulnerabilities	B-15
	Political Activities	B-15
	Violent Activities	B-16
	Insurgent Organizational Structure and Key Personalities	B-17
	Section V – Determine Threat Courses of Action	**B-18**
	Insurgent Strategies	B-18
	Tactical Courses of Action	B-19
Appendix C	**LINGUISTIC SUPPORT**	**C-1**
	Selecting Interpreters	C-1
	Establishing Rapport	C-3
	Orienting Interpreters	C-4
	Preparing for Presentations	C-4

Conducting Presentations C-4

Speaking Techniques... C-5

Summary C-6

Appendix D   **LEGAL CONSIDERATIONS** **D-1**

Authority to Assist a Foreign Government D-1

Authorization to Use Military Force D-2

Rules of Engagement.. D-2

The Law of War D-3

Internal Armed Conflict.. D-3

Detention and Interrogation.. D-4

Enforcing Discipline of U.S. Forces D-5

Humanitarian Relief and Reconstruction D-6

Training and Equipping Foreign Forces... D-7

Claims and Solatia D-7

Establishing the Rule of Law D-8

Appendix E   **SOCIAL NETWORK ANALYSIS**..**E-1**

Networks and Insurgents E-1

Performing Social Network Analysis E-2

Social Network Data Collection.. E-4

Social Network Graphs and Insurgent Organization...................................... E-5

Social Network Measures E-7

The Network Perspective E-10

Appendix F   **AIRPOWER IN COUNTERINSURGENCY**...**F-1**

Overview F-1

The Advantages of Airpower... F-1

Airpower in the Strike Role... F-2

Airpower in Intelligence Collection F-2

The Role of High-Tech Assets F-3

The Role of Low-Tech Assets... F-3

The Airpower Command Structure F-3

Building the Host-nation's Airpower Capability F-4

Appendix G   **LEARNING COUNTERINSURGENCY** **G-1**

GLOSSARY **Glossary-1**

ANNOTATED BIBLIOGRAPHY...**Annotated Bibliography-1**

MILITARY REFERENCES...**References-1**

Figures

Figure 1-1. Possible counterinsurgency phases 1-3

Figure 1-2. Support for an insurgency 1-16

Figure 1-3. Successful and unsuccessful counterinsurgency operational practices 1-24

Figure 2-1. Sample country team 2-10

Figure 4-1. Design planning continuum 4-2

Figure 4-2. Iterative campaign design 4-5

Figure 4-3: 1st Marine Division's operational design for Operation Iraq Freedom II 4-8

Figure 5-1. Common counterinsurgency operational-level logical lines of operations 5-4

Figure 5-2. The effect of proper application of LLOs in counterinsurgency 5-5

Figure 5-3. The strengthening effect of interrelated LLOs 5-6

Figure 5-4. Unit application of the essential services logical lines of operations 5-13

Figure 6-1. Matrix of security force development 6-9

Figure 8-1. Conventional and counterinsurgency operations contrasted 8-2

Figure 8-2. Comparison of essential services availability to insurgency effectiveness 8-14

Figure 8-3. Tactical financial management organizations 8-19

Figure E-1. Simple network E-3

Figure E-2. Large complex network E-3

Figure E-3. Dyad examples E-4

Figure E-4. Activities matrix E-5

Figure E-5. Network organization with high connections E-6

Figure E-6. Fragmented network E-6

Figure E-7. Assessments E-7

Figure E-8. Hypothetical regional insurgency E-7

Figure E-9. Comparison of subgroup densities E-8

Figure E-10. Density shift E-9

Preface

If the individual members of the organizations were of the same mind, if every organiza-
tion worked according to a standard pattern, the problem would be solved. Is this not
precisely what a coherent, well-understood, and accepted doctrine would tend to
achieve? More than anything else, a doctrine appears to be the practical answer to the
problem of how to channel efforts in a single direction.

David Galula

This field manual/Marine Corps reference publication (FM/MCRP) establishes doctrine (fundamental princi-ples) for military operations in a counterinsurgency environment. It is based on lessons learned from previous counterinsurgencies and relevant combat operations. It is also based on existing interim doctrine, and doctrine recently developed. Many people who articulate and apply U.S. policy and those who conduct operations in counterinsurgency environments are neither conversant nor familiar with the principles of insurgencies and counterinsurgencies. This is not surprising, considering that those subjects have been generally neglected in broader American military doctrine and national security policies since the end of the Vietnam War over 40 years ago. This manual is designed to reverse that trend. It is also designed to merge traditional approaches to counterinsurgency with the realities of a new international arena shaped by technological advances and global-ization.

To make this text useful to leaders involved in counterinsurgency operations regardless of where these opera-tions may occur, the doctrine contained herein is broad in scope and involves principles applicable to various areas of operation. This FM/MCRP is not focused on any region or country. Insurgencies have many common characteristics and patterns, while their ideological basis may vary widely. Fundamental to all counterinsurgen-cies is the need to assist local authorities to secure the populace and thereby separate the people from the insur-gents while enhancing the legitimacy of the government. Insurgents thrive on terrorizing and intimidating the population to gain control over them. Thus, creating chaos to undercut and reduce governmental legitimacy and authority, and fostering overall instability. But American military forces must also appreciate the importance of other agencies and other missions in achieving a lasting victory. They must be prepared to render required as-sistance as well as perform traditionally nonmilitary tasks.

The primary audience for this manual is leaders and planners at the battalion level and above. This manual ap-plies to the United States Marine Corps, the Active Army, the Army National Guard, and the United States Army Reserve unless otherwise stated.

Terms that have joint or Army definitions are identified in both the glossary and the text. FM 3-24 is not the proponent field manual (the authority) for any Army term. For definitions in the text, the term is italicized and the number of the proponent manual follows the definition.

Headquarters, U.S. Army Training and Doctrine Command (TRADOC) is the proponent for this publication. The preparing agency is the Combined Arms Doctrine Directorate, U.S. Army Combined Arms Center. Send written comments and recommendations on Department of the Army (DA) Form 2028 (Recommended Changes to Publications and Blank Forms) directly to: Commander, U.S. Army Combined Arms Center and Fort Leavenworth, ATTN: ATZL-CD (FM 3-24), 201 Reynolds Avenue (Building 285), Fort Leavenworth, Kansas 66027-1352.

Chapter 1

Insurgency and Counterinsurgency

Counterinsurgency is not just thinking man's warfare—it is the graduate level of war.
Special Forces Officer in Iraq, 2005

This chapter provides background information on insurgency and counterinsurgency (COIN). The first half of this chapter describes insurgency, while the second part examines the much more complex challenge of countering it. The chapter concludes with a set of principles and imperatives necessary for success in COIN.

OVERVIEW

1-1.   Insurgency and counterinsurgency are subsets of war. Though globalization and technological advancement have influenced contemporary conflict, the nature of war in the 21st century is the same as it has been since ancient times, "…a violent clash of interests between or among organized groups characterized by the use of military force." Success in war still depends on a group's ability to mobilize support for its political interests and generate sufficient violence to achieve political consequences. Means to achieve these goals are not limited to regular armies employed by a nation-state. At its core, war is a violent struggle between hostile, independent, and irreconcilable wills attempting to impose their desires on another. It is a complex interaction between human beings and is played out in a continuous process of action, reaction, and adaptation. As an extension of both policy and politics with the addition of military force, war can take different forms across the spectrum of conflict. It may range from large-scale forces engaged in conventional warfare to subtler forms of conflict that barely reach the threshold of violence. It is within this spectrum that insurgency and counterinsurgency exist.

1-2.   Insurgency and its tactics are as old as warfare itself. Joint doctrine defines an *insurgency* as an organized movement aimed at the overthrow of a constituted government through the use of subversion and armed conflict (JP 1-02). *Counterinsurgency* is those political, economic, military, paramilitary, psychological, and civic actions taken by a government to defeat an insurgency (JP 1-02).

1-3.   Those definitions are a good starting point, but they do not properly highlight the key paradox that insurgency and counterinsurgency are distinctly different types of operations. However, they are related, though opposing, and are two sides of a phenomenon that has sometimes been called revolutionary war or internal war.

1-4.   Political power is the central issue in an insurgency, and each side has this as its aim. The insurgent attempts to overthrow or subvert an established government or authority; the counterinsurgent uses all of the instruments of national power to support the government in restoring and enforcing the rule of law. Counterinsurgency thus involves the controlled application of national power in political, information, economic, social, military, and diplomatic fields and disciplines. Its scale and complexity should never be underestimated by leaders and planners; indeed, the possible scale and complexity must be understood before the beginning of any such operation.

ASPECTS OF INSURGENCY

1-5.   Governments can be overthrown in a number of ways. A revolution is an unplanned, spontaneous explosion of popular will, such as the French Revolution of 1789. At another extreme is a coup d'etat, where a small group of plotters replace state leaders with little support from the people at large. An insurgency is an organized, protracted politico-military struggle designed to weaken government control and le-

42   gitimacy while increasing insurgent control. Insurgencies normally seek to either overthrow the existing
43   social order and reallocate power within the country, or to break away from state control and form an
44   autonomous area. Insurgency is always a form of internal war, while coups and revolutions can become
45   such a war if they do not come to an immediate resolution. As the name "internal wars" implies, these are
46   primarily conflicts within states, not conflicts between states, and they all contain at least some element of
47   civil war.

48   1-6.   The one possible exception to this rule involves what can be termed a "liberation insurgency," where
49   indigenous elements seek to expel or overthrow what they perceive to be a foreign or occupation govern-
50   ment. Such a resistance movement could be mounted by a legitimate government-in-exile as well as by ele-
51   ments competing for that role.

52   1-7.   Even in internal wars, the involvement of outside actors is to be expected. During the Cold War, the
53   Soviets and the United States played roles in many such conflicts. Today there is a growing global aspect
54   in the form of transnational extremist organizations linked to and exploiting the internal conditions that
55   plague failed and failing states and result in conditions ripe for insurgency. In all cases, however, the long-
56   term objective for all sides remains acceptance of the legitimacy of one side's claim to political power by
57   the people of the state or region'.

58   1-8.   Calling the terrorist or guerrilla tactics common to insurgency "unconventional" or "irregular" can
59   be very misleading, since they have been among the most common approaches to warfare throughout his-
60   tory. Any combatant prefers a quick, cheap, overwhelming victory to a long, bloody, protracted struggle.
61   But to achieve success in the face of superior resources and technology, weaker actors have had to adapt.
62   The recent dominant performance of American military forces in major combat operations may lead many
63   future opponents to pursue asymmetric approaches. Because America retains significant advantages in fires
64   and surveillance, a thinking enemy is unlikely to choose to fight U.S. forces in open battle. Opponents who
65   have attempted to do so, such as in Panama in 1989 or Iraq in 1991 and 2003, have been destroyed in con-
66   flicts that have been measured in hours or days. Conversely, opponents who have offset America's fire and
67   surveillance advantages by operating close to civilians and news media, such as Somali clans in 1993 and
68   Iraqi insurgents in 2005, have been more successful in achieving their aims. This does not mean that coun-
69   terinsurgents do not face open warfare. Insurgents resort to conventional military operations if conditions
70   seem right, in addition to using milder means such as nonviolent political mobilization of people, legal po-
71   litical action, and strikes.

72   1-9.   The contest of internal war is not "fair," as most of the rules favor the insurgent. That is why insur-
73   gency has been a common approach used by the weak to combat the strong. At the beginning of a conflict,
74   insurgents have the strategic initiative. Though they may be prodded into violence by regime changes or
75   government actions, the insurgents generally initiate the war. They may strive to disguise their intentions,
76   and the potential counterinsurgent will be at a great disadvantage until political and military leaders recog-
77   nize that an insurgency exists and are able to determine its makeup and characteristics to facilitate a coor-
78   dinated reaction. While the government prepares to respond, the insurgent is gaining strength and creating
79   increasing disruptions throughout the state. The existing government normally has an initial advantage in
80   resources, but that edge is counterbalanced by the requirement to maintain order. The insurgent succeeds
81   by sowing chaos and disorder anywhere; the government fails unless it maintains order everywhere.

82   1-10. The vast and necessary resources expended for security by the counterinsurgent always dwarf those
83   of their opponents, and successful COIN often requires a very high force ratio. That is a major reason why
84   protracted wars are hard for a counterinsurgent to sustain. The scale and complexity of COIN should never
85   be underestimated. The effort requires a firm political will and extreme patience on the part of the govern-
86   ment, its people, and countries providing support. The widespread security requirements also limit flexibil-
87   ity, again ceding initiative to the insurgents.

88   1-11. Insurgents have an additional advantage in shaping the information environment. While the counter-
89   insurgent seeking to preserve legitimacy must stick to the truth and make sure that words are backed up by
90   appropriate deeds, the insurgent can make exorbitant promises and point out governmental shortcomings,
91   many caused by the insurgency. Ironically, as the insurgent achieves more success and begins to control
92   larger portions of the populace, many of these asymmetries diminish. That may produce new vulnerabilities
93   to be exploited by an adaptive counterinsurgent.

94   1-12. COIN is not an approach to war that can be classified simply as foreign internal defense. It features
95 full spectrum operations, including stability operations, like any other campaign. (See figure 1-1.) The
96 course of an insurgency involves significant variations in the proportion of effort devoted to the different
97 types of operations by region and time. In all cases, however, insurgencies will not be defeated by simply
98 killing insurgents. Since the insurgent begins with the strategic initiative, the counterinsurgent is usually
99 involved initially in more defensive than offensive operations. To regain the initiative, the counterinsurgent
100 strategy focuses on stability operations, addressing the root causes of societal discontent through reforms
101 or reconstruction projects, performing other measures to positively influence the support of the people, and
102 conducting combat operations against insurgent forces. As the counterinsurgent gains success, offensive
103 and defensive operations become more in balance and eventually diminish in importance compared to sta-
104 bility operations. Victory cannot be gained until the people accept the legitimacy of the government mount-
105 ing COIN and stop actively and passively supporting the insurgents.

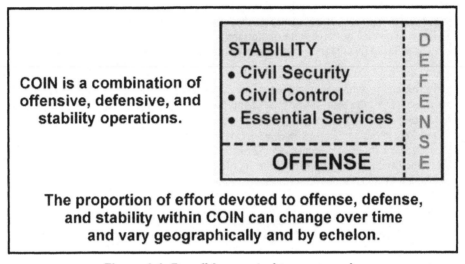

106   **Figure 1-1. Possible counterinsurgency phases**

107 ## THE EVOLUTION OF INSURGENCY

108   1-13. Revolutionary situations may result from regime changes, external interventions, and grievances
109 carefully nurtured and manipulated by unscrupulous leaders. Sometimes societies are most prone to unrest
110 not when conditions are the worst, but when the situation begins to improve and people's expectations rise.
111 Influences of globalization and the international media may create a sense of relative deprivation, contrib-
112 uting to increased discontent as well.

113   1-14. Insurgency has taken many forms over time, from struggles for independence against colonial pow-
114 ers, to the rising of disadvantaged ethnic or religious groups against their rivals, to resistance to foreign in-
115 vaders. The student of insurgency and practitioner of counterinsurgency must begin by understanding the
116 specific circumstances of their particular situation; while studying the history of this mode of warfare, they
117 must also understand how varied it can be.

118   1-15. Insurgencies and counterinsurgencies have been common forms of warfare, especially since the be-
119 ginning of the 20th century. The United States began the era by defeating the Philippine Insurrection. The
120 turmoil of World War I and its aftermath produced numerous internal wars. Trotsky and Lenin seized
121 power in Russia and then defended the new regime against counterrevolutionaries. T.E. Lawrence and
122 Arab raiders bedeviled the Turks during the Arab Revolt. The British developed and practiced concepts of
123 imperial policing worldwide. Before the World War I, insurgencies tended to be conservative and were
124 usually based on defense of hearth, home, traditional monarchies, and traditional religion. For example, the
125 19th century Spanish uprising against Napoleon sapped French strength and attracted the attention of writ-
126 ers like Clausewitz. Since World War I, insurgencies have generally had more revolutionary purposes. The
127 Bolshevik takeover of Russia demonstrated a conspiratorial approach to overthrowing a government and

128 spawned a Communist movement that supported further "wars of national liberation." Lawrence's experi-
129 ences in the Arab Revolt made him a hero and provide rich insights for today.

130 1-16. World War II and its upheavals really launched the modern era of insurgencies and internal wars.
131 The many resistance movements against German and Japanese occupation generated a momentum that
132 continued after the Axis defeat. As nationalism rose, imperial powers declined. Along with the suprana-
133 tional philosophy of communism, nationalism became an important motivation for people to form govern-
134 ments responsive to their needs. They were assisted by the development of portable and effective killing
135 technologies that dramatically increased the firepower available to insurgent groups. Just as important has
136 been the increase in the media's ability to get close to conflict and transmit imagery to the local population
137 and around the world. T.E. Lawrence noted "The printing press is the greatest weapon in the armory of the
138 modern commander." He might have added "and the modern insurgent." The changes in the ends of insur-
139 gency to communist or nationalist domination of a government and in the means available to achieve such
140 ends transformed insurgency. Where Clausewitz envisioned that wars by the armed populace could only
141 serve as a strategic defensive, insurgent theorists of the post–World War II era realized that insurgency
142 could now become a decisive form of warfare in its own right. This era spawned the Maoist, *foco*, and ur-
143 ban approaches to insurgency, which are described below.

144 1-17. The end of the Cold War has brought a new wave of insurgencies. Many are fueled by more tradi-
145 tional religious or ethnic motivations. They are often based on clan or tribal affiliations, and they differ
146 considerably from the post–World War II approaches. With the collapse of empires in the 20th century and
147 the resulting wave of decolonization, weak and failed states have proliferated, now no longer propped up
148 by Cold War rivalries. These power vacuums breed insurgencies. Similar conditions exist when regimes
149 are changed by force or circumstance. American forces supporting a counterinsurgency often find them-
150 selves allied with a struggling nation in its bid to reestablish a functioning government. And the chaotic
151 environment of failed states includes many groups of "spoilers" that counterinsurgents must sort out to de-
152 velop appropriate responses.

153 1-18. Interconnectedness and information technology are new aspects of this contemporary wave of insur-
154 gencies. Insurgents can now link with allied groups, including criminal organizations, throughout the state,
155 region, and world, joining in loose organizations with common objectives but different motivations. Break-
156 ing up these networks can be difficult.

157 1-19. The contemporary environment also features a new kind of globalized insurgency, represented by Al
158 Qaeda, which seeks to transform the Islamic world and reorder its relationship with the rest of the globe.
159 Such groups feed on local grievances, integrate them into broader ideologies, and link disparate conflicts
160 through globalized communications, finances, and technology. While the scale of the effort is new, the
161 grievances and methods that sustain it are not. As in other insurgencies, terrorism, subversion, propaganda,
162 and open warfare are its tools. But defeating such an enemy requires a similarly globalized response to deal
163 with the array of linked resources and conflicts that sustain it.

164 ## INSURGENTS AND THEIR APPROACHES

165 1-20. Each insurgency is unique, although there may be similarities among them. In all cases, the insur-
166 gents' aim is to force political change; any military action is secondary and subordinate, a means to an end.
167 Few insurgencies fit neatly into any rigid classification. Examining the specific type of insurgency one
168 faces enables commanders and staffs to build a more accurate picture of the insurgents and the thinking
169 behind their overall campaign plan. Such an examination identifies the following:

170 - Root cause or causes of the insurgency.
171 - Extent to which it enjoys support, both internally and externally.
172 - Bases on which insurgents appeal to the target population.
173 - Insurgents' motivation and depth of commitment.
174 - Likely insurgent weapons and tactics.
175 - Operational environment in which insurgents seek to initiate and develop their campaign.

176 Counterinsurgent commanders may face a confusing and shifting coalition of many kinds of opponents,
177 who may very well be at odds with one another.

1-21. There are several types of insurgents. These first four seek to completely change the existing political system:

- **Anarchists** seek disorder as their desired end state and consider any political authority illegitimate.
- **Egalitarians** seek to impose a centrally controlled political system to ensure equitable distribution of resources and a radically transformed social structure.
- **Traditionalists** desire a return to some golden age or religious-based value system. Often their goals are regional or international, and their rigid ideological structure leaves little room for compromise or negotiation. Some fringe religious groups, such as Aum Shinrikyo in Japan, envision creating a new world order by precipitating an apocalypse through terrorism.
- **Pluralists** profess traditional Western values, such as freedom and liberty, and aim to establish liberal democracies.

1-22. The remaining types of insurgents do not seek total political power in a state:

- **Secessionists** seek to withdraw altogether to pursue their own independent destiny or to join a different state.
- **Reformists** aim to use violence to make changes within a state to create a more equitable distribution of political and economic power.
- **Preservationists** use their own illegal violence against anyone trying to make changes or institute reforms.
- **Commercialists** pursue economic gain, a behavior seen in some clans and warlords in less-developed societies.

INSURGENT STRATEGIES

1-23. Counterinsurgents not only have to be able to determine what motivates their opponents, which will influence the development of programs designed to attack the root causes of the insurgency, but also what sort of strategy is being used to advance the insurgent cause. Analysis of the insurgent approach shapes counterinsurgent military options. Insurgent strategies include the following:

- Conspiratorial.
- Military-focus.
- Urban warfare.
- Protracted popular war.
- Traditional.

Conspiratorial

1-24. A conspiratorial strategy involves a few leaders and a militant cadre or activist party seizing control of government structures or exploiting a revolutionary situation, as Lenin did in 1917 with the Bolshevik Revolution. Such insurgents remain secretive as long as possible, emerging only when success can be achieved quickly. This approach usually involves the creation of a small, secretive, "vanguard" party or force. Like the Bolsheviks, successful insurgents using this approach may have to create security forces and mass support later to maintain power after seizing it.

Military-focus

1-25. Other strategies aim to create their own revolutionary possibilities or force the issue primarily through the application of military force. For example, the militarily dominant *foco* strategy, popularized by figures like Che Guevera, posited that an insurrection itself could create the conditions to overthrow a government. A small group of guerrillas taking up arms in a rural environment where grievances exist could eventually gather enough support to achieve their aims. Some secessionist insurgencies have relied on major conventional forces to try to secure their independence. Militarily focused insurgencies conducted by Islamic extremist groups or insurgents in Africa or Latin America have little or no political structure and spread their control through movement of combat forces rather than political subversion.

225 **Urban**

226 1-26. Organizations like the Irish Republican Army and Latin American groups have pursued an urban
227 warfare strategy. This approach uses terrorist tactics in cities to sow disorder and create government re-
228 pression, though such activities have not generated much success without wider rural support. However, as
229 societies become more urbanized and insurgent networks more sophisticated, this approach may become
230 more successful. When facing adequately run internal security forces, urban insurgencies typically have a
231 conspiratorial cellular structure recruited along lines of close association—familial, religious affiliation,
232 political party, or social group.

233 **Protracted Popular War**

234 1-27. Protracted conflicts favor insurgents, and no strategy makes better use of that asymmetry than the
235 People's War conducted so well by the Chinese Communists and adapted by the North Vietnamese and
236 Algerians. While this complex approach is relatively uncommon, it is the most difficult to counter. Mao
237 Zedong's model divides structurally into parallel political and military hierarchies, with the political hierar-
238 chy dominant. The overall strategic direction of the conflict comes from a centralized command element.
239 The bulk of the effort is political subversion through united fronts and mass movements. Subversion often
240 precedes the introduction of military forces into a region, and elements of the political cadre often remain
241 behind after government forces have driven out the military elements. The guerrilla force expects to even-
242 tually wage a conventional war and seeks to form large conventional (what Mao calls "main force") units.

243 1-28. Mao's "Theory of Protracted War" outlines a three-phased politico-military approach:

244 - **Strategic Defensive.** The enemy has the much stronger correlation of forces, and the insurgent
245   must concentrate on survival and building support. Bases are established, local leaders recruited,
246   and cellular networks and parallel governments established. The primary military activity is se-
247   lective terrorist strikes to gain popular support and influence recalcitrant individuals.
248 - **Strategic Stalemate.** Guerrilla warfare becomes the most important activity, as force correla-
249   tions approach equilibrium. In the political arena, the insurgent concentrates on separating the
250   people from the government and expanding areas of control.
251 - **Strategic Offensive.** The insurgent now has superior strength. Military forces move to more
252   conventional operations to destroy government military capability, while political actions are
253   designed to completely displace all government authorities.

254 1-29. Maoist strategy does not require a sequential or complete application of all three stages. The aim is
255 the seizure of political power within the state, and if the government's will and capability collapses early in
256 the process, so much the better. Later insurgent practitioners of this strategy have added new twists, to in-
257 clude rejecting the notion of an eventual switch to large-scale conventional operations. The Algerian insur-
258 gents could not manage much military success at all, but they were able to garner decisive popular support
259 through superior organizational skills and astute propaganda that leveraged French mistakes. The North
260 Vietnamese developed a detailed variant known as *dau tranh*, "the struggle," which featured a variant of
261 the logical lines of operations that will be explained in chapter 5. Besides modifying the three phases of
262 military activity developed by Mao, they delineated specific political lines of operations to be conducted
263 among the enemy population, enemy soldiers, and friendly forces. Though the North Vietnamese approach
264 envisioned a culminating "general offensive–general uprising," which never occurred as they expected, it
265 was also designed to achieve victory by whatever means were effective. It did not attack one specific en-
266 emy center of gravity but instead put pressure on several, believing that over time victory would result
267 from the activities of one or many lines of operation.

268 1-30. These protracted People's War approaches are not only conducted nonsequentially along multiple
269 politico-military lines of operations, but also are locally configured. One province could be in a guerrilla
270 war while another was experiencing terrorist attacks. There may be differences in political activities be-
271 tween villages in the same province. The result is more than just a "three-block war"; it is a shifting "mo-
272 saic war" that is very difficult for a counterinsurgent to envision as a coherent whole, let alone centralize.
273 This pattern can exist for any single insurgency, Maoist or not, and the COIN environment can become in-
274 credibly complex.

Traditional

1-31. Many contemporary insurgencies fit a more traditional approach based on clan, tribe, or ethnic group. Such an insurgency typically has entire communities join the effort as a whole, bringing with them their existing social/military hierarchy. There is not necessarily a dual Maoist military/political hierarchy, and insurgent mobilization strategies are often aimed at the leadership of other clans and tribes.

MOBILIZATION MEANS AND CAUSES

1-32. The primary struggle in an internal war is to mobilize people in a battle for legitimacy. The insurgent gathers the resources to sustain that struggle while discouraging support for the government. Two aspects of this effort are mobilization means and causes.

Mobilization Means

1-33. There are five means to mobilize popular support:

- Persuasion.
- Coercion.
- Reaction to abuses.
- Foreign support.
- Apolitical motivations.

A mixture of them may motivate any one individual.

Persuasion

1-34. Political, social, security, and economic benefits entice people to support one side or the other. Ideology/religion is one means of persuasion, especially for the elites/leadership. In this case, legitimacy derives from the consent of the governed, though leaders and led can have very different motivations. In Iraq, the primary issue motivating fighters in some Baghdad neighborhoods was provision of adequate sewer, water, electricity, and trash services. Their concerns were totally disconnected from the overall Baathist goal of expelling the Americans.

Coercion

1-35. The struggle in Iraq has produced many examples of how insurgent coercion can block government success. In the eyes of some, a government that cannot protect its people forfeits the right to rule. Legitimacy in this instance is determined by the old adage "Might makes right," as citizens seek to ally with groups who can guarantee safety. In some areas of Iraq and Afghanistan, for instance, militias established themselves as extragovernmental arbiters of the physical security of the population. Insurgents may use coercive force to provide security for people or else to intimidate them into active or passive support. Kidnapping or killing local leaders or their families is a common insurgent tactic to discourage anyone from working with the government. Militias sometimes use the promise of security, or the threat to take it away, to maintain control of cities and towns.

Reaction to Abuses

1-36. A government that abuses its people or is tyrannical generates resistance to its rule. People who have been injured, maltreated, victimized, dishonored, or had close friends or relatives killed by the government, particularly by its security forces, may lash back at their attackers. Security force abuses and the social upheaval caused by collateral damage from combat can be major escalating factors for insurgencies.

Foreign Support

1-37. Foreign governments or nongovernmental organizations can provide expertise, international legitimacy, and money to buy or exacerbate a conflict. In Chechnya in 1999, there was no popular support for the renewal of fighting, but the war resumed because foreign supporters and warlords had enough money to hire a guerrilla army.

319 *Apolitical Motivations*

320 1-38. Criminals, mercenaries, people attracted by the holy warrior's romantic status, and others whose
321 self-image is being a fighter for a cause might also join an insurgency. Political solutions might not satisfy
322 them enough to end their participation. Fighters who have joined for a paycheck will probably become
323 bandits once the formal conflict ends unless there are jobs for them. This category includes opportunists
324 who exploit the absence of security to engage in economically lucrative criminal activity, such as kidnap-
325 ping and theft.

326 ## Causes

327 1-39. A *cause* is a principle or movement militantly defended or supported. Insurgent leaders often seek to
328 adopt attractive and persuasive causes to mobilize support. These causes often stem from what Mao de-
329 scribed as the unresolved contradictions that exist within any society or culture. Often, contradictions are
330 based on real problems. However, insurgents may create artificial contradictions using propaganda and
331 misinformation. Insurgents have much to gain by not limiting themselves to a single cause. By selecting an
332 assortment of causes and tailoring them for the various groups within the society, insurgents increase their
333 base of sympathetic and complicit support.

334 1-40. Insurgents employ deep-seated strategic causes as well as temporary local causes, adding or deleting
335 them as circumstances demand. Leaders often use a bait-and-switch approach to draw in supporters, ap-
336 pealing initially to local grievances to lure followers into a broader movement. Without an attractive cause,
337 an insurgency might not be able to sustain itself. But a carefully chosen cause is a formidable and intangi-
338 ble asset that can provide a fledgling movement a long-term, concrete base of support. The ideal cause at-
339 tracts the most people while alienating the fewest. It is one the counterinsurgent cannot undermine the in-
340 surgency by advocating. Insurgents must be able to fully identify themselves with the cause and those
341 attracted to it.

342 1-41. Potential insurgents can capitalize on a wealth of potential causes. Any country ruled by a small
343 group without broad popular participation provides a political cause for insurgents. Exploited or repressed
344 social groups—be they entire classes, ethnic groups, or small elites—may support larger causes in reaction
345 to their own, narrower grievances. Economic inequities can nurture revolutionary unrest, as can racial or
346 ethnic persecution. As previously noted, an efficient insurgent propaganda campaign can also turn an arti-
347 ficial problem into a real one.

348 1-42. A skillful counterinsurgent can deal a significant blow to an insurgency by appropriating its cause,
349 such as when the Philippine government adopted land reform to help defuse the Huk Rebellion in 1951. In-
350 surgents often exploit multiple causes, making the counterinsurgents' problems more acute. In the end, any
351 successful counterinsurgency expecting lasting results against a serious insurgency must address in some
352 way the underlying conditions that produced the revolutionary situation. These may be very different in
353 each local area, requiring a complex set of solutions.

354 1-43. As was noted earlier, insurgents now use communications technology, including the Internet, to link
355 with allied groups within and outside the country, joining in loose organizations with a common objective
356 but very different motivations. For many groups involved with the current radical Islamic insurgency, de-
357 crying the very globalization that allows them to coordinate their activities and achieve synergy is one of
358 their main causes. Thus, even though the old adage "all politics is local" remains applicable to modern in-
359 surgencies, they may have global dimensions.

360 ## Mobilizing Resources

361 1-44. Insurgents resort to such tactics as guerrilla warfare and terrorism not only because of their disadvan-
362 tages in manpower or organization, but also because they usually begin with limited or inadequate re-
363 sources compared to the counterinsurgent. To strengthen and sustain their effort, insurgents require money,
364 supplies, and weapons.

365 1-45. Weapons are especially important. In some parts of the world, lack of an access to weapons may
366 forestall insurgency as an option for pursuing political change. Unfortunately, there is widespread avail-
367 ability of weapons, with especially large surpluses in the most violent areas of the world. Skillful counter-

368
369
370
insurgents cut off the flow of arms into the theater and eliminate their source. Insurgents can obtain weapons through legal or illegal purchases or from foreign sources. A common tactic is to capture them from government forces.

371
372
373
374
375
1-46. Income is essential not just to purchase weapons, but also to pay recruits and buy the compliance of corrupt government officials. Money and supplies can be obtained through many sources. Foreign support has already been mentioned. Local supporters or international front organizations may provide donations. Sometimes legitimate businesses can be established to furnish funding. In areas controlled by insurgent forces, confiscation or taxation might be utilized. Another common source of funding is criminal activity.

376
INSURGENCY AND CRIME

377
378
379
380
381
382
383
384
385
1-47. Sustainment requirements often drive insurgents into relationships with organized crime or into criminal activity themselves. Funding has a great influence on the character of an insurgency and its vulnerabilities. Insurgent doctrine is critical in determining materiel requirements. Maoist approaches emphasizing mobilization of the masses require a greater level of resources, both human and material, than a *focoist* emphasis primarily upon armed action. The first requires the resources necessary to construct and maintain a true counterstate. (A counterstate is a competing structure set up by an insurgent to replace the government in power. It includes the administrative and bureaucratic trappings of political power, and performs the normal functions of government). The second requires only what is necessary to sustain a military campaign. A conspiratorial approach requires even less.

386
387
388
389
390
391
392
393
394
395
396
397
1-48. Reaping windfall profits and avoiding the costs and difficulties involved in securing external support makes illegal activity attractive to insurgents. While taxation of a mass base is inherently low-return, kidnapping, extortion, and drugs—to cite three prominent examples of activities favored by insurgents—are very lucrative. The activities of the *Fuerzas Armadas Revolucionarias de Colombia* (FARC) in Colombia serve to illustrate this point: profits from single kidnappings often totaled millions of U.S. dollars. In the case of the Maoist Communist Party of Nepal, taxing the mass base directly proved inferior to other criminal forms of "revolutionary taxation," such as extortion and kidnapping. Drugs retain the highest potential for garnering large profits from relatively small investments. Questioned in the 1990s concerning their willingness to sell gold at half the market price, insurgents in Suriname, South America, responded that the quick profits they made provided seed money for investment in the drug trade, from which they "could make real money." Failed and failing states with rich natural resources like oil are particularly lucrative areas for criminal activity.

398
399
400
401
402
403
404
405
406
407
1-49. It stands to reason that any insurgent movement that increasingly devotes exceptional amounts of time and effort to fund-raising must of necessity shortchange ideological or even armed action. Just where this leads in the case of any particular movement is at the heart of debate in characterizing movements as diverse at the Provisional Irish Republican Army in Ulster or the FARC in Colombia. The first has been involved in all manner of criminal activity for many years, yet certainly remains committed to its ideological aims; the latter, through its involvement in the drug trade, became the richest self-sustaining insurgent group in history while continuing to claim pursuance of "Bolivarian" and "socialist" or "Marxist-Leninist" ends. FARC activities, though, have increasingly been labeled "narco-terrorist" or simply criminal by a variety of critics. Some criminals and organized crime organizations may also begin to develop ideologies and counterstate attributes without actually becoming insurgencies.

408
409
410
411
412
413
414
415
1-50. Throughout history there has been no shortage of insurgencies that have degenerated to criminality. This often occurred as the primary movements disintegrated and remaining elements were cast adrift. Doctrinal thinking has long held that such disintegration is desirable because it takes a truly dangerous ideologically inspired body of disaffiliated individuals and replaces it with a less dangerous but more diverse body, normally of very uneven character. The first is a security threat, the second a law-and-order concern. This should not be interpreted, of course, as denigrating the armed capacity of a law-and-order threat. A successful counterinsurgent is prepared to deal with this disintegration. The best solution is eliminating both the insurgency and any criminal threats its elimination produces.

416 **ELEMENTS OF INSURGENCY**

417   1-51. Though the previous passages highlight the many varieties of insurgency, they share some common
418   attributes. An insurgent organization normally consists of five elements:
419   • Leaders.
420   • Combatants (main forces, regional forces, local forces).
421   • Political cadre (also called militants or the party).
422   • Auxiliaries (active followers who provide important support services).
423   • Mass base (the bulk of the membership).

424   1-52. The proportions of these elements relative to the larger movement depend on the strategic approach
425   adopted by the insurgency. A conspiratorial approach does not pay much attention to combatants or a mass
426   base. The *focoists* downplay the importance of a political cadre and emphasize military action to generate
427   popular support. The People's War approach is the most complex: If the state presence has been elimi-
428   nated, the elements can exist openly. If the state remains a continuous or occasional presence, the elements
429   must maintain a clandestine existence.

430 ## Leaders

431   1-53. The leaders control the insurgent movement. They are the "idea people" and the planners. They usu-
432   ally exercise leadership through force of personality, the power of revolutionary ideas, or personal cha-
433   risma. In some insurgencies, they may hold their position through clan or tribal authority.

434 ## Combatants

435   1-54. The combatants do the actual fighting and provide security. They are often mistaken for the move-
436   ment itself. They exist only to support the insurgency's broader political agenda. The combatants maintain
437   local control. They also protect and expand the shadow government and counterstate, if the insurgency
438   seeks to set up such institutions.

439 ## Political Cadre

440   1-55. The cadre is the political core of the insurgency. They are actively engaged in the struggle to accom-
441   plish insurgent goals. They may also be designated as a formal party to signify their political importance.
442   They implement guidance and procedures provided by the leadership. Modern noncommunist insurgents
443   rarely, if ever, use the term "cadre," but usually include a group that performs similar functions.

444   1-56. The cadre assesses the grievances in local areas and carries out activities that satisfy them. They then
445   attribute the solutions they have provided to the insurgent movement itself. As the insurgency matures,
446   deeds become more important to make insurgent slogans meaningful to the population. Larger societal is-
447   sues, such as foreign presence, facilitate such political activism, because these larger issues may be blamed
448   for life's smaller problems. Destroying the state bureaucracy or preventing national reconstruction after a
449   conflict to sow disorder and sever legitimate links with the people is a common insurgent tactic. In time,
450   the cadre may seek to replace that bureaucracy and assume functions in a shadow government.

451 ## Auxiliaries

452   1-57. Auxiliaries are active sympathizers who provide important support services. They may run safe
453   houses or store weapons and supplies but do not participate in direct action.

454 ## Mass Base

455   1-58. The mass base consists of the followers of the insurgent movement, the supporting populace. Mass
456   base members are often recruited and indoctrinated by the cadre, though in many politically charged situa-
457   tions or traditional insurgencies such active pursuit is not necessary. Mass base members may continue in
458   their normal positions in society. Many, however, lead clandestine lives for the insurgent movement. They
459   may even pursue full-time positions within the insurgency. For example, combatants normally begin as

460 members of the mass base before becoming armed manpower. Such roles are particularly hard to define in
461 a tribal or clan-based insurgency, where there is no clear cadre, and people will drift between combatant,
462 auxiliary, and follower status as needed.

Employing the Elements

464 1-59. The insurgent leaders provide organizational and managerial skills to transform mobilized individu-
465 als and communities into an effective force for armed political action. What results, as in any conflict, is a
466 contest of resource mobilization and force deployment. A state is challenged by a revolutionary or resis-
467 tance movement that may also seek to create a counterstate. No objective force level guarantees victory for
468 either side. It is frequently stated that a 10 to 1 or 20 to 1 combatant ratio of counterinsurgents to insurgents
469 is necessary for counterinsurgency victory. In reality, no such fixed ratios exist. As in any war, all correla-
470 tions of forces to defeat an insurgency depend on the situation. Of necessity, however, counterinsurgency
471 is manpower intensive because of the requirements to maintain widespread order and security. And coun-
472 terinsurgents might have to adopt different approaches to deal with each element of the insurgency. Auxil-
473 iaries might be co-opted by economic reforms. Fanatic combatants will have to be killed or captured.

DYNAMICS OF AN INSURGENCY

475 1-60. Insurgencies also contain several common dynamics:

476 - Leadership.
477 - Objectives.
478 - Ideology.
479 - Environment and geography.
480 - External support and sanctuaries.
481 - Phasing and timing.

482 These provide a framework for analysis that can reveal the insurgency's strengths and weaknesses. Al-
483 though analysts can examine the following dynamics separately, they must study their interaction to fully
484 understand an insurgency.

Leadership

486 1-61. Leadership is critical to any insurgency. Insurgency is not simply random violence. It is directed and
487 focused violence aimed at achieving an object beyond the war that is political in nature. It requires leader-
488 ship to provide vision, direction, guidance, coordination, and organizational coherence. Successful insur-
489 gent leaders make their cause known to the people and gain popular support. Their key tasks are to break
490 the ties between the people and the government and to establish credibility for their movement. Their edu-
491 cation, background, family, social connections, social positions and experiences contribute to their ability
492 to organize and inspire resistance and insurgency.

493 1-62. Revolutionary insurgent movements usually begin as the tangible manifestation of political es-
494 trangement. Alienated elite members, however defined (for example, a school teacher is an "elite member"
495 in most of the world), advance alternatives to existing conditions. As their movement grows, leaders decide
496 which approach to adopt. The level of decentralization of responsibility and authority drives the precise
497 structure and operational procedures of the insurgency. Extreme decentralization results in a movement
498 that rarely functions as a coherent body. It is, however, capable of inflicting substantial casualties and dam-
499 age. Loose networks find it difficult to create a viable counterstate and therefore have great difficulty seiz-
500 ing political power. However, they are also very hard to destroy and can continue to sow disorder, even
501 when degraded. It takes very little coordination to disrupt most states.

502 1-63. Many contemporary insurgencies involve mobilization based on clan, tribe, or ethnic group. These
503 insurgencies are often led by traditional authority figures, such as sheikhs and religious leaders. Rather
504 than being codified in impersonal rules, traditional authority is usually invested in a hereditary line or in-
505 vested in a particular office by a higher power. As the Indonesian *Dar 'ul Islam* rebellions of 1948 and
506 1961 demonstrate, traditional authority figures often wield enough power, especially in rural areas, to sin-

507 gle-handedly drive an insurgency. This type of insurgency can be defeated by co-opting the responsible
508 traditional authority figure.

Objectives

1-64. Effective analysis of an insurgency requires identifying its strategic, operational, and tactical objectives. The strategic objective is the insurgents' desired end state, or how the insurgents will use power. Operational objectives are those that the insurgents pursue as part of the total process of destroying government legitimacy and progressively establishing their desired end state. Tactical objectives are the immediate aims of insurgent acts, and can be psychological and physical in nature. One example of a psychological objective is gaining support for the insurgency by the assassination of local government officials. An example of a physical objective is the disruption of government services by the attack and seizure of a key facility. These tactical acts are linked to higher purposes. In fact, the levels of war often become so collapsed and nested that they are difficult to discern separately. Commanders and staffs must prepare to deal with that complexity. In both insurgency and counterinsurgency, tactical actions may have immediate strategic effects.

Ideology and Narrative

1-65. Ideas are a motivating factor in insurgent violence. Insurgencies gather recruits and amass popular support through ideological appeal. Recruits are often young men suffering from frustrated hopes and unable to improve their lot in life. The insurgent group provides them identity, purpose, and community in addition to physical, economic, and psychological security. The movement's ideology explains its followers' tribulations and provides a course of action to remedy those ills. The most powerful ideologies tap latent, emotive concerns of the populace, such as the desire for justice, religious beliefs, or liberation from foreign occupation. Ideology provides a prism, including a vocabulary and analytical categories, through which the situation is assessed. Thus ideology can shape the movement's organization and operational methods.

1-66. The central mechanism through which ideologies are expressed and absorbed is the narrative. A *narrative* is an organizational scheme expressed in story form. Narratives are central to the representation of identity, particularly the collective identity of groups such as religions, nations, and cultures. Stories about a community's history provide models of how actions and consequences are linked and are often the basis for strategies, actions, and interpretation of the intentions of other actors. Insurgent organizations such as Al Qaeda use narratives very efficiently in the development of a legitimating ideology. In the jihadist narrative, Osama Bin Laden's depiction of himself as a man purified in the mountains of Afghanistan who begins converting followers and punishing the infidels, resonates powerfully with the historic figure of Mohammed. In the collective imagination of Bin Laden and his followers, Islamic history is a story about the decline of the *umma* and the inevitable triumph against Western imperialism. Only through *jihad* can Islam be renewed both politically and theologically.

1-67. Though insurgencies have historically been phenomena specific to nation-states, there have been a number of transnational insurgencies. Likewise, there have been efforts by external powers to tap or create general upheaval by coordinating national insurgencies so that they take on a transnational character. The effort of the Moscow-directed Communist International between the two world wars is possibly the best historical illustration. The recent activities of Al Qaeda are another attempt to create and support such a transnational array of insurgencies. Operational-level commanders address elements of the transnational movement within their theater of operations, while other government agencies and higher-level officials must deal with the national-strategic response to such threats.

Environment and Geography

1-68. Environment and geography, including cultural and demographic factors, affect all participants in a conflict. The manner in which insurgents and counterinsurgents adapt to these realities creates advantages and disadvantages for each. The effects of these factors are immediately visible at the tactical level, where they are perhaps the predominant influence on decisions regarding force structure and doctrine (including tactics, techniques, and procedures). Insurgencies in an urban environment present a different set of planning considerations than those in rural environments. Border areas contiguous to states that may wittingly or unwittingly provide external support and sanctuary to the insurgents are a particularly vulnerable area for counterinsurgents. Chapter 3 discusses his dynamic.

External Support and Sanctuaries

1-69. Access to external resources and sanctuaries has always been a factor influencing the effectiveness of insurgencies. External support can provide political, psychological, and material resources that might otherwise be limited or totally unavailable. Such assistance does not need to come just from neighboring states, however. Insurgents can be supported from countries outside the region looking for political or economic influence. Insurgencies may turn to criminal elements for funding or take advantage of the interconnected nature of the Internet to create a network of support among nongovernmental organizations, even in the United States. Ethnic diaspora communities also provide a form of external support and sanctuary, particularly in a transnational insurgency.

1-70. The meaning of sanctuary is evolving. Sanctuaries were traditionally physical safe havens, such as base areas, and this form of safe haven still exists. But insurgents today can also draw on "virtual" sanctuaries in the Internet, global financial systems, and international media that make their actions seem acceptable or laudable to the population.

1-71. Historically, sanctuary in neighboring countries has provided insurgents with a sheltered zone to rebuild and reorganize without fear of counterinsurgent interference. The impressive targeting abilities of modern weaponry and intelligence gathering technology make insurgents in isolation, even in neighboring states, more vulnerable than those hidden among the population. Thus, contemporary insurgencies often develop in urban environments and try to avoid concentration. This dispersion limits effectiveness in seizing power, but increases survivability.

1-72. The changed security environment following the end of the Cold War and the terrorist attacks on 11 September 2001 have resulted in greater concern for the role of nonstate actors in insurgencies. Nonstate actors, such as transnational terrorist organizations, often represent a security threat beyond the areas they inhabit—even to the point that they can pose a concern for the United States and its multinational partners. The same nonstate actors often team with insurgents and, in this sense, profit from the conflict. Insurgency can open up sanctuaries within a state over which the host nation's forces cannot extend control or significant influence. In these sanctuaries, nonstate actors with intentions hostile to the United States can develop unimpaired. When it is to their advantage, they provide support for insurgencies and, in turn, derive physical support for their purposes in much the same fashion a parasite does from its host. For these reasons, the issue of sanctuary cannot be ignored in campaign strategy. Effective counterinsurgency operations work to reduce and destroy such sanctuaries.

Phasing and Timing

1-73. Insurgencies often pass through common phases of development.**Error! Bookmark not defined.** Not all insurgencies experience such phased development, and progression through all phases is not a requirement for success. The same insurgent movement may be in different phases in different regions of a country. Insurgencies can also revert to an earlier phase when under pressure. Indeed, this is the key strength of a phased approach, which provides fallback positions for the insurgents when threatened. They then resume development when favorable conditions return. While the traditional Maoist phases may not provide a complete template for understanding contemporary insurgencies, they do explain a shifting mosaic of activities that are usually present in some form.

1-74. Versions of the Maoist concept have been used by movements as diverse as communist and Islamist insurgencies because it is logical and based on mass mobilization. Strategic movement from one phase to another incorporates the operational and tactical activity typical of earlier phases. It does not end them. The Vietnamese explicitly recognized this in their "war of interlocking" doctrine. This held that all forms of warfare occur simultaneously, even as a particular form is paramount. Debates about Vietnam that focus on whether U.S. forces should have concentrated on guerrilla or conventional operations ignore this complexity. In fact, forces that win a mosaic war can conduct both types, rapidly transitioning between them as required.

606 1-75. If the insurgents adopt a conspiratorial or *focoist* approach emphasizing quick or armed action and
607 deemphasizing political organization, these phases do not necessarily apply. In many ways these are less
608 complex strategies to counter, though the counterinsurgent must not lose sight of the overall importance of
609 long-term political objectives in dealing with any insurgent approach.

610 ## INSURGENT NETWORKS

611 1-76. Insurgents use the technological, economic, and social aspects of globalization to create networks
612 involving a wide array of partners. Networking is a tool available to traditional territorially-rooted insur-
613 gencies, such as the FARC in Colombia and the Zapatistas in Chiapas, and extends the range and variety of
614 both their military and political actions. Other groups exist almost entirely in networks with little physical
615 presence in their target countries. Networked organizations do not have a center of gravity in the traditional
616 sense—rather the network's links themselves may become centers of gravity. Networked organizations are
617 very hard to destroy and tend to heal, adapt, and learn rapidly. However, they have a limited chance of at-
618 taining strategic success themselves because they have a difficult time mustering and focusing power. The
619 best they can hope for is to create a security vacuum leading to a collapse of the targeted regime's will and
620 then to gain in the competition for the spoils. However, their enhanced abilities to sow disorder and survive
621 present particularly difficult problems for counterinsurgents.

622 ## INSURGENT VULNERABILITIES

623 1-77. While this chapter so far has stressed the difficulties insurgencies present, they do have vulnerabili-
624 ties that may be exploited by a skilled counterinsurgent. This process is discussed in more detail in chapters
625 4 and 5 However, a few of these potential vulnerabilities are worth highlighting here.

626 ### Secrecy

627 1-78. Any group beginning from a position of weakness that intends to use force and violence to prosecute
628 its political aims must initially adopt a covert approach for their planning and activities. This practice can
629 become counterproductive once an active insurgency begins. Excessive secrecy can limit insurgent free-
630 dom of action, lessen or distort information about insurgent goals and ideals, and restrict communication
631 within the insurgency. One of the ways insurgent groups attempt to avoid the effects of too much secrecy is
632 splitting into political and military wings, as in the case of Sinn Fein and the Irish Republican Army, to
633 deal separately with the public (political) requirements of an insurgency while still conducting clandestine
634 (military) actions.

635 ### Mobilization

636 1-79. In the early stages of an insurgency especially, movements may be tempted to go to almost any ex-
637 tremes to attract followers. To mobilize their base of support, insurgent groups use a combination of propa-
638 ganda and intimidation, and they may overreach in both. Counterinsurgents should be able to use informa-
639 tion operations to effectively exploit inconsistencies in the insurgents' message as well as their excessive
640 uses of force or intimidation.

641 ### Base of Operations

642 1-80. Insurgents can experience serious difficulty finding a viable base of operations. If the chosen loca-
643 tion is too far away from the major centers of activity, it is potentially secure but out of touch with the peo-
644 ple and vulnerable to isolation. If the base is too near to centers of activity, the insurgency can be open to
645 observation and perhaps infiltration. It is also closer to the machinery of state control. Bases close to na-
646 tional borders can be attractive when they are beyond the authority or reach of the counterinsurgent force
647 yet safe enough to avoid the unwanted suspicions of the neighboring authority or population. Timely, reso-
648 lute counterinsurgent action to exploit poor enemy base locations and eliminate or disrupt good ones can
649 do much to weaken an insurgency.

Financial Weakness

1-81. All insurgencies require funding to a greater or lesser extent. Working with criminal organizations brings insurgents into contact with unreliable and vulnerable groups who could attract undue attention from the authorities. It also creates vulnerability to counterinsurgent intelligence operations. Funding from outside donors may come with a political price that could distort and affect the overall aim of an insurgency, weakening its popular appeal. Insurgent financial weakness can also be exploited by counterinsurgents and exacerbated by controls and regulations to limit the movement and exchange of goods and funds, particularly if an insurgency is being funded from outside the state.

Internal Divisions

1-82. Counterinsurgents remain alert for signs of divisions within an insurgent movement. A series of successes by the counterinsurgent force or some errors made by insurgent groups could cause some members to question their cause or even challenge the leadership of the insurgency. In addition, relations within an insurgent movement, like other social networks, will not remain harmonious if cabals form to vie for power. Rifts that exist between insurgent leaders, if identified, can be exploited. An offer of amnesty or a seemingly generous compromise offered by the counterinsurgent could also cause division within an insurgency and present opportunities to split or weaken the movement.

Maintaining Momentum

1-83. Controlling the pace and timing of operations is vital to the success of any insurgency. Insurgents, if they have planned properly, control when the war begins and have some measure of control over subsequent activity. But many insurgencies have failed to capitalize on their initial opportunities or have allowed the pace of events and scope of activities to be dictated by a skilled counterinsurgent force. If insurgent momentum is lost, the strategic initiative can be regained by the counterinsurgent.

Informants

1-84. There is nothing more demoralizing to insurgents than the realization that people inside their movement or trusted supporters among the public are providing information to the state authorities. While informers have sometimes been infiltrated into insurgencies, it is more common to achieve success by turning someone who is already in the organization or is an auxiliary who has contact with key insurgents (for example, the couriers or suppliers who are the links between clandestine cells and their base of supporters). *Turning* is the intelligence term for persuading such a person to become an informer. This may be best achieved by discovering a participant who has become dissatisfied with the insurgency or who, for personal or family reasons, wants to leave it. Counterinsurgents may further exercise pressure to turn by arousing fear of prosecution or by offering rewards. Informers must be confident that they and their families will be protected against retribution.

ASPECTS OF COUNTERINSURGENCY

1-85. In almost every case, the counterinsurgent faces a populace containing an active minority supporting a government and a similar militant faction opposing it. To be successful, the government must be accepted as legitimate by most of that uncommitted middle, which also includes passive supporters of both sides. (See figure 1-2.) Because of the ease of sowing disorder, it is usually not enough for a counterinsurgent to get 51 percent of popular support; a solid majority is often essential. However, for an insurgent with strong support from external agencies or criminal elements, a passive populace may be all that is necessary to facilitate seizing political power.

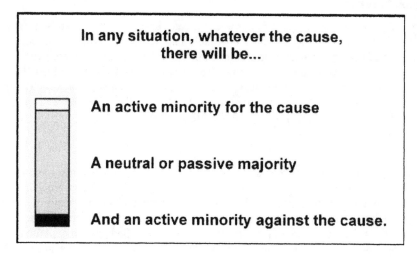

Figure 1-2. Support for an insurgency

691

692   1-86. Upon assuming a COIN mission, Soldiers and Marines must not only be prepared to identify their
693 opponents and the approaches to insurgency they are applying; counterinsurgents must also understand the
694 broader strategic context within which they are operating. Options are different when the United States is
695 invited to assist a functioning government than if no such viable entity exists or if a regime has been
696 changed by conflict. The last two situations add complex sovereignty and national reconstruction issues to
697 what is already a complicated mission. The level of violence and state of infrastructure influences the type
698 and amount of resources from sources within and outside the Department of Defense that can be drawn
699 upon. The number and type of outside agencies shape the amount of interagency and international coordi-
700 nation required.

701   1-87. The rest of this manual presents detailed discussions about conducting counterinsurgency operations.
702 The following discussion addresses some general themes shape the chapters that follow. It addresses the
703 following:

704   ●   Principles derived from past insurgencies.
705   ●   Imperatives based on current COIN operations.
706   ●   Paradoxes present in a COIN environment.
707   ●   Successful and unsuccessful COIN practices.

708   1-88. .The principles and imperatives of modern counterinsurgency derived from that historical record and
709 detailed below provide some guideposts for forces engaged in counterinsurgency operations. However,
710 counterinsurgency is complicated, and even following the principles and imperatives does not guarantee
711 success. This paradox is present in all forms of warfare but is most obvious in counterinsurgency. The fol-
712 lowing principles and imperatives are presented in the belief that understanding them helps to illuminate
713 the challenges inherent in defeating an insurgency.

714 **HISTORICAL PRINCIPLES**

715   1-89. The following principles are derived from past insurgencies.

716 **Legitimacy is the Main Objective**

717   1-90. The primary objective of any counterinsurgent is to foster the development of effective governance
718 by a legitimate government. All governments rule through a combination of consent and coercion. Gov-
719 ernments described as "legitimate" rule primarily with the consent of the governed, while those described
720 as "illegitimate" tend to rely mainly or entirely on coercion. Their citizens obey the state for fear of the

721
722
723
consequences of doing otherwise, rather than because they voluntarily accept its rule. A government that derives its powers from the governed tends to be accepted by its citizens as legitimate. It still uses coercion—for example, against criminals—but the bulk of the population voluntarily accepts its governance.

724
725
726
727
728
729
730
1-91. In Western, liberal tradition, a government that derives its just powers from the people and responds to their desires while looking out for their welfare is accepted as legitimate. In theocratic societies, political and religious authorities are fused, and political figures are accepted as legitimate because the populace views them as implementing the will of God. Medieval monarchies claimed "the divine right of kings" and imperial China governed with "the mandate of heaven." Since the 1979 revolution, Iran has operated on the theocratic "rule of the jurists." In other societies, "might makes right" and security is the prime determinants of legitimacy.

731
732
733
734
735
736
737
738
739
740
741
1-92. Legitimacy makes it easier for the state to carry out its key functions, which include the capability to regulate social relationships, extract resources and appropriate or use resources in determined ways. A legitimate government can develop these capabilities more easily, allowing it to competently manage, coordinate, and sustain collective security as well as political, economic, and social development. Conversely, illegitimate states typically fail to regulate society, or can only do so by the application of overwhelming coercion ("police states"). Legitimate governance is inherently stable because it engenders societal support to adequately manage the internal problems, change, and conflict that invariably affect individual and collective well-being. Conversely, governance that is not legitimate is inherently unstable because as soon as the state's coercive power is disrupted the population ceases to obey it. Thus legitimate governments tend to be resilient and exercise better governance, while illegitimate states tend to be fragile and poorly administered.

742
1-93. Five indicators of legitimacy that can be used to analyze threats to stability include—

743
744
- Frequent selection of leaders in a manner considered just and fair by a substantial majority of the population.
745
- A high level of popular participation in or support for the political process.
746
- A low level of corruption.
747
- A culturally acceptable level or rate of political, economic, and social development.
748
- A high level of regime acceptance by major social institutions.

749
750
751
752
753
1-94. Governments that score high in these categories probably have the support of an adequate majority of the population. Every culture, however, has varying concepts of acceptable levels of development, corruption, and participation. The first two indicators above reflect Western concepts. For some societies, it may be enough for a government to only provide security and some basic services. However, that concept of security could be very broad, and different groups could have widely divergent expectations.

754
755
756
757
758
759
1-95. During mission analysis for a COIN effort, commanders and staffs determine what the host population defines as effective and legitimate governance, since this influences all ensuing operations. Additionally, planners may also consider perceptions of legitimacy held by outside supporters of the host government or the insurgents. The differences between American, local, and international visions of legitimacy will further complicate operations. But the most important attitude remains that of the indigenous population, who in the end will decide the ultimate victor of the conflict.

760
761
762
763
764
765
1-96. The rule of law is a major factor in assuring voluntary acceptance of a government's authority and, therefore, its legitimacy. Because power may be exercised illegitimately by individuals who use the instruments of a state for personal ends, a government's respect for pre-existing and impersonal legal rules can provide the key to gaining it widespread and enduring societal support. Such governmental respect for rules—ideally ones recorded in a constitution and in laws adopted through a credible and democratic process—is the essence of rule of law. As such, it is a powerful potential force in counterinsurgency.

766
767
768
1-97. Military action can address the symptoms of a loss of legitimacy. However, restoring legitimacy can only be accomplished using all instruments of national power. Without the host-nation government achieving legitimacy, COIN cannot succeed.

Unity of Effort is Essential

1-98. Unity of effort must pervade every echelon. Otherwise, well-intentioned but uncoordinated actions can cancel each other out or provide a competent insurgent many vulnerabilities to exploit. Ideally a counterinsurgent should have authority over all government agencies involved in operations. However, the best situation that military commanders can generally hope for is to be able to achieve unity of effort through communication and liaison with those responsible for the nonmilitary agencies. There are many U.S., international, and indigenous organizations needing coordination. The U. S. Ambassador and country team must be key players in higher-level planning, while similar connections are needed throughout the chain of command.

1-99. Nongovernmental organizations often play an important role at the local level. Many such agencies resist being overtly involved with military forces, but efforts to establish some kind of liaison are needed. The most important connections are those with joint, interagency, multinational, and host-nation organizations to ensure, as much as possible, that objectives are shared and actions and messages synchronized. Achieving synergy is another essential element for effective COIN.

Political Factors are Primary

1-100. General Chang Ting-chen of Mao Zedong's Central Committee once stated that revolutionary war was 80 percent political action and only 20 percent military. At the time he was involved with establishing special schools to train the extensive political cadres that accompanied the victorious communist forces that completed their conquest of mainland China in 1949. While in the initial stages of COIN, military actions appear especially predominant, political objectives must retain primacy. Commanders conduct all operations with consideration of their contribution toward strengthening the legitimacy of the host government and achieving the political goals set by the U.S. government. This means that the country team and U.S. government political-military staff must be active participants throughout the conduct (planning, preparation, execution, and assessment) of COIN operations. The political and military aspects of insurgencies are so bound together as to usually be inseparable, and most insurgent approaches recognize that fact. Military actions conducted without proper assessment of political effects at best result in reduced effectiveness and at worst are counterproductive. Resolving most insurgencies requires a political solution, and it is imperative that the actions of the counterinsurgent do not make achieving that political solution more difficult.

Understand the Environment

1-101. The local population is a critical center of gravity of an insurgency. Successful conduct of counterinsurgency operations depends on thoroughly understanding the society and culture within which they are being conducted. Soldiers and Marines must understand the following about the population in the area of operations (AO):

- How key groups in the society are organized.
- Relationships and tensions among them.
- Ideologies and narratives that resonate with the groups.
- Group interests and motivations.
- Means by which groups communicate.
- The society's leadership system.

1-102. In the vast majority of counterinsurgency operations since 1945, the insurgents have held a distinct advantage in their level of local knowledge. They speak the language, move easily within the society, and are more likely to understand the interests of the population. Thus, effective counterinsurgency requires a leap of imagination and a peculiar skill set not encountered in conventional warfare. The interconnected, politico-military nature of insurgency and COIN requires immersion in the people and their lives to achieve victory. Successful COIN operations require Soldiers and Marines at every echelon to possess the following within the cultural context of the AO:

- A clear, nuanced, and empathetic appreciation of the essential nature of the conflict.
- An understanding of the motivation, strengths, and weaknesses of the insurgents.

817   ●   Knowledge of the roles of other actors in the AO.

818   Without this understanding of the environment, intelligence cannot be understood and properly applied.

Intelligence Drives Operations
819

820   1-103. Without good intelligence, a counterinsurgent is like a blind boxer, wasting energy flailing at an
821 unseen opponent and perhaps causing unintended harm. With good intelligence, a counterinsurgent is like
822 a surgeon, cutting out cancerous tissue while keeping other vital organs intact. Effective operations are
823 shaped by carefully considered actionable intelligence, gathered and analyzed at the lowest possible levels
824 and disseminated and distributed throughout the force.

825   1-104. Because of the dispersed nature of COIN, counterinsurgents' own operations are a key generator
826 of intelligence. A cycle develops where operations produce intelligence that drives subsequent operations.
827 Tactical reporting by units, members of the country team, and associated civilian agencies is of equal or
828 greater importance than reporting by specialized intelligence assets. These factors, along with the need to
829 generate a superior tempo, drive the requirement to produce and disseminate intelligence at the lowest
830 practical level.

Isolate Insurgents from Their Cause and Support
831

832   1-105. It is easier to cut an insurgency off and let it die than to kill every insurgent. Attempting to kill
833 every insurgent is normally impossible. It can also be counterproductive, generating popular resentment,
834 creating martyrs that motivate new recruits, and producing cycles of revenge. Dynamic insurgencies also
835 replace losses quickly. A skillful counterinsurgent cuts off the sources of that recuperative power. Some
836 can be reduced by redressing the social, political, and economic grievances that fuel the insurgency. Physi-
837 cal support can be cut off by population control or border security. International or local legal action might
838 be required to limit financial support. As the host government increases its own legitimacy, the people be-
839 gin to more actively assist it, eventually marginalizing and stigmatizing insurgents to the point where their
840 legitimacy is destroyed. Victory is gained not when this isolation is achieved, but when it is permanently
841 maintained by and with the active support of the populace.

Establish Security Under the Rule of Law
842

843   1-106. The cornerstone of any COIN effort is security for the civilian population. Without that, no per-
844 manent reforms can be implemented, and disorder spreads. To establish legitimacy, security activities
845 move as quickly as feasible from major combat operations to law enforcement. When insurgents are seen
846 as criminals they lose public support; if they are dealt with by an established legal system in line with local
847 culture and practices, the legitimacy of the host government is enhanced. This process takes time, but Sol-
848 diers and Marines remain aware of the legal procedures applicable to their conduct and work to apply
849 them. They also assist in the establishment of the indigenous institutions that sustain that legal regime, in-
850 cluding police forces, court systems, and penal facilities.

851   1-107. Illegitimate actions by government officials, security forces, and multinational partners are those
852 involving the use of power without authority. Efforts to build a legitimate government though illegitimate
853 action—including unjustified or excessive use of force, unlawful detention, torture, or punishment without
854 trial—are self-defeating, even against insurgents who conceal themselves amid noncombatants and flout
855 the law. Moreover, participation in counterinsurgency operations by United States forces must be pursuant
856 to United States law, which includes domestic laws and international treaties to which the United States is
857 party as well as certain laws of the host nation. (See Appendix D.) Any human rights abuses or legal viola-
858 tions committed by U.S. forces quickly become known throughout the local population and eventually
859 around the world because of the globalized media and work to undermine the COIN effort.

860   1-108. Every action by counterinsurgents leaves a "forensic trace" that may need to be used in a court of
861 law. Counterinsurgents document all their activities to preserve, wherever possible, a chain of evidence—
862 and often more importantly, as a means to counter insurgent propaganda.

Prepare for a Long-Term Commitment

1-109. By its nature, insurgency is protracted. The conduct of counterinsurgency always demands considerable expenditures of time and resources. Even if people prefer the host-nation government to the insurgents, they do not actively support that government unless they are convinced the counterinsurgent forces have the means, ability, stamina, and will to win. The insurgent's primary battle is against the indigenous government, not the United States, but American support can be crucial to building public faith in that government's viability. Insurgents and local populations often believe that a few casualties or a few years will cause the United States to abandon a COIN effort. Constant reaffirmations of commitment, backed by deeds, can overcome that perception and bolster faith in the steadfastness of American support.

1-110. Preparing for the necessarily protracted support of COIN requires putting in place purpose-dedicated headquarters and support structures designed for long-term operations. Planning and commitments should be based on sustainable operating tempo and personnel tempo limits for the various components of the force. Even in situations where the American goal is reducing its military force levels as quickly as possible, some support for host-nation institutions usually remains for a long time.

1-111. At the strategic level, winning and maintaining the support of the American people for a protracted deployment is arguably the critical COIN activity. Military commanders are almost never directly involved in this process, which is properly a political activity. They take care to ensure that their actions and statements, and the conduct of the operation neither make it more difficult for elected leaders to maintain public support nor undermine public confidence.

CONTEMPORARY IMPERATIVES OF COUNTERINSURGENCY

1-112. Recent experiences with counterinsurgency have highlighted an important set of additional imperatives to keep in mind for success.

Manage Information and Expectations

1-113. Information and expectations are related, and both are carefully managed by a skillful counterinsurgent. To limit discontent and build support, a host government and any counterinsurgents assisting it create and maintain a realistic set of expectations among the populace, friendly military forces, and the international community. Information operations (including its related activities of public affairs, and civil-military operations) are key tools to accomplish this. Achieving steady progress toward a set of reasonable expectations can increase the population's tolerance for the inevitable inconveniences entailed by ongoing counterinsurgency operations. Where large American forces are present to help establish a regime, such progress can extend the period before an army of liberation becomes perceived as an army of occupation.

1-114. Americans start with an automatic disadvantage because of their reputation for accomplishment, what some call the "man on the moon syndrome." This refers to the expressed disbelief that a nation that can put a man on the moon cannot quickly restore basic services. U.S. agencies trying to fan enthusiasm for their efforts should also avoid making exorbitant promises. In some cultures, failure to deliver promised results is automatically interpreted as deliberate deception, not good intentions gone awry. In other cultures, exorbitant promises are the norm, and people do not expect them to be kept. So counterinsurgents must understand these local norms and employ locally tailored approaches to ensure expectations are controlled. Managing expectations also involves demonstrating economic and political progress to show the populace how life is improving. Increasing the number of people who feel they have a stake in the success of the state and its government is a key to successful COIN. In the final judgment, victory comes by convincing the people that their life will be better under the government than under the insurgent.

1-115. Both the counterinsurgent and the host government ensure that their deeds match their words. They also understand that any action has an information reaction, carefully consider that impact on the many audiences involved in the war and on the sidelines, and work actively to shape the responses that further their ends. In particular, messages to different audiences must be consistent. In the worldwide information environment, local people in a counterinsurgency theater can access the Internet and satellite television to find out what messages the counterinsurgent is sending to the international community or the American people. Any perceived inconsistency destroys credibility and undermines COIN efforts.

Use Measured Force

1-116. Any use of force generates a series of reactions. There may be times when an overwhelming effort is necessary to intimidate an opponent or reassure the populace. But the type and amount of force to be applied, and who wields it, should be carefully calculated by a counterinsurgent for any operation. An operation that kills five insurgents is counterproductive if the collateral damage or the creation of blood feuds leads to the recruitment of fifty more.

1-117. In a COIN environment, it is vital for commanders to adopt appropriate and measured levels of force that accomplish the mission without causing unnecessary loss of life or suffering. Normally, the counterinsurgent can minimize potential loss of life by employing escalation of force procedures, especially at checkpoints and roadblocks, and during convoy operations. The concept of escalation of force refers to the use of lesser means of force when such use is likely to achieve the desired effects and Soldiers and Marines can do so without endangering themselves, others, or the mission. Escalation of force procedures are not limitations on the right of self-defense, including deadly force, when such force is necessary to defend against a hostile act or demonstrated hostile intent. Commanders must be familiar with the concept of force escalation and ensure that their Soldiers and Marines are properly trained in its use

1-118. Who wields force is also important. Providing the police have a reasonable reputation for competence and impartiality, it is better to let police handle urban raids than Soldiers or Marines, even if the police are not as well armed or capable as military units, since the populace is likely to view that application of force as more legitimate. In other circumstances, police may be seen as part of a particular ethnic or sectarian group that oppresses the general population, and hence their use may be counterproductive. So effective counterinsurgents understand the character of the local police and popular perceptions of both police and military units to ensure that the application of force is measured and reinforces the rule of law.

Learn and Adapt

1-119. An effective counterinsurgent force is a learning organization. Insurgents constantly shift between military and political phases and approaches. In addition, networked insurgents constantly exchange information about their enemy's vulnerabilities—including with other insurgents in distant theaters. A skillful counterinsurgent is able to adapt at least as fast as the insurgents. Every unit needs to be able to make observations, draw lessons, apply them, and assess results. Headquarters must develop an effective system to circulate best practices throughout the command quickly. Combatant commanders might also need to seek new laws or policies to authorize or resource necessary changes. Insurgents will shift their areas of operations looking for weak links, so widespread competence is required throughout the counterinsurgent force.

Empower the Lowest Levels.

1-120. *Mission command* is the conduct of military operations through decentralized execution based upon mission orders for effective mission accomplishment. Successful mission command results from subordinate leaders at all echelons exercising disciplined initiative within the commander's intent to accomplish missions. It requires an environment of trust and mutual understanding (FM 6-0). It is the Army's preferred method for commanding and controlling forces during all types of operations. Under mission command, commanders provide subordinates with a mission, their commander's intent, a concept of operations, and resources adequate to accomplish the mission. Higher commanders empower subordinates to make decisions within the commander's intent. They leave details of execution to their subordinates and require them to use initiative and judgment to accomplish the mission.

1-121. Mission command is ideally suited to the mosaic nature of insurgency and COIN. Local commanders have the best grasp of their situations. Under mission command, they are given access to or control of the assets needed to produce actionable intelligence and manage information operations and civil-military operations. Effective COIN operations are decentralized. Higher commanders owe it to their subordinates to push as many capabilities as possible down to their level. Mission command encourages subordinates' initiative within legal limits. It facilitates the learning process that must occur at every level. Mission command is a major characteristic of a counterinsurgency force that can adapt and react as quickly as the insurgents.

961 **Support the Host Nation**

962 1-122. American forces committed to COIN are there to assist a host government. The long-term goal is
963 to leave a host that is capable of standing on its own. In the end, the host nation has to win its own war.
964 Achieving this requires the development of viable local leaders and institutions. U.S. forces and agencies
965 can help, but host-nation elements must be able to accept responsibilities to achieve real victory. While it
966 may be easier for American military units to conduct operations themselves, it is better to work to
967 strengthen local forces and then assist them. Host governments have the final responsibility to solve their
968 own problems. Eventually all foreign armies are seen as interlopers or occupiers, and the sooner the main
969 effort can be turned over to host-nation institutions, the better.

970 PARADOXES OF COUNTERINSURGENCY

971 1-123. The principles and imperatives discussed above reveal that COIN presents a complex and often
972 unfamiliar set of missions and considerations for a military commander. In many ways, the conduct of
973 counterinsurgency is counterintuitive to the traditional American view of war—although it has actually
974 formed a substantial part of America's actual experience. Some representative paradoxes of COIN are pre-
975 sented here as examples of the different mindset required.

976 **The More You Protect Your Force, The Less Secure You Are**

977 1-124. Ultimate success in COIN is gained by protecting the populace, not the COIN force. If military
978 forces stay locked up in compounds, they lose touch with the people, appear to be running scared, and cede
979 the initiative to the insurgents. Patrols must be conducted, risk must be shared, and contact maintained.
980 This ensures access to the intelligence needed to drive operations and reinforces the connections with the
981 people that establish real legitimacy.

982 **The More Force Used, the Less Effective It Is**

983 1-125. Any use of force produces many effects, not all of which can be foreseen. The more force applied,
984 the greater the chance of collateral damage and mistakes. It also increases the opportunity for insurgent
985 propaganda to portray lethal military activities as brutal. The precise and discriminate use of force also
986 strengthens the rule of law that needs to be established.

987 **The More Successful COIN is, the Less Force That Can be Used and the More Risk That Must**
988 **be Accepted**

989 1-126. This is really a corollary to the previous paradox. As the level of insurgent violence drops, the re-
990 quirements of international law and the expectations of the populace allow less use of military actions by
991 the counterinsurgent. More reliance is placed on police work. Rules of engagement get stricter, and troops
992 have to exercise increased restraint. Soldiers and Marines may also have to accept more risk to maintain
993 involvement with the people.

994 **Sometimes Doing Nothing is the Best Reaction**

995 1-127. Often an insurgent carries out a terrorist act or guerrilla raid with the primary purpose of enticing
996 the counterinsurgent to overreact, or at least to react in a way that can then be exploited. If a careful as-
997 sessment of the effects of a course of action concludes that more negative than positive effects may result,
998 an alternative should be considered—potentially including a decision not to act.

999 **The Best Weapons for COIN Do Not Shoot**

1000 1-128. Counterinsurgents achieve the most meaningful success by gaining popular support and legitimacy
1001 for the host government, not by killing insurgents. Security plays an important role in setting the stage for
1002 other progress, but lasting victory comes from a vibrant economy, political participation, and restored
1003 hope. Often dollars and ballots have a more important impact than bombs and bullets. Soldiers and Marines

1004 prepare to engage in a host of nonmilitary missions to support COIN. Everyone has a role in nation-
1005 building, not just the State Department or civil affairs soldiers.

The Host Nation Doing Something Tolerably is Sometimes Better Than Us Doing It Well

1007 1-129. It is just as important to consider who performs an operation as to assess how well it is done. In
1008 cases where the United States is supporting a host nation, long-term success requires the establishment of
1009 viable indigenous leaders and institutions that can carry on without significant American support. The
1010 longer that process takes, the more popular support in the United States will wane, and the more the local
1011 populace will question the legitimacy of their own forces and government. T.E. Lawrence said of his ex-
1012 perience leading the Arab Revolt against the Ottoman Empire, "Do not try and do too much with your own
1013 hands. Better the Arabs do it tolerably than you do it perfectly. It is their war, and you are to help them, not
1014 win it for them." However, a key word in Lawrence's advice is "tolerably." If the host nation cannot per-
1015 form tolerably, the COIN force may have to act. Experience, knowledge of the AO, and cultural sensitivity
1016 are essential to deciding when such action is necessary.

If A Tactic Works This Week, It Might Not Work Next Week; If It Works In This Province, It Might Not Work In The Next

1019 1-130. Competent insurgents are adaptive and today are often part of a widespread network that con-
1020 stantly and instantly communicates. Insurgents quickly disseminate information about successful COIN
1021 practices throughout the insurgency and adapt to them. Indeed, the more effective a COIN tactic is, the
1022 faster it becomes out of date because the insurgents have a greater need to counter it. Effective leaders at
1023 all levels avoid complacency and are at least as adaptive as their enemies. There is no "silver bullet" set of
1024 procedures for COIN. Constantly developing new practices is essential.

Tactical Success Guarantees Nothing

1026 1-131. When COL Harry Summers allegedly told a North Vietnamese counterpart in 1975 that "You
1027 know you never defeated us on the battlefield," the reply supposedly was, "That may be so, but it is also ir-
1028 relevant." Military actions by themselves cannot achieve success in COIN. Tactical actions must not only
1029 be linked to operational and strategic military objectives, but also to the essential political goals of COIN.
1030 Without those connections, lives and resources may be wasted for no real gain.

Most of the Important Decisions are Not Made By Generals

1032 1-132. Successful COIN relies on the competence and judgment of Soldiers and Marines at all levels.
1033 Senior leaders set the proper tone for actions by their organizations with thorough training and clear guid-
1034 ance, and then trust their subordinates to do the right thing. Preparation for tactical-level leaders requires
1035 more than Service doctrine; they must also be trained and educated to adapt to their local situations, under-
1036 stand the legal and ethical implications of their actions, and exercise subordinates' initiative and sound
1037 judgment to meet their senior commanders' intent.

SUCCESSFUL AND UNSUCCESSFUL PRACTICES

1039 1-133. Figure 1-3 (below) lists some practices that have contributed significantly to success or failure in
1040 past counterinsurgencies.

1041

Successful Practices	Unsuccessful Practices
• Emphasize intelligence. • Focus on the population, their needs, and security. • Establish and expand secure areas. • Isolate insurgents from the population (population control). • Appoint a single authority, usually a dynamic, charismatic leader. • Conduct effective, pervasive psychological operations. • Provide amnesty and rehabilitation for insurgents. • Place police in the lead with military support. • Expand and diversify the police force. • Train military forces to conduct counterinsurgency operations. • Embed special operations forces and advisors with indigenous forces. • Deny the insurgents sanctuary.	• Place priority on killing and capturing the enemy, not on engaging the population. • Conduct battalion-sized operations as the norm. • Concentrate military forces in large bases for protection. • Focus special operations forces primarily on raiding. • Place a low priority on assigning quality advisors to host-nation forces. • Build and train host-nation security forces in the of the U.S. Army's image. • Ignore peacetime government processes, including legal procedures. • Allow open borders, airspace, and coastlines.

1042   **Figure 1-3. Successful and unsuccessful counterinsurgency operational practices**

1043   # SUMMARY

1044   1-134.   Counterinsurgency is an extremely complex form of warfare—truly "war at the graduate level." At
1045   its core, counterinsurgency warfare is a struggle for the support of the population. Their protection and
1046   welfare is the center of gravity for friendly forces.

1047   1-135.   Gaining and maintaining that support is a formidable challenge. There are a host of non-military
1048   agencies whose efforts must be synchronized and coordinated with those of the military to achieve these
1049   aims. Popular support allows counterinsurgents to develop the intelligence necessary to identify and defeat
1050   the insurgents. Intelligence drives operations in successful counterinsurgency operations; those operations
1051   in turn provide additional intelligence, driving further operations. Designing and then executing a cam-
1052   paign to garner the support of the population requires careful coordination of several logical lines of opera-
1053   tions to produce success over time. One of these lines is developing host nation security forces who can
1054   take over primary responsibility for combating the insurgency. COIN also places distinct burdens on lead-
1055   ers and logisticians. All of these aspects of COIN are described and analyzed in the chapters that follow.

1056   1-136.   Insurgents are also fighting for the support of the population, but are constrained by neither the
1057   law of war nor the bounds of human decency. They do anything to preserve their greatest advantage, the
1058   ability to hide among the people. These amoral enemies survive by their wits, constantly adapting to the
1059   situation. To defeat them, it is essential that counterinsurgent forces develop the ability to learn and adapt
1060   rapidly and continuously. This manual focuses on this "learn and adapt" counterinsurgency imperative as it
1061   discusses ways to gain and maintain the support of the people.

Chapter 2

Unity of Effort: Integrating Civilian and Military Activities

Essential though it is, the military action is secondary to the political one, its primary purpose being to afford the political power enough freedom to work safely with the population.

David Galula, *Counterinsurgency Warfare*, 1964

This chapter begins by discussing the principles involved in integrating the activities of military and civilian organizations. It then describes the categories of organizations usually involved in counterinsurgency operations. After that, it discusses how responsibilities may be assigned and the mechanisms used to integrate civilian and military activities. It concludes by listing information commanders need to know about civilian agencies operating in their area of operations.

INTEGRATION

2-1.   Although military efforts are necessary and important, they are only effective if integrated into a comprehensive strategy employing all instruments of national power. A successful counterinsurgency (COIN) meets the contested population's needs while protecting it from the insurgents. Effective COIN operations ultimately render the insurgents irrelevant. This requires military forces engaged in COIN to—

- Be acutely aware of the roles and capabilities of U.S., international, and host-nation (HN) partners.
- Include other participants in planning at every level.
- Be supportive of civilian efforts.
- If necessary, conduct or participate in political, social, informational, and economic programs.

2-2.   The integration of civilian and military efforts is crucial in COIN and must be focused on supporting the local population and the HN government. Political, social, and economic programs are usually more valuable than conventional military operations as a means to address root causes of conflict and undermine an insurgency. In COIN, military personnel, diplomats, police, politicians, humanitarian aid workers, contractors, and local leaders are faced with making decisions and solving problems in a complex and acutely challenging environment. Controlling the level of violence is a key element of the struggle. The insurgents often benefit from a high level of violence and societal insecurity that discourages or precludes nonmilitary participants from helping the local population. The higher the level of violence that defines the operational environment, the less likely it is that nonmilitary organizations, particularly external agencies, can work with the local population to address social, political, economic and other challenges. The more benign the security environment, the more likely it is that civilian agencies can provide their resources and expertise, and relieve the burden on the military forces.

2-3.   Organized combat is an expertise that only military forces are required to possess. Military forces do apply their combat skills to fighting insurgents; however, they can and should be engaged in using their capabilities to meet the local population's fundamental needs as well. Only regaining active and continued support of the HN government by the local population can deprive an insurgency of its power and appeal. The military forces' primary function in COIN is protecting that population. However, military force is not the sole means to provide security or defeat insurgents. Indeed, a dilemma for military forces engaged in COIN is that they frequently have greater potential to undermine policy objectives through excessive emphasis on military force than to achieve the overarching political goals that define success. This dilemma places tremendous importance on the measured application of coercive force by COIN forces.

2-4. Durable policy success requires balancing the measured use of force with an emphasis on nonmilitary programs. Although political, social, and economic programs are most commonly and appropriately associated with civilian organizations and expertise, the salient aspect of such programs is their effective implementation, not who performs the tasks. If adequate civilian capacity is not available, members of the military forces must be prepared to fill the gap. COIN programs for political, social, and economic well-being are essential elements for developing local capacity that will command popular support when properly perceived. COIN is also a battle of ideas, and insurgents exploit actions and programs to create misperceptions and further their cause. Comprehensive informational programs are necessary to amplify the messages of positive deeds and to counter insurgent propaganda.

2-5. COIN is fought among the population, and the counterinsurgents bear responsibility for the people's well-being in all its manifestations. These include the following:

- Security from violence and crime.
- Provision of basic economic needs.
- Maintenance of infrastructure.
- Sustainment of key social and cultural institutions.
- Other aspects that contribute to a society's basic quality of life.

The COIN program must address all aspects of the local population's concerns in a unified fashion. Insurgents succeed by maintaining turbulence and highlighting local costs due to gaps in the COIN effort. COIN forces success by eliminating turbulence and meeting the population's basic needs.

2-6. When the United States commits to assisting a host nation against an insurgency, success requires the application of national resources along multiple lines of operations. (See chapter 5.) The fact that efforts along one line of operations can easily affect progress in others means that uncoordinated actions are frequently counterproductive.

2-7. Lines of operations in COIN focus primarily on the population. Each line is dependent on the others. Their interdependence is similar to factors in a multiplication equation: if the value of one of the lines of operations is zero, the overall product is zero. Many of these lines of operations require the application of expertise usually found in civilian organizations. These include—

- U.S. government agencies other than the Department of Defense (DOD).
- International organizations (such as the United Nations and its many suborganizations).
- Nongovernmental organizations (NGOs).
- Private corporations.
- Other organizations that wield diplomatic, informational, and economic power.

These civilian organizations bring expertise and capabilities that complement those of military forces engaged in COIN operations. At the same time, civilian capabilities cannot be brought to bear without the security provided by military forces. The interdependent relationship of all these groups must be understood and orchestrated to achieve harmony of action and coherent results.

UNITY OF COMMAND

2-8. Where possible, formal relationships should be established and maintained for unity of command. For all elements of the U.S. government engaged in a particular COIN mission, formal command and control using established command relationships within a clear hierarchy should be axiomatic. Unity of command should also extend to all military forces supporting a host nation. The ultimate objective of these arrangements is for local military forces, police, and other security forces to establish effective command and control while attaining a monopoly on the legitimate use of violence within the society.

2-9. As important as the principle of unity of command is to military operations, it is one of the most difficult and sensitive issues to resolve in COIN. U.S. and other external military participation in COIN are inherently problematic, as it influences perceptions of the capacity and legitimacy of local security forces. Although unity of command of military forces may be desirable, it may be impractical due to political considerations. Political sensitivities about the perceived subordination of national forces to those of other states or international organizations often preclude strong command relationships. The differing goals and

93 fundamental independence of NGOs and local organizations frequently prevent formal relationships. In the
94 absence of formal relationships governed by command authority, military leaders seek to persuade and in-
95 fluence other participants to contribute to attaining COIN objectives. Informal or less authoritative rela-
96 tionships include coordination and liaison with other participants. In some cases, direct interaction among
97 various organizations may be impractical or undesirable. Basic awareness and general information sharing
98 may be the most that can be accomplished.

99 ## UNITY OF EFFORT

100 2-10. Informed and strong leadership is a foundation of successful COIN operations. The focus of leader-
101 ship must be on the central problems that affect the local population. A clear vision of resolution should in-
102 fuse all efforts, regardless of the specific agencies or individuals charged with their execution. (See chapter
103 4.) All elements supporting the COIN should strive for maximum unity of effort. Given the primacy of po-
104 litical considerations, military forces often support civilian efforts. However, the mosaic nature of COIN
105 operations means that lead responsibility shifts among military, civilian, and HN authorities. Regardless,
106 military leaders must be prepared to assume local leadership for COIN efforts and remember that the orga-
107 nizing imperative is to focus on what needs to be done, not on who does it.

108 2-11. Countering an insurgency begins with understanding the complex environment and the numerous
109 competing forces acting upon it. Gaining an understanding of the environment—to include the insurgents,
110 affected populace, and disparate organizations attempting to counter the insurgency—is essential to an in-
111 tegrated COIN operation. The complexity of resolving the causes of the insurgency and integrating actions
112 across multiple and interrelated lines of operations requires an understanding of the civilian and military
113 capabilities, activities, and vision of resolution. Just as Soldiers and Marines use different tactics to achieve
114 an objective, so the various agencies acting to re-establish stability may differ in goals and approaches.
115 When their actions are allowed to adversely impact each other, the population suffers and insurgents iden-
116 tify gaps to exploit. Integrated actions are essential to defeat the ideologies professed by insurgents. A
117 shared understanding of the operation's purpose provides a unifying theme for COIN efforts. Through a
118 common understanding of that purpose, the COIN team can design an operation that promotes effective
119 collaboration and coordination among all agencies and the affected population.

120 ## COORDINATION AND LIAISON

121 2-12. A vast array of organizations can influence successful COIN operations. Given the complex diplo-
122 matic, informational, military, and economic context of an insurgency, there is no way for military leaders
123 to assert command over all elements—nor should they try to do so. Among interagency partners, NGOs,
124 and private organizations, there are many interests and agendas that military forces will be unable to con-
125 trol. Additionally, local legitimacy is frequently affected by the degree to which local institutions are per-
126 ceived as independent and capable without external support. Nevertheless, military leaders should make
127 every effort to ensure that actions in support of the COIN are as well integrated as possible. Active leader-
128 ship by military leaders is imperative to conduct coordination, establish liaison (formal and informal) and
129 share information among various groups working on behalf of the local population. Influencing and per-
130 suading groups beyond a commander's direct control requires great skill and often great subtlety. As ac-
131 tively as commanders may pursue unity of effort, they should also be mindful of the visibility of their role
132 and recognize the wisdom of acting indirectly and in ways that allow credit for success to go to others—
133 particularly local individuals and organizations.

134 2-13. Local leaders, informal associations, families, tribes, some private enterprises, some humanitarian
135 groups, and the media often play critical roles in influencing the outcome of a COIN but are beyond the
136 control of military forces or civilian governing institutions. Commanders remain aware of the influence of
137 such groups and are prepared to work with, through, or around them.

138
139

"Hand Shake Con" in Provide Comfort

140
141
142
143
144
145
146
147
148

[Regarding relationships with other multinational contingents, t]he Chairman of the Joint Chiefs of Staff asked me..."The lines in your command chart, the command relationships, what are they? OPCON? TACON? Command?" "Sir, we don't ask, because no one can sign up to any of that stuff." "Well, how do you do business?" "Hand Shake Con. That's it." No memoranda of agreement. No memoranda of understanding...the relationships are worked out on the scene, and they aren't pretty. And you don't really want to try to capture them...distill them, and say as you go off in the future, you're going to have this sort of command relationship...it is Hand Shake Con and that's the way it works. It is consultative. It is behind-the-scene.

149

General Anthony Zinni, 1994

KEY COUNTERINSURGENCY PARTICIPANTS AND THEIR LIKELY ROLES

150
151

152   2-14. Likely participants in COIN operations include the following:
153   ● U.S. military forces.
154   ● Multinational forces.
155   ● U.S. government agencies.
156   ● Nongovernmental organizations.
157   ● International organizations.
158   ● Multinational corporations and contractors.
159   ● HN civil authorities.

160   ### U.S. MILITARY FORCES

161   2-15. The role of military forces in COIN operations is extensive. As one of the most demanding and com-
162   plex forms of warfare, COIN draws heavily on the broad range of capabilities of the joint force. Military
163   forces must be prepared to conduct a different mix of offensive, defensive and stability operations from
164   that expected in major combat operations. Air, land, and maritime elements all contribute to successful op-
165   erations and to the vital effort to cut off and isolate insurgents from the populations they seek to control.
166   Nonetheless, the Army and Marine Corps usually furnish the principal U.S. military contributions to
167   COIN. Within the land forces, special operations forces (SOF) are particularly valuable to COIN due to
168   their specialized capabilities: civil affairs, psychological operations, intelligence, language skills, and re-
169   gional-specific knowledge. SOF can also provide very light, agile, high-capability teams that can operate
170   discreetly in local communities.

171   2-16. U.S. military forces are vastly capable. Designed predominantly for conventional warfare, they none-
172   theless have the essential components to successfully prosecute COIN. The most important assets in COIN
173   are disciplined Soldiers and Marines with adaptive, self-aware, and intelligent leaders. There are also ele-
174   ments of the military forces that are particularly relevant to common COIN challenges. For example, COIN
175   often requires dismounted infantry, human intelligence, language specialists, military police, civil affairs,
176   engineers, medical units, logistical support, legal affairs, and contracting elements. All of these elements
177   are found in the Army; most are be found in the Marine Corps as well. To a limited degree, they are also
178   found in the Air Force and Navy.

179   2-17. U.S. forces help HN military, paramilitary, and police forces conduct COIN operations, including
180   area security and local security operations. U.S. forces advise and assist in finding, dispersing, capturing,
181   and defeating the insurgent force, while emphasizing the training of HN forces to perform essential defense
182   functions. These are the central tasks of foreign internal defense, which until recently was primarily the re-

183 sponsibility of SOF. The current and more extensive national security demands for such efforts require that
184 all Services be prepared to contribute to the establishment and training of local security forces.

185 2-18. Land combat forces conduct full spectrum operations to disrupt or destroy insurgent military capa-
186 bilities. Land forces use offensive combat operations to disrupt insurgent efforts to establish base areas and
187 consolidate their forces. They conduct defensive operations to provide area and local security. They con-
188 duct stability operations to thwart insurgent efforts to control or disrupt people's lives and routine activi-
189 ties. In all applications of combat power, U.S. forces first ensure that likely costs do not outweigh or un-
190 dermine other more important COIN efforts.

191 2-19. Most valuable to long-term success in winning the support of the population are the contributions
192 land forces can make by conducting stability operations. A *stability operation* is an operation to establish,
193 preserve, and exploit security and control over areas, populations, and resources (FM 3-0). Forces engaged
194 in stability operations establish, safeguard, or restore basic civil services. They act directly and in support
195 of governmental agencies. Success in stability operations enables the local population and government
196 agencies of the host nation to resume or develop the capabilities needed to conduct COIN operations and
197 create the conditions that will permit U.S. military forces to disengage.

198 2-20. Military forces can also use their capabilities to enable the efforts of nonmilitary participants. Logis-
199 tics, transportation, equipment, personnel, and other assets can support interagency partners and other civil-
200 ian organizations.

MULTINATIONAL MILITARY FORCES

202 2-21. U.S. military forces rarely operate alone. They normally function as part of a multinational force. In
203 a COIN operation, U.S. forces usually operate with the security forces of the local population or host na-
204 tion. Few multinational military forces have the comprehensive capabilities possessed by the United States,
205 but many possess significant similar capabilities that can contribute to COIN. Many other countries' mili-
206 tary forces bring cultural backgrounds, historical experiences, and other capabilities that can be particularly
207 valuable to COIN efforts.

208 2-22. However, nations join coalitions for varied policy aims. Although missions may be ostensibly simi-
209 lar, rules of engagement, home-country policies, and sensitivities will differ among multinational partners.
210 U.S. military leaders require a strong cultural and political awareness of HN and other multinational mili-
211 tary partners.

NONMILITARY COUNTERINSURGENCY PARTICIPANTS

213 2-23. Many nonmilitary organizations may be engaged in supporting a host nation as it confronts an insur-
214 gency. Some of these organizations are discussed below. JP 3-08 contains a detailed description of all or-
215 ganizations listed here as well as many others.

U.S. Government Organizations

217 2-24. Commanders must be familiar with other U.S. government organizations and aware of the capabili-
218 ties they can provide for COIN. During planning, all forces should determine which organizations are sup-
219 porting their function or operating in their AO. Commanders and civilian leaders of U.S. government or-
220 ganizations should collaboratively plan and coordinate actions to avoid conflict or duplication of effort.
221 Within the U.S. government, key organizations include—

222 • Department of State.
223 • U.S. Agency for International Development (USAID).
224 • Central Intelligence Agency.
225 • Department of Justice.
226 • Department of the Treasury.
227 • Department of Homeland Security.
228 • Department of Agriculture.

229   ●   Department of Commerce.

230   ●   Department of Transportation.

231   ●   The U.S. Coast Guard (under Department of Homeland Security).

232   ●   The Federal Bureau of Investigation (under Department of Justice).

233   ●   Immigration Customs Enforcement (under Department of Homeland Security).

234   Nongovernmental Organizations

235   2-25. Joint doctrine defines a *nongovernmental organization* as a private, self-governing, not-for-profit or-
236   ganization dedicated to alleviating human suffering; and/or promoting education, health care, economic
237   development, environmental protection, human rights and conflict resolution; and/or encouraging the es-
238   tablishment of democratic institutions and civil society (JP 1-02). There are several thousand NGOs of
239   many different types. Their activities are governed by their organizing charters and the motivations of their
240   members. Some NGOs receive at least part of their funding from national governments or international
241   governmental organizations and may become implementing partners in accordance with specific grants or
242   contracts. (For example, USAID provides some NGO funding.) In these cases, the funding organization of-
243   ten gains oversight and authority over how the funds are used.

244   2-26. Some NGOs maintain strict independence from governments and other belligerents in a conflict and
245   do not want to be seen directly associating with military forces. Gaining their support and coordinating op-
246   erations with them can be difficult. While establishing basic awareness of these groups and their activities
247   may be the most commanders can achieve, it is important to note the role of NGOs in resolving an insur-
248   gency. Many of these NGOs arrive before military forces and remain long afterwards. They can support
249   lasting stability. To the greatest extent possible, the military should complement and not override their ca-
250   pabilities. Building a complementary and trust-based relationship is vital. Examples of NGOs include—

251   ●   International Committee of the Red Cross.

252   ●   World Vision.

253   ●   Médecins sans Frontières (Doctors Without Borders).

254   ●   Cooperative for Assistance and Relief Everywhere (CARE).

255   ●   Oxford Committee for Famine Relief (OXFAM).

256   ●   Save the Children.

257   ●   Mercy Corps.

258   ●   Academy for Educational Development.

259   International Organizations

260   2-27. The most notable international organization is the United Nations (UN). Regional organizations like
261   the Organization of American States and the European Union may be involved in some COIN operations.
262   The UN in particular has many subordinate or affiliated agencies that are active around the world. Depend-
263   ing on the nature of the situation and the needs of the country involved, any of several UN organizations
264   may be present, such as the following:

265   ●   Office of the Chief of Humanitarian Affairs.

266   ●   Department of Peacekeeping Operations.

267   ●   World Food Program.

268   ●   UN Refugee Agency (known by the acronym for its director, the UN High Commissioner for
269   Refugees—UNHCR).

270   ●   UN High Commissioner for Human Rights.

271   ●   UN Development Program.

272   Multinational Corporations and Contractors

273   2-28. Multinational corporations often engage in reconstruction, economic development, and governance
274   activities. At a minimum, commanders should know which companies are present in their AO and where

275
276

those companies are conducting business. Such information can prevent fratricide or destruction of private property.

277
278
279
280
281
282

2-29. Recently, private contractors from firms providing military-related services have become more prominent in theaters of operations. This category includes armed contractors providing an array of security services to the U.S. government, NGOs, and private businesses. There are also many businesses that market expertise in areas related to supporting governance, economics, education, and other aspects of civil society. Providing capabilities similar to some NGOs, these firms often attain contracts through government agencies.

283
284
285
286
287
288

2-30. When contractors or other businesses are being paid to support U.S. military or other government agencies, the principle of unity of command should apply. Commanders should be able to influence contractors' performance through U.S. government contract supervisors. When under contract to the United States, many such contractors should behave as an extension of the organizations or agencies for which they work. Commanders should identify contractors operating in their AO and determine the nature of their contract, existing accountability mechanisms, and appropriate coordination relationships.

289

Host-nation Civil Authorities

290
291
292
293
294
295
296
297
298

2-31. Sovereignty issues are among the most difficult for commanders conducting COIN operations, both in regard to forces contributed by other nations and by the host nation. Often, commanders are required to lead through coordination, communication, and consensus, in addition to traditional command practices. Political sensitivities must be acknowledged. Commanders and subordinates often act as diplomats as well as warriors. Within military units, legal officers and their staffs are particularly valuable for clarifying legal arrangements with the host nation. To avoid adverse effects on operations, commanders should address all sovereignty issues through the chain of command to the U.S. Ambassador. As much as possible, sovereignty issues should be addressed before executing operations. Examples of key sovereignty issues include the following:

299
300
301
302
303
304
305
306
307
308
309

- Collecting and sharing information.
- Basing.
- Overflight rights.
- Aerial ports of debarkation.
- Seaports of debarkation.
- Location and access.
- Railheads.
- Border crossings.
- Force protection.
- Jurisdiction over members of the U.S. and multinational forces.
- Operations in the territorial sea and internal waters.

310
311
312
313
314
315
316
317

2-32. Commanders should create coordinating mechanisms, such as committees or liaison elements, to facilitate cooperation and build trust with HN authorities. Military or nonmilitary representatives of the host nation should have leading roles in such mechanisms. These organizations facilitate operations by reducing sensitivities and misunderstandings while removing impediments. Sovereignty issues can be formally resolved with the host nation through the development of appropriate technical agreements to augment existing or recently developed status of forces agreements. In many cases, security assistance organizations, NGOs, and international organizations have detailed local knowledge and reservoirs of good will that can facilitate the establishment of a positive and constructive relationship with the host nation.

318
319
320
321
322
323

2-33. Coordination and support should exist down to local levels (such as villages and neighborhoods). Soldiers and Marines should be aware of the political and societal structures in their AOs. Political structures usually have designated leaders responsible to the government and people. However, the societal structure may include informal leaders who operate outside the political structure. These leaders may be economic (such as businessmen), theological (such as clerics and lay leaders), informational (such as newspaper publishers or journalists), or family-based (such as elders or patriarchs). Some societal leaders may

324
325

emerge due to charisma or other intangible influences. Commanders should identify the key leaders and the manner in which they are likely to influence COIN efforts.

326

KEY RESPONSIBILITIES IN COUNTERINSURGENCY

327
328
329
330

2-34. David Galula wisely notes, "To confine soldiers to purely military functions while urgent and vital tasks have to be done, and nobody else is available to undertake them, would be senseless. The soldier must then be prepared to become...a social worker, a civil engineer, a schoolteacher, a nurse, a boy scout. But only for as long as he cannot be replaced, for it is better to entrust civilian tasks to civilians."

331

PREFERRED DIVISION OF LABOR

332
333
334
335
336
337
338
339
340

2-35. The preference in COIN is always to have civilians carry out civilian tasks. Civilian agencies or individuals with the greatest expertise for a given task should perform it—with special preference for legitimate local civil authorities. Although there are many U.S. and other international civilian agencies that possess greater expertise than military forces for meeting the fundamental needs of a population under assault, the ability of such agencies to deploy to foreign countries in sustainable numbers and with ready access to necessary resources is usually limited. The degree of violence in the AO also affects the ability of civilian agencies to operate. The more violent the environment, the more difficult it is for civilians to operate effectively. Hence, the preferred or ideal division of labor is frequently unattainable. The more violent the insurgency, the more unrealistic is this preferred division of labor.

341

REALISTIC DIVISION OF LABOR

342
343
344
345
346
347

2-36. By default, U.S. and other military forces often possess the only readily available capability to meet many of the fundamental needs of local populations. Human decency and the law of war require land forces to assist populations in their AOs. Leaders at all levels prepare to address civilian needs. Leaders identify people in their units with regional expertise, interagency know-how, civil-military competence, and other critical skills that can usefully support a local population and HN government. Useful skill sets may include the following:

348
349
350
351
352
353
354

- Knowledge, cultural understanding, and appreciation of the host nation and its region.
- Functional skills for interagency and HN coordination (for example, liaison, negotiation, and appropriate social or political relationships).
- Language skills enabling more effective coordination with the host nation, NGOs, and multinational partners.
- Knowledge of the civil foundations for infrastructure, economy, governance, or other lines of operation being pursued as part of the COIN effort.

355
356
357
358
359
360
361

2-37. U.S. government and international agencies rarely have the resources and capabilities needed to address all tasks required in a COIN environment. Successfully accomplishing a COIN mission requires leaders who are adaptive in their thinking and understand that they must be prepared to address the required tasks with available resources. They must understand that the overall purpose is always guided by the fact that long-term security does not come solely or even primarily from a gun; it comes from an integrated and balanced application of efforts along all lines of operations, with the goal of supporting the local population and achieving legitimacy for the HN government.

362

TRANSITIONS

363
364
365
366
367
368
369
370

2-38. Regardless of the division of labor, an important recurring feature of COIN is the transition of responsibility and participation in key lines of operations. As consistently and conscientiously as possible, military leaders ensure continuity in meeting the needs of the HN government and the local population. The same general guidelines that govern battle handovers apply to transitions between participants in COIN. Whether the transition is from one military unit to another or from a military unit to civilian agencies (external or local), everyone involved requires a clear understanding of tasks and responsibilities to be passed from one element to the next. Maintaining unity of effort is particularly important during transitions, especially between elements of different capabilities and capacity. Relationships tend to break down during transitions. Transition is not a single event, where all activity happens at once. It is a rolling process of little handoffs between different actors along several streams of activities, and there are usually multiple transitions for any one stream of activity over time. Using the coordination mechanisms in the next section can help to create and sustain the linkages that support effective transitions without compromising unity of effort.

CIVILIAN AND MILITARY INTEGRATION MECHANISMS

2-39. Applying the principle of unity of effort is possible in many organizational forms. The first choice should be to identify existing coordination mechanisms and incorporate them into comprehensive COIN efforts. This includes existing U.S. government, multinational, and HN mechanisms. Context is extremely important. Although many of these structures exist and are often employed in other types of missions (such as peacekeeping or humanitarian relief), there is an acute and fundamental difference in an insurgency environment. The nature of the conflict and its focus on the population makes civilian and military unity a decisive point for the insurgents. The following discussion highlights some of the well established, general mechanisms for civilian and military integration. Many civil-military organizations and mechanisms have been created for specific missions. Although the names and acronyms differ, in their general outlines they usually reflect the concepts portrayed in the following sections.

2-40. The U.S. government influences events worldwide through the effective employment of the instruments of national power. These instruments are diplomatic, informational, military and economic and are coordinated by the appropriate executive branch officials, often with National Security Council assistance.

2-41. The National Security Council is the President's principal forum for considering national security and foreign policy matters. It serves as the President's principal means for coordinating policy among various interagency organizations. At the strategic level, the National Security Council creates the interagency political-military plan for COIN. (See JP 1.)

JOINT INTERAGENCY COORDINATION GROUP

2-42. Joint interagency coordination groups (JIACGs) help combatant commanders in the conduct COIN operations by providing interagency support of plans, operations, contingencies, and initiatives. The goal of a JIACG is to provide timely, usable information and advice from an interagency perspective to the combatant commander by information sharing, integration, synchronization, training, and exercising. JIACGs may include representatives from other federal departments and agencies, state and local authorities, and liaison officers from other commands and DOD components. The interagency representatives and liaison officers are the subject matter experts for their respective agencies and commands, and provide the critical bridge between the combatant commander and interagency organizations. (See JP 3-08, Vol I.)

COUNTRY TEAM

2-43. At the HN level, the U.S. country team is the primary interagency coordinating structure for COIN. (See figure 2-1.) The country team is the senior in-country coordinating and supervising body, headed by the U.S. chief of mission—usually the Ambassador—and composed of the senior member of each represented department or agency. In a foreign country, the chief of mission is the highest U.S. civil authority. The Foreign Service Act states, "The Chief of Mission to a foreign country has responsibility for the direction, coordination, and supervision of all Government executive branch employees in that country except for service-members and employees under the command of a United States area military commander. The chief of mission is the senior United States Government official permanently assigned in the host country, and as such is responsible to the president for policy oversight of all United States government programs." The chief of mission leads the country team and is responsible for integrating U.S. efforts in support of the host nation. As permanently established interagency organizations, country teams represent a priceless COIN resource, with deep reservoirs of local knowledge and interaction with the HN government and population.

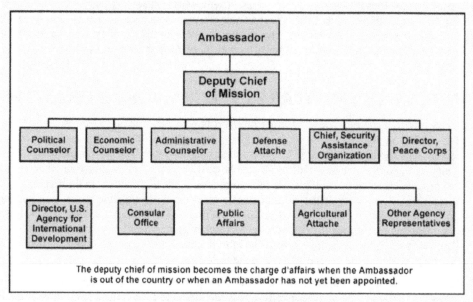

The deputy chief of mission becomes the charge d'affairs when the Ambassador is out of the country or when an Ambassador has not yet been appointed.

417   **Figure 2-1. Sample country team**

418 2-44. The more extensive the U.S. participation is in a COIN and the more dispersed U.S. forces are
419 throughout a country, the greater the need for additional mechanisms to extend civilian oversight and assis-
420 tance to match the scope of the challenge. However, given the limited resources of the Department of State
421 and the other U.S. government agencies, military forces often represent the country team in decentralized
422 and diffuse conflict environments. Operating with a clear understanding of the guiding political aims,
423 members of the military at all levels must be prepared to exercise judgment and act without the benefit of
424 immediate civilian oversight and control. At each subordinate political level of the HN government, mili-
425 tary and civilian leaders should establish a coordinating structure, such as a civil-military operations center
426 (CMOC), that includes representatives of the HN government and security forces, as well as U.S. and mul-
427 tinational forces and agencies. CMOCs facilitate the integration of military and political actions. Below the
428 national level, additional structures may be established where military commanders and civilian leaders can
429 meet directly with local leaders to discuss issues. Where possible, international organizations and NGOs
430 should be encouraged to participate in coordination meetings to ensure their actions are integrated with
431 military and HN plans.

432 2-45. In practice, the makeup of country team varies widely, depending on the U.S. departments and agen-
433 cies represented in country, the desires of the Ambassador and the HN situation. During COIN, members
434 of the country team meet regularly to coordinate U.S. government diplomatic, informational, military, and
435 economic activities in the host nation to ensure unity of effort. The interagency representatives usually in-
436 clude at least the Departments of State, Defense, Justice and Treasury; USAID; Central Intelligence
437 Agency; and Drug Enforcement Administration. Participation of other U.S. government organizations de-
438 pends on the situation.

439 2-46. In almost all bilateral missions, DOD is represented on the country team by the U.S. defense atta-
440 ché's office or the security assistance organization. They are key military sources of information for inter-
441 agency coordination in foreign countries. (Security assistance organizations are called by various names,
442 such as the Office of Defense Cooperation, Security Assistance Office, or Military Group. The choice is
443 largely governed by the preference of the host nation.)

444
445

Provincial Reconstruction Teams in Afghanistan

446
447
448
449
450
451
452
453
454
455
456
457
458
459
460
461

A model for civil-military cooperation is the provincial reconstruction teams (PRTs) first fielded in 2003 in Afghanistan. PRTs were conceived as a means to extend the reach and enhance the legitimacy of the central government into the provinces of Afghanistan at a time when most assistance was limited to the nation's capital. Though PRTs were staffed by a number of coalition and NATO allied countries, they generally consisted of 50 to 300 troops as well as representatives from multinational development and diplomatic agencies. Within U.S. PRTs, USAID and State Department leaders and the PRT commander formed a senior team that coordinated the policies, strategies, and activities of each agency towards a common goal. In secure areas, PRTs maintained a low profile. In areas where coalition combat operations were underway, PRTs worked closely with maneuver units and local government entities to ensure that shaping operations achieved their desired effects. Each PRT leadership team received tremendous latitude to determine its own strategy. However, each PRT used its significant funding and diverse expertise to pursue activities that fell into one of three general lines of operations: pursue security sector reform, build local governance, or execute reconstruction and development.

462
463
464
465
466
467
468

2-47. The country team determines how the United States can effectively apply interagency capabilities to assist a HN government in creating a complementary institutional capacity to deal with an insurgency. Efforts to support local officials and build HN capacity must be integrated with information operations so HN citizens are aware of their government's efforts. In addition, interagency capabilities must be applied down to the tactical level so commanders have access to the options that such capabilities make available. The CORDS approach developed during the Vietnam War provides a positive example of integrated civilian and military structures that reached every level of the COIN effort.

469
470
471

Civil Operations and Rural Development Support and Accelerated Pacification in Vietnam (1967-72)

472
473
474
475
476
477
478

During the Vietnam War, one of the most valuable and successful elements of COIN was the Civil Operations and Revolutionary—later Rural—Development Support (CORDS) program. CORDS was created in 1967 to integrate U.S. civilian and military support of the South Vietnamese government and people. CORDS achieved great success in supporting and protecting the South Vietnamese population and undermining the communist insurgents' influence and appeal, particularly after the implementation of accelerated pacification in 1968.

479
480
481
482
483
484
485
486
487
488
489

Pacification was the process by which the government asserted its influence and control into an area beset by insurgents. This included local security efforts, programs to distribute food and medical supplies, and lasting reforms (like land redistribution). In 1965, U.S. civilian contributions to pacification consisted of several civilian agencies (among others, the Central Intelligence Agency, Agency for International Development, U.S. Information Service, and State Department). Each developed its own programs. Coordination was uneven. The U.S. military contribution to pacification consisted of thousands of advisors. By early 1966, there were military advisory teams in all of South Vietnam's 44 provinces and most of its 243 districts. But there were two separate chains of command for military and civilian pacification efforts, making it particularly difficult for the civilian-run pacification program to function.

490
491
492

To establish closer integration of civilian and military efforts, in 1967 President Lyndon B. Johnson established CORDS within General William Westmoreland's Military Assistance Command, Vietnam (MACV). Robert Komer was appointed to run the

493
494
495
496
497
498

program, with a three-star-equivalent rank. Civilians, including an assistant chief of staff for CORDS, were integrated into military staffs at all levels. This placed civilians in charge of military personnel and resources. Komer was energetic, strong-willed, and persistent in getting the program started. Nicknamed "Blowtorch" for his aggressive style, Komer was modestly successful in leading improvements in pacification before the 1968 Tet offensive.

499
500
501
502
503
504
505

In mid-1968 the new MACV Commander, General Creighton Abrams, and his new civilian deputy, William Colby, used CORDS as the implementing mechanism for an accelerated pacification program that became the priority effort for the United States. Significant allocations of personnel helped make CORDS effective. In this, the military's involvement was key. In September 1969—the highpoint of the pacification effort in terms of total manpower—there were 7,601 advisors assigned to province and district pacification teams. Of these 6,464 were military.

506
507
508
509
510
511
512
513
514
515
516
517
518
519
520

The effectiveness of CORDS was a function of integrated civilian and military teams at every level of society in Vietnam. From district to province to national level, U.S. advisors and U.S. interagency partners worked closely with their Vietnamese counterparts. The entire effort was well established under the direction of the country team, led by Ambassador Ellsworth Bunker. General Abrams and his civilian deputy were clear in their focus on pacification as the priority and ensured the military and civilian agencies worked closely together. Keen attention was given to the ultimate objective of serving the needs of the local population. Success in meeting basic needs of the population led, in turn, to improved intelligence that facilitated an assault on the Viet Cong political infrastructure. By early 1970, statistics indicated that 93 percent of South Vietnamese lived in "relatively secure" villages, an increase of almost 20 percent from the middle of 1968. By 1972, pacification had largely uprooted the insurgency from among the South Vietnamese population and forced the communists to rely more heavily on infiltrating conventional forces from North Vietnam and employing them in irregular and conventional operations.

521
522
523
524
525

In 1972, South Vietnamese forces operating with significant support from U.S. airpower, defeated large-scale North Vietnamese conventional attacks. Unfortunately, a North Vietnamese conventional assault succeeded in 1975 after the withdrawal of U.S. forces, ending of U.S. air support, and curtailment of U.S. funding to South Vietnam.

526
527
528

Pacification, once it was integrated under CORDS, was generally led, planned, and executed well. CORDS was a successful synthesis of military and civilian efforts. It is a useful model to consider for other COIN operations.

529 ## CIVIL-MILITARY OPERATIONS CENTER

530
531
532
533
534

2-48. Another mechanism for bringing elements together for coordination is the CMOC. CMOCs can be established at all levels of command. CMOCs coordinate the interaction of U.S. and multinational military forces with a wide variety of civilian agencies. A CMOC is not designed, nor should it be used as, a command and control element. However, it does provide a useful venue to transmit the commander's guidance to other agencies and exchange information, and to facilitate complementary and unified efforts.

535
536
537
538
539
540
541

2-49. Overall management of a CMOC may be assigned to a multinational force commander, shared by a U.S. and a multinational commander, or shared by a U.S. commander and a civilian agency head. A CMOC is an operations center that can be used to build on-site, interagency coordination to achieve unity of effort. There is no established structure for a CMOC; its size and composition depend on the situation. Senior civil affairs officers normally serve as the CMOC director and deputy director. Other military participants usually include civil affairs, legal, operations, logistics, engineering, and medical sections of the supported headquarters. Civilian members of a CMOC may include U.S. government organizations, multi-

542   national partners, international organizations, HN or other local organizations, and NGOs. (For more in-
543   formation on CMOCs, see FM 41-10.)

544   # TACTICAL-LEVEL INTERAGENCY CONSIDERATIONS

545   2-50. Tactical units may find interagency expertise pushed to their level when they are responsible for
546   large AOs in a COIN environment. Tactical units down to company level must be prepared to integrate
547   their efforts with civilian organizations.

548   2-51. To ensure integration of interagency capabilities, units should conduct coordination with all inter-
549   agency representatives and organizations that enter their AO. Despite the best efforts to coordinate, often
550   the fog and friction inherent in COIN will lead to civilian organizations entering an AO without prior coor-
551   dination. Below is a suggested list for conducting coordination with interagency and other nonmilitary or-
552   ganizations who may operate in a unit's AO:

553   ● Identify organizational structures and leadership.
554   ● Identify key objectives, responsibilities, capabilities, and programs.
555   ● Develop common courses of action or options for inclusion in planning, movement coordina-
556   tion, and security briefings.
557   ● Develop relationships that enable the greatest possible integration.
558   ● Assign liaison officers to the most important civilian organizations.
559   ● Define problems in clear and unambiguous terms.

560   2-52. Other useful information about the organizations includes the following:
561   ● Intended duration of operations.
562   ● Location of bases of operations.
563   ● Number of personnel (names, descriptions).
564   ● Type, color, number and license numbers of civilian vehicles.
565   ● Other resources in the AO.
566   ● Local groups and agencies with whom they are working.
567   ● Terms of reference or operating procedures.

568   # SUMMARY

569   2-53. As President John F. Kennedy eloquently noted, "You [military professionals] must know something
570   about strategy and tactics and logistics, but also economics and politics and diplomacy and history. You
571   must know everything you can know about military power, and you must also understand the limits of
572   military power. You must understand that few of the important problems of our time have...been finally
573   solved by military power alone." Nowhere is this insight more relevant than in COIN. Successful COIN
574   requires unity of effort in bringing all instruments of national power to bear. Civilian agencies can contrib-
575   ute directly to military operations, particularly by providing intelligence. That theme will be developed in
576   more detail in the next chapter.
577
578
579

Chapter 3

Intelligence in Counterinsurgency

Everything good that happens seems to come from good intelligence.

General Creighton Abrams

Effective, accurate, and timely intelligence is essential to the conduct of any form of warfare. This maxim applies especially to counterinsurgency operations, as the ultimate success or failure of the mission depends on the effectiveness of the intelligence effort. This chapter builds upon previous concepts to further describe insurgencies, requirements for intelligence preparation of the battlefield and predeployment planning, collection and analysis of intelligence in counterinsurgency, intelligence fusion, and general methodology for integrating intelligence to support operations. This chapter does not supersede processes in U.S. military doctrine (see FM 2-0, FM 34-130/FMFRP 3-23-2, and FMI 2-91.4) but instead provides specific guidance for counterinsurgency.

SECTION I – INTELLIGENCE CHARACTERISTICS IN COUNTERINSURGENCY

3-1.   Counterinsurgency (COIN) is an intelligence war. The function of intelligence in COIN is to facilitate understanding of the operational environment, with emphasis on the populace, host nation, and insurgents, so commanders can best address the issues driving the insurgency. Both insurgents and counterinsurgents require an effective intelligence capability to be successful. They attempt to create and maintain intelligence networks while trying to neutralize their opponent's intelligence capabilities.

3-2.   Intelligence in COIN is about people. U.S. forces must understand the people of the host nation, the insurgents, and the host-nation (HN) government. Commanders and planners must have insight into cultures, perceptions, values, beliefs, interests and decision-making processes of individuals and groups. These requirements are the basis for collection and analytical efforts.

3-3.   Intelligence and operations feed back on one another. Effective intelligence drives effective operations, which produce more intelligence. Similarly, ineffective or inaccurate intelligence produces ineffective operations, which reduce the availability of intelligence.

3-4.   All operations have an intelligence component to them. This is because all Soldiers and Marines are potential intelligence collectors whenever they interact with the people. Operations should therefore always include intelligence collection requirements.

3-5.   Insurgencies are local. They vary greatly in time and space. The insurgency one battalion faces will often be different from that of an adjacent battalion. This shifting mosaic nature of insurgencies coupled with the fact that all Soldiers and Marines are intelligence collectors means all echelons are both intelligence producers and consumers, resulting in a bottom-up flow of intelligence. This pattern also means that tactical units at brigade and below will require a great deal of support for both intelligence collection and analysis, as their organic intelligence structure will often be inadequate.

3-6.   Counterinsurgency occurs in a joint, interagency, and multinational environment at all echelons. Soldiers and Marines must coordinate intelligence collection and analysis with foreign militaries, foreign intelligence services, and many different U.S. intelligence organizations.

40
41

SECTION II – PREDEPLOYMENT PLANNING AND INTELLIGENCE PREPARATION OF THE BATTLEFIELD

42 3-7. The purpose of predeployment planning and intelligence preparation of the battlefield (IPB) is to de-
43 velop an understanding of the operating environment to drive planning and predeployment training. Prede-
44 ployment intelligence must be as detailed as possible and focus on the host nation, its people, and insur-
45 gents in the AO. Commanders and staffs use predeployment intelligence to establish a plan for addressing
46 the underlying causes of the insurgency and to prepare their units to interact with the populace in an appro-
47 priate manner. The desired end state is that commanders and their subordinates are not surprised by what
48 they encounter in theater.

49 3-8. Preparation for deployment begins with IPB. This is the systematic, continuous process of analyzing
50 the threat and environment in a specific geographic area. It is accomplished via the four steps described in
51 FM 34-130/FMFRP 3-23-2:

52 • Define the operational environment.
53 • Describe the effects of the environment.
54 • Evaluate the threat.
55 • Determine threat courses of action.

56 3-9. IPB in COIN operations follows the methodology described in FM 34-130/FMFRP 3-23-2, but it
57 varies in that it places greater emphasis on people and leaders in an area of operations (AO). (IPB is cov-
58 ered in depth in appendix B.) Though IPB is a continuous process and products resulting from it will be re-
59 vised throughout the mission, the initial products created prior to deployment are of particular importance
60 for the reasons explained above. Whenever possible, personnel must conduct a very thorough and detailed
61 IPB in preparation for deployment. IPB in COIN requires intelligence personnel to work in areas such as
62 economics, anthropology, and governance that may be outside their expertise. Therefore, integrating staffs
63 and drawing upon the knowledge of nonintelligence personnel and external subject matter experts with lo-
64 cal and regional knowledge are critical to effective unit preparation.

65 3-10. If there are already units deployed in theater, these are the best potential sources of intelligence
66 available. Deploying units should make an effort to reach forward to units in theater. Governmental agen-
67 cies such as the Department of State, U.S. Agency for International Development, and intelligence agen-
68 cies are often able to provide country studies and other background information on a theater. Open-source
69 intelligence is also very important to predeployment IPB. In many cases, background information on the
70 populace, cultures, languages, history, and government of an AO may be found in open sources. These in-
71 clude books, magazines, encyclopedias, Web sites, tourist maps, and atlases. Academic sources, such as
72 journal articles and university professors, can also be of great benefit.

73 # DEFINE THE OPERATIONAL ENVIRONMENT

74 3-11. The *operational environment* consists of the air, land, sea, space, and associated adversary, friendly,
75 and neutral systems (political, military, economic, social, informational, infrastructure, legal, and others)
76 that are relevant to a specific joint operation (JP 1-02). At the tactical and operational levels, defining the
77 operational environment involves defining a unit's AO and then establishing an area of interest. AOs may
78 be relatively static, but people and information flow through the AO continuously. Therefore, when defin-
79 ing an area of interest, both physical and human geography are taken into account. This is because an AO
80 will often cut across not just physical lines of communication, such as roads, but also areas that are tribally,
81 economically, or culturally interrelated. For instance, tribal and family groups in both Iraq and Afghanistan
82 cross national borders into neighboring countries. The cross-border ties allow insurgents safe haven outside
83 of their country and aid them in cross-border smuggling. The area of interest can be large relative to the
84 AO, as it must often account for the various influences that affect the AO such as—

85 • Cultural geography: family, tribal, ethnic, religious, or other links that go beyond the AO.
86 • Communications links to other regions.
87 • Economic links to other regions.
88 • Media influence on the local populace, U.S. populace, and multinational partners.

89   ●   External financial, moral, and logistic support of the enemy.

90 3-12. At the theater level, this can mean the area of interest is global if enemy forces have an international
91 financial network or are able to affect popular support within the U.S. or multinational partners. At the tac-
92 tical level, this means commands must be aware of activities in neighboring regions and population centers
93 that affect the population in their AO.

94 3-13. As explained in chapter 2, another consideration for defining the operational environment is to learn
95 the lower, adjacent, and higher units of the COIN force. In addition, determine the non–Department of De-
96 fense (DOD) agencies, multinational forces, nongovernmental organizations, and HN organizations in the
97 AO. Knowledge of these units and organizations allows for establishment of working relationships and
98 procedures for sharing intelligence. These relationships and procedures are critical to developing a com-
99 prehensive common operational picture (COP) and enabling unity of effort.

DESCRIBE THE EFFECTS OF THE OPERATIONAL ENVIRONMENT

101 3-14. This step involves developing an understanding of the operational environment and is critical to the
102 success of operations. It includes—
103   ●   Civil considerations, with emphasis on the people, history, and HN government in the AO.
104   ●   Terrain analysis (physical geography): with emphasis on complex terrain, suburban and urban
105   terrain, key infrastructure, and lines of communication.
106   ●   Weather analysis, with attention given to the weather's effects on activities of the population,
107   such as agriculture, smuggling, or insurgent actions.

108 3-15. All staff members are involved in this step. This improves the knowledge base used to develop an
109 understanding of the AO. For instance, civil affairs personnel receive training in analysis of populations,
110 cultures, and economic development, and can contribute greatly to understanding the civil considerations
111 of an AO. As another example, foreign area officers have linguistic, historical, and cultural knowledge
112 about particular regions and have often lived there for extended periods.

113 3-16. The products that result from describing the effects of the operational environment influence opera-
114 tions from the theater level down to the fire team. The description informs political activities and economic
115 policies of theater-level commanders, and drives information operations and civil-military operations plan-
116 ning. The knowledge gained affects the way Soldiers and Marines interact with the populace.

CIVIL CONSIDERATIONS: SOCIO-CULTURAL FACTORS ANALYSIS

118 3-17. *Civil considerations* are how the manmade infrastructure, civilian institutions, and attitudes and ac-
119 tivities of the civilian leaders, populations, and organizations within an area of operations influence the
120 conduct of military operations (FM 6-0). The memory aid ASCOPE expresses the six characteristics of
121 civil considerations: areas, structures, capabilities, organizations, people, and events.

122 3-18. While all characteristics of civil considerations are important, understanding the *people* characteristic
123 (cultural awareness) is particularly key in COIN. Cultural awareness involves understanding of socio-
124 cultural factors and their impact on military operations. Socio-cultural factors analysis is used to develop
125 cultural awareness. There are six socio-cultural considerations:
126   ●   Society.
127   ●   Social structure.
128   ●   Culture.
129   ●   Power and authority.
130   ●   Interests.
131   ●   Economics.

Society

3-19. A *society* can be defined as a population whose members are subject to the same political authority, occupy a common territory, have a common culture, and share a sense of identity. A society is not easily created or destroyed, but it is possible to do so through genocide or war.

3-20. No society is homogeneous. A society usually has a dominant culture, but can also have a vast number of secondary cultures. Different societies may share similar cultures, such as Canada and the United States do. Societies are not static, but change over time.

3-21. Understanding the societies in the AO allows COIN forces to achieve objectives and gain support. Commanders also consider societies outside the AO whose actions, opinions, or political influence can affect the mission.

Social Structure

3-22. Each society is composed of both social structure and culture. Social structure refers to the relations among groups of persons within a system of groups. Social structure is persistent over time, meaning that it is regular and continuous despite disturbances, and the relation between the parts holds steady even as groups expand or contract. In an army, for example, the structure consists of the arrangement into groups like divisions, battalions, and companies, and the arrangement into ranks. In a society, the social structure includes groups, institutions, organizations, and networks. Social structure involves the arrangement of the parts that constitute society, the organization of social positions, and the distribution of people within those positions.

Groups

3-23. A *group* is two or more people regularly interacting on the basis of shared expectations of others' behavior and who have interrelated statuses and roles. In an AO, the social structure will include a variety of groups, including racial groups, ethnic groups, religious groups and/or tribal groups and other kinship-based groups.

3-24. A *race* is a human group that defines itself or is defined by other groups as different by virtue of innate physical characteristics. Biologically, there is no such thing as race among human beings; race is a social category.

3-25. An *ethnic group* is a human community whose learned cultural practices, language, history, ancestry, or religion, distinguish them from others. Members of ethnic groups see themselves as different from other groups in a society, and are recognized as such by others. Religious groups may be subsets of larger ethnic groups, such as Sri Lankan Buddhists, or religious groups may have members from many different ethnicities, such as the Roman Catholic faith.

3-26. *Tribes* are generally defined as autonomous, genealogically structured groups in which the rights of individuals are largely determined by their ancestry and belonging to a particular lineage. Tribes are essentially adaptive social networks, organized by extended kinship and descent, with common needs for physical and economic security.

3-27. Understanding the composition of groups in the AO is of vital importance for COIN operations, especially because insurgents may organize around racial, ethnic, religious, or tribal identities. Furthermore, tensions or hostilities between groups in an AO may destabilize a society and provide opportunities for insurgents. Commanders should thus identify powerful groups both inside and outside of their AO, and identify the formal relationships (such as treaties or alliances) and the informal relationships (such as tolerance or friction) between groups. Commanders should identify cleavages between groups and cross-cutting ties (for example, religious alignments that cut across ethnic differences). Insurgent leadership and their rank and file may belong to separate groups, or the bulk of the population may differ from the insurgents. These characteristics may suggest courses of action aimed at reinforcing or widening seams among insurgents or between insurgents and the population.

178 *Networks*

179   3-28. *Networks*, which are a series of direct and indirect ties from one actor to a collection of others, may
180 be an important element within a social structure. Common types of networks include elite networks,
181 prison networks, transnational diasporas, and neighborhood networks. Networks can have many purposes:
182 economic, criminal, and emotional. Network analysis should consider both the structure of the network and
183 the nature of the interactions between network actors.

184 *Institutions*

185   3-29. Groups engaged in patterned activity to complete a common task are called *institutions*. Educational
186 institutions bring together groups and individuals who statuses and roles concern teaching and learning.
187 Military institutions bring together groups and individuals whose statuses and roles concern defense and
188 security. Institutions, the basic building blocks of societies, are continuous through many generations and
189 continue to exist even when the individuals that compose them are replaced.

190 *Organizations*

191   3-30. *Organizations*, both formal and informal, are institutions that have bounded membership, defined
192 goals, established operations, fixed facilities or meeting places, and a means of financial or logistic support.
193 In an AO, organizations may control, direct, restrain, or regulate the local populace. Thus, commanders
194 should identify influential organizations both inside and outside of their AO, which members of what
195 groups belong to the organization, and how their activities may affect military operations. The next step is
196 to determine how these organizations affect the local populace, whose interests they fulfill, and what role
197 they play in influencing local perceptions. Planners can generally group these organizations into communi-
198 cating, religious, economic, governance, and social.

199   3-31. *Communicating* organizations have the power to influence a population's perceptions.

200   3-32. *Religious* organizations regulate norms, restrain or empower activities, reaffirm worldviews, and
201 provide social support. A religious organization differs from a religious group: whereas a religious group is
202 a general category, such as Christian, a religious organization is a specific community, such as the Episco-
203 pal Church.

204   3-33. *Economic* organizations provide employment, help regulate and stabilize monetary flow, assist in
205 development, and create social networks.

206   3-34. *Social* institutions, such as schools, civil society groups, and sports teams, provide support to the
207 population, create social networks, and can influence ideologies. Organizations may belong to more than
208 one category. For instance, an influential religious organization may also be a communicating organization.

209 *Roles and Statuses*

210   3-35. To facilitate COIN operations, the most common roles, statuses, and institutions within the society
211 must be identified. Individuals in any given society interact as members with social positions, referred to as
212 a status. Most societies associate particular statuses with particular social groups, such as family, lineage,
213 ethnicity, or religion. Statuses may be achieved by meeting certain criteria, or may be ascribed by birth.
214 Statuses are often reciprocal, such as that of husband and wife or teacher and student. Every status carries a
215 cluster of expected behaviors, such as how a person in that status is expected to think, feel, and act, as well
216 as expectations about how they should be treated by others. Thus, in American society parents (status) have
217 the obligation to care for their children (role) and the right to discipline them (role).

218 *Social Norms*

219   3-36. Violation of a role prescribed by a given status, such as failing to feed one's children, results in so-
220 cial disapproval. The standard of conduct for social roles is known as a 'social norm.' A social norm is
221 what people are expected to do or should do, rather than what people actually do. Norms may be either
222 moral (incest prohibition, homicide prohibition) or customary (prayer before a meal, removing shoes be-
223 fore entering a house). When a person's behavior does not conform to social norms, the person may be

224   sanctioned. Understanding roles, statuses, and social norms within an AO can clarify and provide guidance
225   to COIN forces about expected behavior.

Culture

3-37. Once the social structure has been thoroughly mapped out, COIN forces should identify and analyze the culture of the society as a whole, and of each major group within the society. Whereas social structure (groups, institutions, organizations) comprises the relationships among groups, institutions, and individuals within a society, culture (ideas, norms, rituals, codes of behavior) provides the meaning to individuals within the society. For example, families are a core institutional building block of social structure, found universally. However, marital monogamy, expectations of a certain number of children, and willingness to live with in-laws are all matters of culture, and are highly variable in different societies. Social structure can be thought of as a skeleton, and culture is like the muscle on the bones. The two are mutually dependent and reinforcing, and a change in one results in a change in the other.

3-38. Culture is "web of meaning" shared by members of a particular society or group within a society. (See FM 3-05.301/MCRP 3-40.6A.) Culture has the following characteristics:

- Culture is a system of shared beliefs, values, customs, behaviors, and artifacts that the members of society use to cope with their world and with one another.
- Culture is learned, though a process called enculturation.
- Culture is shared by members of a society; there is no "culture of one."
- Culture is patterned, meaning that people in a society live and think in ways that form definite, repeating patterns.
- Culture is changeable, meaning it is constantly changing through social interactions between people and groups.
- Culture is arbitrary, meaning that Soldiers and Marines should make no assumptions regarding what a society considers right and wrong, good and bad.
- Culture is internalized, in the sense that is habitual, taken-for-granted, and perceived as "natural" by people within the society.

3-39. Culture might also be described as an operational code that is valid for an entire group of people. Culture conditions the individual's range of action and ideas, including what to do (and what not to do), how to do it (or not do it), whom to do it with (or whom not to do it with), and under what circumstances the "rules" shift and change. Culture influences how people make judgments about what is right and wrong, assess what is important and unimportant, categorize things, and deal with things that do not fit into existing categories. Cultural rules are flexible in practice. For example, the kinship system of the Yanomamo, an Amazonian Indian tribe, requires that individuals marry their cross cousin. However, the definition of cross cousin is often changed to make people eligible for marriage.

Identity

3-40. Each individual belongs to multiple groups, through birth, assimilation, or achievement. Each group to which the individual belongs influences his or her beliefs, values, attitudes, and perceptions. Individuals rank their identities consciously or unconsciously into primary (national, racial, religious) identities and secondary identities (hunter, blogger, coffee drinker). Frequently, individuals' identities may be in conflict, and COIN forces can use these conflicts to influence key leaders' decision-making processes.

Beliefs

3-41. Beliefs are concepts and ideas accepted as true. Beliefs can be core, intermediate, or peripheral.

3-42. *Core* beliefs are those views that are part of a person's deep identity, and are not easily changed. Examples include belief in the existence of God, the value of democratic government, and the role of the family. Core beliefs are unstated, taken for granted, resistant to change, and not consciously considered. Attempts to change the central beliefs of a culture may result in significant unintended second- and third-order consequences. Decisions to do so are made at the national-strategic level.

271
272

3-43. Beliefs in the *intermediate* region are predicated on reference to authority figures or authoritative texts. Thus, intermediate beliefs can be influenced through co-optation of the opinion leaders.

273
274
275
276

3-44. From intermediate beliefs flow *peripheral* beliefs, which are open to debate, consciously considered, and easiest to change. For example, peripheral beliefs about birth control, the New Deal, or the theory of sexual repression derive from beliefs about the Roman Catholic Church, President Franklin Roosevelt, or Sigmund Freud respectively.

277

Values

278
279
280
281
282
283
284
285
286
287

3-45. A *value* is an enduring belief that a specific mode of conduct or end state of existence is preferable to an opposite or converse mode of conduct or end state of existence. Each group to which an individual belongs inculcates that individual with its values and their ranking of importance. Individuals do not unquestioningly absorb all the values of the groups to which they belong, but accept some and reject others. The values of each culture to which individuals simultaneously belong are often in conflict: religious values may conflict with generational values, gender values with organizational practices. Commanders should evaluate the values of each group, such as toleration, stability, prosperity, social change, and self-determination. Commanders should also consider whether the values promoted by the insurgency correspond to the values of other social groups in the AO or to those of the government, and whether these differences in values can be exploited by U.S. forces.

288

Attitudes and Perceptions

289
290
291
292

3-46. Attitudes are affinities for and aversions to groups, persons, and objects. Commanders should consider groups' attitudes toward other groups, outsiders, their government, the United States, the U.S. military, globalization, and so forth. Attitudes affect perception, which is the process by which an individual selects, evaluates, and organizes information from the external environment.

293

Belief Systems

294
295
296
297
298

3-47. The totality of the identities, beliefs, values, attitudes, and perceptions that an individual holds—and the ranking of their importance—is that person's belief system. Religions, ideologies, and all types of "-isms" fall in this category. As a belief system, a religion may include such things as a concept of God, a view of the afterlife, ideas about the sacred and the profane, funeral practices, rules of conduct, and modes of worship.

299
300
301

3-48. The belief system acts as a filter for new information: it is the lens through which people perceive the world. Understanding the belief systems of various groups in an AO allows COIN forces to more effectively influence the population.

302
303
304
305
306

3-49. The belief systems of insurgents and other groups within the AO should be given careful attention. The insurgency may frame its objectives in terms of a belief system or may use a belief system to mobilize and recruit followers from among the population. Where there are differences between the belief system of the insurgents and civilian groups, an opportunity exists for counterinsurgents to engineer a split between the insurgency and the larger population.

307

Cultural Forms

308
309
310
311
312

3-50. *Cultural forms* are the concrete expression of the belief systems shared by members of a particular culture. Cultural forms, such as rituals, symbols and narratives, are the medium for communicating ideologies, values, and norms that influence thought and behavior. Each culture constructs or invents its own cultural forms, through which cultural meanings are transmitted and reproduced. COIN forces can identify the belief systems of a culture by observing and analyzing its cultural forms.

313
314
315
316
317

3-51. The most important cultural form for COIN forces to understand is the narrative. A *cultural narrative* is a story recounted in the form of a causally linked set of events that explains an event in group's history and expresses the values, character, or self-identity of the group. Narratives are the means through which ideologies are expressed and absorbed by individuals in a society. For example, at the Boston Tea Party in 1773, Sam Adams and the Sons of Liberty dumped five tons of tea in the Boston Harbor as a pro-

318
319
320
321
322
323
324
test against what they considered unfair British taxation. This narrative explains in part why the Revolutionary War began but also tells Americans something about themselves each time they hear the story: that fairness, independence, and justice are worth fighting for. As this example indicates, narratives may not conform to historical facts, or they may drastically simplify facts to more clearly express basic cultural values. (For example, Americans in 1773 were taxed less than their British counterparts, and most British attempts to raise revenues from the colonies were designed to help reduce the crushing national debt incurred in their defense.) By listening to narratives, COIN forces can identify the basic core values of the society.

325
326
327
328
329
330
331
332
3-52. Other cultural forms include ritual and symbols. *Ritual* is a stereotyped sequence of activities involving gestures, words, and objects, performed to influence supernatural entities or forces on behalf of the actors' goals and interests. Rituals can be either sacred or secular. A *symbol* is the smallest unit of cultural meaning and is filled with a vast amount of information that can be decoded by a knowledgeable observer. Symbols can be objects, activities, words, relationships, events, or gestures. Institutions and organizations often use cultural symbols to amass political power or to generate resistance against external groups. Careful attention should be paid to the meaning of the common symbols and how various groups in the AO use them.

333 Power and Authority

334
335
336
3-53. Once the social structure has been mapped and the culture is understood, COIN forces must understand how power is apportioned and used within a society. Understanding power is the key to manipulating the interests of groups within the society.

337
338
339
340
341
3-54. There may be many power holders in a society, both formal and informal. Understanding the formal political system (including central governments, local governments, political interest groups, political parties, unions, government agencies, and regional and international political bodies) is necessary but not sufficient for COIN operations. Informal power holders are often more important for COIN operations, and may include ethno-religious groups, social elites, or religious figures.

342
343
344
345
346
347
3-55. For each group within the AO, COIN forces should identify what type of power the group has, what they use their power for (such as amassing resources, protecting followers, and so forth) and how they acquire and maintain power. Commanders should also identify which leaders have power within particular groups, what type of power they have, what they use their power for, and how their power is acquired and maintained. There are four major forms of power in a society: coercive force, social capital, economic resources, and authority.

348 *Coercive Force*

349
350
351
352
353
354
355
356
357
358
359
360
361
3-56. *Coercion* is the ability to compel a person to act through threat of harm or by use of physical force. Coercive force can be positive, in the sense that a group may provide security to its members (such as policing and defense of territory), or it may be negative, in the sense that a group may intimidate or threaten group members or outsiders. One essential role of government is to monopolize coercive force within its territory in order to provide physical security to its citizens. Where the state fails to provide security to its citizens or becomes a threat to them, citizens may seek alternative security guarantees from an ethnic, political, religious, or tribal group in the AO. Insurgents, and other nongovernmental groups, may possess considerable coercive force, in the form of paramilitary units, tribal militias, gangs, or organizational security personnel that they use to gain power over a civilian population. These groups may be using their coercive force for a variety of purposes unrelated to the insurgency, such as protecting their own community members, carrying out vendettas, or engaging in criminal activities. For example, what may appear to be insurgent violence against "innocent civilians" could in fact be a tribal blood feud unrelated to the insurgency.

362 *Social Capital*

363
364
365
366
3-57. *Social capital* refers to the power of individuals and groups to utilize social networks of reciprocity and exchange to accomplish their goals. In many non-Western societies, patron-client relationships are an important form of social capital. In a system based on patron-client relationships, an individual in a powerful position provides goods, services, security, or other resources to followers in exchange for political sup-

367
368

port or loyalty, thereby amassing power. COIN forces must identify, where possible, which groups and individuals have social capital and how they attract and maintain followers.

369 *Economic Power*

370
371
372
373
374
375
376
377
378
379

3-58. *Economic power* is the power of groups or individuals to use economic incentives and disincentives to change people's behavior. Economic systems can be formal or informal, or a mixture of both. In weak or failed states, where the formal economy may be functioning in a diminished capacity, the informal economy plays a central role in people's daily lives. For example, in many societies, monies and other economic goods are distributed though the tribal or clan networks and are connected to indigenous patronage systems. Those groups able to provide their members with economic resources through an informal economy (whether smuggling, black market activities, barter, or exchange) gain followers and may amass considerable political power. Counterinsurgents must therefore be aware of the local informal economy and evaluate the role played by various groups and individuals within it. Insurgent organizations may also attract followers through criminal activities that provide income.

380 *Authority*

381
382

3-59. *Authority* is power that is attached to positions and is justified by the beliefs of the obedient. There are three primary types of authority:

383
384

- Rational-legal authority, which is grounded in law and contract, codified in impersonal rules, and commonly found in developed, capitalist societies.

385
386
387

- Charismatic authority, which is exercised by leaders who develop allegiance among their followers because of their unique, individual charismatic appeal, whether ethical, religious, political, or social.

388
389

- Traditional authority, which is usually invested in a hereditary line or in a particular office by a higher power.

390
391
392
393
394
395

3-60. Traditional authority, which relies on the precedent of history, is a common type of authority in non-Western societies. In particular, tribal and religious forms of organization rely heavily on traditional authority. Traditional authority figures often wield enough power, especially in rural areas, to single-handedly drive an insurgency. Understanding the types of authority at work in the formal and informal political systems of the AO helps COIN forces identify the agents of influence who can help or hinder the accomplishment of objectives.

396 **Interests and Economics**

397
398
399
400

3-61. After mapping the social structure, evaluating the culture of the society as a whole and that of each group within the society, and identifying who holds formal and informal power and why, COIN forces should consider how to use the capabilities at their disposal to reduce support for insurgents and gain support for the legitimate government.

401
402
403
404
405
406
407

3-62. To accomplish this task, requires COIN forces to understand the population's *interests*, meaning the core motivations that drive behavior, such as physical security, basic necessities, economic well-being, political participation, and social identity. During times of instability when the government cannot function, the groups and organizations to which an individual belongs provide physical security, basic necessities, economic resources, and political identity. Understanding a group's interests illuminates the various opportunities available to a commander to meet or frustrate those interests with the capabilities at his or her disposal. A group's interests may become grievances if the government does not fulfill the group's needs.

408 *Physical Security*

409
410
411
412
413

3-63. During an insurgency or any period of political instability, the primary interest of the civilian population is physical security for themselves and their families. When government forces fail to provide security or threaten the security of civilians, the civilian population is much more likely to seek alternative security guarantees from insurgents, militias, or other armed groups. This process can drive an insurgency. However, when government forces provide physical security, civilians are more likely to support the govern-

414 ment against the insurgents. Commanders should therefore identify who is providing security to each group
415 within the civilian population. The provision of security must occur in conjunction with political and eco-
416 nomic reform.

Essential Services

418 3-64. Other interests of a population include that which is necessary to sustain life, such as food, water,
419 clothing, shelter, and medical treatment. Stabilizing a populace requires meeting all these essential needs.
420 People pursue these needs until they are met, at any cost and from any source. If an insurgent source pro-
421 vides for the populace's needs, the populace will support that source. This may impede mission accom-
422 plishment. If the government provides that which is necessary to sustain life, the population is more likely
423 to support the government. Commanders should therefore identify who is providing basic life necessities to
424 each group within the civilian population.

Economy

426 3-65. Individuals and groups within a social system satisfy their economic interests through the produc-
427 tion, distribution, and consumption of goods and services. How individuals within a society satisfy their
428 economic needs depends on the level and type of economic development. For instance, in a rural-based so-
429 ciety, land ownership may be a chief component of any economic development plan. For a more urban so-
430 ciety, jobs in both the public and private sector may be the biggest issue of contention.

431 3-66. Economic disparities between groups may drive political instability, and economic disenfranchise-
432 ment may be a core grievance of insurgent groups. Insurgent leadership or traditional authority figures of-
433 ten use real or perceived societal injustices to drive an insurgency, such as economic disenfranchisement,
434 exploitative economic arrangements, or a significant income disparity that creates (or allows for) intracta-
435 ble class distinctions. The economic system within an AO may be adversely affected by military operations
436 and/or insurgent actions. This can cause the population to resent the government. On the other hand, resto-
437 ration of production and distribution systems can energize the economy, create jobs and growth, and posi-
438 tively influence local perceptions. Commanders should evaluate whether the society has a functioning
439 economy, whether civilians have fair access to land and property, and how to minimize the economic
440 grievances of the civilian population in order to reduce support for insurgency and increase support for the
441 government.

Political Participation

443 3-67. Another interest of the civilian population is political participation. Many insurgencies begin because
444 certain groups within the society perceive that they have been denied political rights. In order to satisfy
445 their political interests, groups may use preexisting cultural narratives and symbols to mobilize for political
446 action. Very often, they will coalesce around traditional or charismatic authority figures. Commanders
447 should investigate whether all members of the civilian population have a guarantee of political participa-
448 tion; whether there is ethnic, religious, or other discrimination; and whether there are legal, social, or other
449 policies that contribute to the insurgency. Commanders should also identify what narratives mobilize po-
450 litical action within the group and who their traditional or charismatic authority figures are.

LANGUAGES

452 3-68. The languages used in an AO have a major impact on operations. Languages must be identified so
453 language training, communication aids such as phrase cards, and requisitioning of translators can be ac-
454 complished. Translators are critical for intelligence collection, interaction with local citizens and commu-
455 nity leaders, and development of products for information operations. The mission cannot proceed without
456 them. (See appendix C.)

457 3-69. Another aspect of language involves transliteration of names not normally written using the English
458 alphabet. This affects all intelligence operations, to include collection, analysis, fusion, and targeting. In
459 countries that do not use the English alphabet, a theaterwide standard should be set for spelling names.
460 Without a spelling standard, it can be difficult to conduct effective analysis. (This problem is compounded
461 in cultures where people use different names depending on the context in which the name is asked.) Unfa-

462   miliar and similar place names can confuse map reading and cause targeting errors. In addition, insurgents
463   may be released from custody if their name is misidentified. To overcome these problems, there must be
464   one spelling standard for a theater. Because of the interagency nature of COIN operations, the standard
465   must also be agreed upon by non-DOD agencies.

466 # EVALUATE THE THREAT

467   3-70. The purpose of evaluating the insurgency and other related threats to the mission is to understand the
468   enemy, enemy capabilities, enemy vulnerabilities, and opportunities commanders may be able to exploit.
469   Evaluating an insurgency is difficult. Neatly arrayed enemy orders of battle are not available and are not
470   the critical information commanders require. Commanders require knowledge of difficult-to-measure char-
471   acteristics of the threat, such as insurgent goals, grievances exploited by the insurgents, means used by the
472   insurgents to generate support, the organization of insurgent forces, and accurate locations of key insurgent
473   leaders. However, insurgents usually look no different from the general populace and do their best to blend
474   with noncombatants. The structure of insurgent groups is often adaptive and flexible, and they may pub-
475   licly claim different motivations and goals than what is truly driving their actions. Further complicating
476   matters, insurgent organizations are often rooted in ethnic or tribal groups and they often take part in
477   criminal activities or link themselves to political parties, charities, or religious organizations. This blurs the
478   lines as to what and who the threat is. The following characteristics of an insurgency provide a basis for
479   evaluating the threat and are discussed in depth in appendix B:

480   ● Insurgent objectives.
481   ● Insurgent motivations.
482   ● Popular support/tolerance.
483   ● Support activities, capabilities, and vulnerabilities.
484   ● Information activities, capabilities, and vulnerabilities.
485   ● Political activities, capabilities, and vulnerabilities.
486   ● Violent activities, capabilities, and vulnerabilities.
487   ● Organization.
488   ● Key leaders and personalities.

489   3-71. Of the preceding characteristics, knowledge of objectives, motivations, and means of generating
490   popular support/tolerance will often be the most important intelligence requirements and the most difficult
491   to ascertain. In particular, generating popular support/tolerance often has the greatest impact on the insur-
492   gency's long-term effectiveness. This is usually the center of gravity of an insurgency. Support/tolerance,
493   provided either willingly or unwillingly by the populace, results in safe havens, freedom of movement, lo-
494   gistic support, financial support, intelligence, and new personnel for the insurgency. It should be noted that
495   support/tolerance is often generated using violent coercion and intimidation of the people. In these cases,
496   even if the people do not favor the insurgent cause, they are forced to tolerate the insurgents or provide
497   them material support.

498   3-72. Generating popular support has a positive effect on an insurgent organization. As the insurgent
499   group gains support, its capabilities grow, which in turn enables it to gain more support. Insurgents gener-
500   ally view popular support as a zero-sum game in which a gain in support for their movement is a loss for
501   the government and a loss of support for the government is a gain for the insurgents.

502   3-73. Although violence is generally the most visible aspect of the insurgency, it can represent only a small
503   portion of the insurgent effort. Support activities, such as recruitment, securing finances, and securing lo-
504   gistics, are more difficult to detect, but often more widespread and important to the insurgency than vio-
505   lence. It is important to track the violent actions of insurgents, but this should not be done at the expense of
506   understanding the other characteristics of the insurgency.

507   3-74. As explained in chapter 1, insurgents are by nature an asymmetric threat. They do not use terrorist
508   and guerrilla tactics because they are cowards afraid of a "fair fight." They use these tactics because they
509   are the best means available for achieving their goals. Terrorist and guerrilla attacks are usually planned to
510   achieve the greatest political and informational impact with the lowest amount of risk to the insurgents.

511
512
513

This means that commanders must not simply understand insurgent tactics and targeting; they must also understand how violence is used to achieve the goals of the insurgent organization and how violent actions are linked to political and informational actions.

514
515
516
517
518

3-75. Attitudes and perceptions play a very important role in understanding the threat. It is important to know how the population perceives the insurgents, the host nation, and U.S. forces. In addition HN and insurgent perceptions of one another and of U.S. forces are also very important. Attitudes and perceptions of different groups and organizations inform decision-making processes and shape popular thinking on the legitimacy of the actors in the war.

519
Insurgency-related Threats

520
521
522

3-76. The presence of an insurgency with widespread support in a state usually means the government of the state is weak and losing control. In such situations, criminal organizations and nongovernmental militia forces can become powerful.

523
Criminal Networks

524
525
526
527
528
529

3-77. Though not necessarily a part of the insurgency, criminal networks can further undermine the authority of the state through banditry, hijackings, kidnappings, smuggling, and other disruptive activities. Insurgent organizations often link themselves to criminal networks in order to secure additional sources of funding and logistics. In some cases, insurgent networks and criminal networks become indistinguishable. As commanders work to reassert government control of an area, they need to know what criminal networks are present, what their activities are, and how they interact with the insurgency.

530
Nongovernmental Militias

531
532
533
534
535
536
537
538

3-78. As the state weakens and violence increases, people look for ways to protect themselves. If the state cannot provide protection, people may organize into armed militias to protect themselves. Examples of this are the Protestant militias in Northern Ireland, the right-wing paramilitary organizations that formed in Colombia to counter the FARC, and the militias of various ethnic and political groups in Iraq during Operation Iraqi Freedom. If militias are outside of the control of the government, they are often disruptive to ending the insurgency. They may become more powerful than the government, particularly at a local level. They may also fuel the insurgency and a downward spiral into full-scale civil war if they conduct their own armed attacks on insurgents.

539
540
541
542

3-79. Although militias may not be an immediate threat to U.S. forces, they constitute a long-term threat to the law and order within the state. They should be tracked by intelligence in the same manner as an insurgent group. The intent of this analysis is to understand what role militias play in the insurgency, what role they play in politics, and how they can be disarmed.

543
Opportunities

544
545
546
547
548
549
550

3-80. As the threat is evaluated and the interests and attitudes of the people are learned, it becomes possible to identify divisions between the insurgents and the people as well as between the HN government and the people. For instance, if the insurgent ideology is not popular, the insurgents may rely on intimidation to generate support. Another example would be discovering that the insurgents gain support by providing social services that the HN government neglects. Determining such divisions provides opportunities for crafting friendly operations that expand splits between the insurgents and the people or lessen the divides between the HN government and the people.

551
DETERMINE THREAT COURSES OF ACTION

552
553
554
555

3-81. The purpose of this IPB step is to understand insurgent strategies and tactics so they may be effectively countered. The initial determination of threat courses of action focuses on two levels of analysis. The first is determining the overall strategy, or combination of strategies, selected by an insurgency to achieve its goals. The second is determining tactical courses of action used in support of the strategy.

INSURGENT STRATEGIES

3-82. As indicated in chapter 1, there are five general strategies insurgents may follow:

- Conspiratorial.
- Protracted popular war.
- Traditional.
- Military-focus.
- Urban warfare.

3-83. These approaches may be combined with one another or occur in parallel in the same AO as different insurgent groups follow different paths. In addition, insurgents may change strategies over time. The type of strategy pursued affects the insurgents' organization, the types of activities they conduct, and the emphasis they place on different activities. It should be noted that insurgents may be inept at the use of a given strategy, or they may misread the operational environment and use an inappropriate strategy. Knowledge of misapplication of strategy or the use of different strategies by different insurgent groups may provide opportunities that can be exploited by counterinsurgents.

Conspiratorial

3-84. A conspiratorial strategy relies on the secret creation of a political body that exploits popular revolution, the actions of other insurgents, or weakness in the government to seize control. Insurgent organizations following this strategy generally conduct very few overt violent or informational actions until they believe the conditions are set for them to seize power. Their cadre is large relative to the number of the combatants in the organization and they may have a small mass base or no mass base at all. The organization will usually be small, secretive, disciplined, and tightly controlled by leadership. Insurgencies following a conspiratorial strategy may be countered effectively by targeting leaders and cadre through direct action. However, information operations and political or economic programs should be used to prevent the government from weakening to the point that conspiratorial insurgents can seize power.

Military-focus

3-85. The military-focus strategy relies almost completely on violence. Followers of this strategy believe the populace shares a general resentment for the government and that violence will create the conditions for overthrowing the government. They further believe that, as the populace sees insurgents fighting the government on their behalf, it will be motivated to support the insurgency. Insurgent organizations following this strategy tend to have leaders and combatants, but little, if any, cadre or mass base. This strategy has rarely been effective historically, although the Cuban revolution is a well-known success. Targeting leaders and combatants through direct action is often the best way to counter an insurgent group using a military-focus strategy. However, political, economic, and social programs can be used to bolster support for the HN government and make it impossible for the insurgents to gain support through violent actions.

Urban

3-86. An urban strategy relies on terrorism in urban areas. The intent of the terrorist activities is to incite repression by COIN forces that bolsters support to the insurgency. In addition, practitioners of the urban strategy may seek to tie-down large numbers of COIN forces in cities so they may gain freedom of movement in rural areas. They may also infiltrate and subvert government and security forces in urban areas. Insurgent organizations using this strategy tend to be composed of small, compartmentalized cells. Like the military-focus strategy, the cadre and mass base in the organization will generally be small relative to the number of combatants. Countering the urban strategy relies on effectively securing cities without being repressive. As in countering other strategies, information operations, political programs, economic programs, and social programs may be used to isolate the insurgents and ensure support to the HN government.

Protracted Popular War

3-87. Protracted popular war is the best-known, most successful insurgent strategy and is most closely associated with Mao Zedong. It relies on the slow development of popular support while simultaneously undermining the will, legitimacy, and popular support of counterinsurgents. This strategy makes the best use of asymmetries between insurgents and counterinsurgents.

3-88. Although Mao ascribed phases to this strategy, it need not follow those phases. The insurgents also need not progress to conventional war. Information and political activities supported by terrorism and/or guerrilla warfare may be sufficient to defeat counterinsurgents. In some cases, information and political activities may be effective without any violent action.

3-89. There are many different ways of structuring an insurgent organization to follow this strategy, and the structure will likely change with both time and location on the battlefield. However, they tend to rely on a large mass base of supporters. Other indicators of this strategy are overt violence, heavy use of information and political activities, and focus on building popular support for the insurgency. Protracted popular war is best countered by winning the "hearts and minds" of the populace and separating the leaders, cadre, and combatants from the mass base through information operations, civil-military operations, economic programs, social programs, and political action.

Traditional Strategy

3-90. A traditional strategy relies on building an insurgency along the lines of ethnic, tribal, or religious groups. Such insurgencies draw heavily upon the identity of a group to differentiate themselves from the government. When this strategy is used effectively, entire communities may join the insurgency together. This strategy is strongly associated with liberation insurgencies as insurgent leaders create an "us-and-them" gap between the government and the ethnic, tribal, or religious group and use this division to encourage others to join the insurgency. Insurgencies following this strategy tend to have a large mass base of passive and active supporters built around pre-existing social networks. There also tend to be many auxiliaries. The number of cadre is generally small, as the insurgency relies on traditional authority structures and is generally not intent on spreading a new political system. The number of combatants varies, but there are often large numbers of part-time combatants. There may or may not be many full-time combatants. Leaders usually come from traditional leadership positions, such as religious leaders, tribal chiefs, or tribal councils. Insurgencies using a traditional strategy can be difficult to counter, particularly if they are fighting to liberate their group from the government. Generally speaking, political or economic deals are necessary to end the conflict.

TACTICAL COURSES OF ACTION

3-91. Insurgents base their tactical courses of action on their capabilities and intentions. The capabilities of an organization are the support, information, political, and violent capabilities evaluated in the third step of IPB. The intentions come from goals, motivations, strategy, culture, perceptions, and leadership personalities. It is important to understand that many different courses of action can happen in any given region at any given time and that tactical courses of action change with both time and location. People and their attitudes, both within the nation and often without, are the ultimate targets of the insurgents in all operations. Therefore, special attention must be paid to what effects insurgent courses of action have on the people and how the insurgents achieve those effects. Finally, it is important to note that tactical actions can have strategic effects. This is because insurgent propaganda and media reporting can reach a global audience, multiplying the effects of insurgent tactical actions. (Tactical courses of action are covered in greater depth in appendix B.)

SECTION III – INTELLIGENCE, SURVEILLANCE, AND RECONNAISSANCE OPERATIONS

3-92. The purpose of intelligence, surveillance, and reconnaissance (ISR) operations is to develop the intelligence necessary for addressing the issues that drive the insurgency. Several factors are particularly important for ISR operations in COIN environments. These include:

648 • A focus on the local populace.

649 • Collection occurring at all echelons.

650 • Localized nature of insurgencies.

651 • All Soldiers and Marines being collectors.

652 • Insurgent use of complex terrain.

653 3-93. Intelligence gaps and information requirements determined during IPB may range from the location
654 of insurgent leaders, to local perceptions of insurgents by the populace, to the status of the HN's political
655 parties. In general, collection focuses on the populace, the insurgents, and the host nation.

656 3-94. The fact that all units collect and report information combined with the localized nature of insurgen-
657 cies means that the flow of intelligence in COIN is more bottom up than top-down. Pushing intelligence
658 collection assets down to the tactical level therefore benefits all echelons. Its advantages include improving
659 the collection capabilities of tactical units, ensuring reports go through appropriate channels to reach
660 higher-echelon audiences, and getting collectors to where the insurgents are.

661 3-95. While collection may occur in any unit and collectors may be pushed down to the lowest levels, the
662 overall intelligence synchronization plan (formerly the collection plan) must remain synchronized so that
663 all echelons receive the information they require. There are several means of accomplishing this. One is to
664 ensure priority intelligence requirements (PIRs) are "nested" at all echelons. They may be tailored to local
665 or regional circumstances, but local and theater collection should support one another. Headquarters moni-
666 tor requests for information from lower echelons and tasking from higher echelons so they can get informa-
667 tion to units that require it in a timely manner. It is important to ensure all Soldiers and Marines understand
668 the PIRs. This enables them to know when they should report something they have seen or heard and what
669 they should report.

670 3-96. Feedback from analysts and operators to collectors is important to synchronize the ISR effort. This
671 response lets collectors know a report is of interest and that there should be follow-up. Such feedback may
672 come from any unit at any echelon.

673 3-97. Also affecting intelligence synchronization is the requirement to work closely with U.S. governmen-
674 tal agencies, HN security and intelligence organizations, and third-country intelligence organizations. Syn-
675 chronization of the efforts of the various agencies and organizations is driven by theater-level ISR plan-
676 ning, but this coordination occurs at all echelons. Communication among collection managers and
677 collectors down to the battalion level is important to ensuring work is not duplicated unnecessarily and cir-
678 cular reporting does not occur. (See the Intelligence Collaboration and Fusion section, below).

679 3-98. Insurgents often seek to use complex terrain to their advantage. Collection managers do not ignore
680 areas of complex terrain. In addition, insurgents use "seams" between maneuver units to their advantage.
681 (Seams are boundaries between units not adequately covered by any unit.) Collection managers must have
682 a means of monitoring seams in order to ensure the enemy cannot establish undetected bases of operation.

THE INTELLIGENCE–OPERATIONS DYNAMIC

684 3-99. Intelligence and operations have a dynamic relationship in all operations. Intelligence drives opera-
685 tions and successful operations generate intelligence. For instance, an operation increasing the security and
686 general happiness of a town often increases the amount of information offered by its inhabitants. This re-
687 sults in more effective operations. The reverse of this is also true. Operations conducted without accurate
688 intelligence may upset the people and cause them to offer less information.

689 3-100. In many cases, tactical units new to an AO have little intelligence on the AO. Operations will be
690 conducted with the purpose of gathering intelligence. Even in permissive environments where a great deal
691 is known about the enemy, there is an intelligence aspect to all operations.

692 3-101. Because intelligence and operations are so closely related, it is important that collectors be linked
693 directly to the analysts and operators they support and remain responsive to their requirements. Further,
694 collectors should not passively wait for requirements, but closely monitor the operational environment and
695 recommend requirements on their own.

696 # HUMAN INTELLIGENCE AND OPERATIONAL REPORTING

697
698
699
700
701
702
703
3-102. *Human intelligence* (HUMINT) is a category of intelligence derived from information collected and provided by human sources (JP 1-02). During COIN operations, much intelligence is based on information gathered from people, to include informants (HUMINT sources) and captured insurgents. HUMINT is often the primary intelligence category in COIN. Because every Soldier and Marine is a potential collector, any interaction with the populace may result in human intelligence reporting. This means that HUMINT will often be based on information from operational reporting as well as dedicated HUMINT collectors.

704
705
706
3-103. The lives of people offering HUMINT are in danger. Insurgents continuously try to defeat ISR collection operations by intimidating or murdering sources. Soldiers and Marines must protect the identity of HUMINT sources.

707
708
709
710
711
3-104. U.S. forces should not expect people to willingly provide information if insurgents have the ability to violently intimidate them. HUMINT reporting increases if COIN forces protect the populace from the insurgents and people begin to believe the insurgents will lose the conflict. To maintain increased cooperation from the people, it is very important that insurgents not be allowed to return and murder or intimidate them. If insurgents do return to an area, COIN forces lose credibility with its inhabitants.

712
713
714
715
716
3-105. Residents may approach patrols and offer them information. Soldiers and Marines should take the information offered and, if appropriate, establish a means of contacting the individual. Patrol members receiving the information should then report it as part of their debriefing. If possible, trained HUMINT personnel should be involved in any contacting of the source. This is to ensure the safety of both U.S. personnel and local residents.

717
718
719
720
721
722
723
3-106. People will often provide inaccurate information to Soldiers and Marines. They may simply be spreading rumors or providing inaccurate information purposefully to use U.S. forces to settle tribal, ethnic, or business disputes. In some cases, insurgents provide bad information to Soldiers and Marines to lead them into ambushes, to get them to undertake an operation that will upset the population, to learn about U.S. planning time and tactical procedures, or to stretch U.S. forces thin by causing them to react to false reports. For these reasons, the accuracy of HUMINT should be verified before being used to support direct action.

724 ## HUMAN INTELLIGENCE NETWORKS

725
726
727
728
729
730
731
732
733
734
735
3-107. An effective means of HUMINT collection is to build a network of HUMINT sources. A HUMINT network provides the COIN equivalent of the reconnaissance and surveillance conducted by scouts in conventional operations. These networks serve as "eyes and ears" on the street and provide an early warning system for tracking insurgent activity. Although Soldiers and Marines will regularly get "walk-in" or "walk-up" intelligence reports from citizens, the only counterintelligence personnel are trained and authorized personnel to operate a HUMINT network. Due to legal considerations, the potential danger informants face if identified, and the potential danger to troops involved, only counterintelligence personnel operate a HUMINT networks and develop HUMINT sources. All Soldiers and Marines may record information given to them by walk-up informants or contacts with whom they have a regular liaison relationship; however, they may not develop HUMINT sources or networks. (When published, FM 2-22.3 will provide complete coverage of HUMINT operations.)

736 ## NONTRADITIONAL HUMAN INTELLIGENCE SOURCES

737
738
739
740
741
742
3-108. As previously stated, the interactions of any Soldier and Marine with the populace can generate HUMINT reporting if troops know PIRs and conduct after-action debriefings. It is therefore imperative that Soldiers and Marines know what information is reportable and that every patrol is debriefed. These debriefings should be as detailed as possible in order to ensure PIRs are accurately answered. In addition, because of their continuous contact with the populace, Soldiers and Marines regularly generate new potential sources for HUMINT personnel to develop. Analysts and HUMINT collectors should work closely

with operations staffs and other personnel to ensure new sources are properly developed. Some valuable sources of HUMINT include the following:

- **Patrol debriefings and after-action reviews**. Patrols regularly encounter individuals offering information and observe new enemy tactics and techniques. Patrol debriefings are especially important to units at brigade level and below, though the information collected can be of higher-echelon significance.

- **Civil affairs action reports**. These are especially useful for gathering information about politics, economy, and infrastructure. Civil affairs personnel also regularly come into contact with individuals offering information.

- **Psychological operations (PSYOP) action reports**. PSYOP personnel conduct opinion polls and gather information on community attitudes, perceptions, and interests/grievances. PSYOP personnel also regularly encounter individuals offering information.

- **Special forces reporting**. Special forces often work closely with local nationals and produce valuable HUMINT reports.

- **Leadership liaison**. Commanders and leaders regularly meet with their counterparts in the HN security forces and with community leaders. These meetings often result in the passing of information or tips.

- **Contracting**. Contracting officers work with theater contractors, both HN and external, performing support functions or building national infrastructure. Contractors may offer information to contracting officers.

- **Multinational operations centers**. These provide a venue for sharing information between the host nation and U.S. personnel.

- **Tips hotlines**. Telephone or e-mail hotlines provide a means for local citizens to provide information without undue exposure to insurgent retribution. They are especially useful for time-sensitive intelligence, such as warning of an attack, or the current location of an insurgent.

- **U.S. persons**. There will be times when U.S. civilians, such as contractors or journalists, offer information to Soldiers and Marines. For legal reasons, it is important to understand regulations regarding intelligence-related information collected from U.S. persons. In addition, military intelligence personnel cannot collect information concerning U.S. persons. (See FM 2-22.3, when published.)

CAPTURED INSURGENTS AND DEFECTORS

3-109. Captured insurgents and defectors from the insurgency are a very important source of HUMINT. The information they provide about the internal workings of an insurgency are usually better than any other HUMINT source can provide. In addition, they have valuable insight into the perceptions, motivations, goals, morale, organization, and tactics of an insurgent organization. Both captured insurgents and defectors should be thoroughly questioned on all aspects of an insurgency discussed in the IPB discussion, above. The answers provided should be compared with captured equipment, pocket litter, and documents to build a better understanding of the insurgency. While properly trained Soldiers and Marines can conduct immediate tactical questioning of captured insurgents or defectors, legally only trained HUMINT personnel can thoroughly interrogate such subjects. (Refer to Chapter 7 and appendix D of this manual, FM 34-52, and FM 2-22.3 [when published] for more information on interrogation of captured personnel.)

SURVEILLANCE AND RECONNAISSANCE CONSIDERATIONS

3-110. Because all Soldiers and Marines are potential collectors, all day-to-day tactical operations are addressed in the ISR plan. This means every patrol or mission should be provided intelligence collection requirements in addition to operations requirements.

3-111. Overt area and zone reconnaissances are excellent means for tactical units to learn more about their AO, to include the terrain, infrastructure, people, government, local leaders, and insurgents. Overt reconnaissance by patrols allows commanders to fill intelligence gaps and develop relationships with local leaders, while simultaneously providing security to the populace.

792
793
794
795
796
797
798
799
800
801
802
803
804

3-112. Covert reconnaissance and surveillance operations employing scouts or concealed observation posts are often ineffective in urban areas, suburban areas, close-knit communities, and other places where the populace is alert and suspicious of outsiders. Under these circumstances, it is very difficult for scouts to conduct reconnaissance or surveillance without being observed by insurgents or people who may tip off insurgents. It is important that commanders and collection managers understand that reconnaissance of a target may be noticed and cause insurgents to leave the area. Likewise, small groups of scouts may be attractive targets for insurgent attacks if the scouts' location is known. For these reasons, the use of a HUMINT network or aerial imagery intelligence platforms is often preferable to ground-based reconnaissance and surveillance. Successful ground reconnaissance in populated areas requires leaders to be creative in how they put personnel in place. For instance, dismounted movements at night, leaving a small "stay behind" observation post while the rest of a patrol moves on, or secretly photographing a place of interest while driving by may all be effective. However, commanders weigh the benefits of these operations with the potential cost of insurgents having early warning of counterinsurgent intentions.

805
CONSIDERATIONS FOR OTHER INTELLIGENCE DISCIPLINES

806
807
808
3-113. An *intelligence discipline* is a well-defined area of intelligence collection, processing, exploitation, and reporting using a specific category of technical or human resources (JP 1-02). The following discussion addresses COIN-specific considerations for selected intelligence disciplines and information types.

809
SIGNALS INTELLIGENCE

810
811
812
813
814
3-114. In conventional environments, signals intelligence (SIGINT) collection is a good source for enemy locations, intentions, capabilities, and morale. The same applies in COIN operations. SIGINT is often helpful for confirming or denying HUMINT reporting and may be the primary source of intelligence in areas under insurgent control. Pushing SIGINT collection platforms down to tactical units can therefore have positive impacts on intelligence collection.

815
OPEN-SOURCE INTELLIGENCE

816
817
818
819
820
821
822
3-115. Open-source intelligence (OSINT) is a valuable tool for understanding the environment. It is often more useful than any other discipline for understanding public attitudes and public support for insurgents and counterinsurgents. OSINT is also an important means of determining the effectiveness of information operations. Monitoring a wide variety of media in multiple languages benefits the COIN effort. If possible, this should occur at every echelon with collection requirements, varying by the kinds of media that are important at each level. For instance, reporting by major news networks often matters a great deal at the theater level, while local newspapers or radio stations may be of importance to a tactical unit.

823
IMAGERY INTELLIGENCE

824
825
826
827
3-116. Imagery intelligence (IMINT): In COIN operations, imagery platforms may be used to conduct surveillance of likely insurgent safe houses and other facilities. Further, aerial IMINT platforms are also effective at detecting unusual movements of personnel and supplies, which can help commanders determine where best to interdict.

828
829
3-117. Static imagery, such as aerial photos of facilities, is useful in operational planning and for detecting long-term changes in structures or activities.

830
831
832
833
3-118. Real-time video, often from aerial surveillance platforms, is critical to assessing whether particular locations are likely sites of insurgent activity and may be used to track insurgents during operations. If flown high enough so insurgents cannot hear the platform, real-time video provides surveillance in areas where it is difficult or impossible to use an observation post.

834
TARGET EXPLOITATION AND DOCUMENT EXPLOITATION

835
836
3-119. Documents and pocket litter, as well as information found in computers and cell phones, can provide critical information analysts need to map insurgent organization, capabilities and intent. Target Ex-

837
838
ploitation (TAREX) and document exploitation (DOCEX) are also of great benefit to interrogators in substantiating what detainees know and whether they are telling the truth.

839
840
841
842
843
844
845
846
847
3-120. TAREX in a COIN environment is like evidence collection in a law enforcement environment. Units must have procedures in place to ensure captured equipment and documents are tracked accurately and attached to the correct insurgents. While this evidence may not need to be sufficient to convict in a court of law (though in some cases it does), it needs to be sufficient to justify the use of operational resources to apprehend the individuals in question. Pushing HUMINT or law enforcement personnel to the battalion level and below can improve TAREX/DOCEX by tactical units. Procedures for ensuring that tactical units get the results of higher-level TAREX/DOCEX are also important. Units must be able to receive intelligence collected from the documents, equipment, and personnel they capture in a timely manner to fully benefit from the operation and adapt to their local situation.

848
PROPERTY OWNERSHIP RECORDS

849
850
851
852
3-121. Property ownership records include census records, deeds, and other means of determining ownership of land and buildings. They help counterinsurgents to determine who should or should not be living in a specific area and aid in securing the populace. In some cases, it may be necessary for Soldiers and Marines to go door to door and collect census data themselves.

853
TECHNICAL INTELLIGENCE

854
855
856
3-122. Insurgents often adapt their tactics, techniques, and procedures rapidly. This may include the use of improvised explosive devices, homemade mortars, and other pieces of customized military equipment. Technical intelligence on insurgent equipment is of benefit for understanding their capabilities.

857
MEASUREMENT AND SIGNATURES INTELLIGENCE

858
859
860
3-123. Measurement and signatures intelligence sensors provide remote monitoring of avenues of approach or border regions for smugglers or insurgents. They can also be used to locate insurgent safe havens and cache sites, aid in targeting, and determining insurgent activities and capabilities.

861
SECTION IV – COUNTERINTELLIGENCE AND COUNTERRECONNAISSANCE

862
863
864
865
866
867
3-124. Insurgents place heavy emphasis on gathering intelligence. They do so using informants, double agents, reconnaissance, surveillance, open-source media, and open-source imagery. Insurgents can potentially use any person interacting with U.S. or multinational personnel as an informant. These include all of the same people that U.S. forces use as potential HUMINT sources. This makes operations security very important. It also means U.S. personnel must carefully screen the contractors, informants, translators, and other personnel working with them.

868
869
870
871
872
3-125. Background screenings should include the collection of personal and biometric data and a search through available reporting databases to determine that the person is not an insurgent. Identification badges may be useful for local nationals working on U.S. and HN government facilities. However, these badges may be forged or stolen, and insurgents can use them to identify people working with the government. Therefore, biometrics is preferable, when available.

873
874
875
876
877
3-126. Insurgents, criminals, political organizations, and disaffected individuals may also provide U.S. personnel with false information. They do so to create animosity between troops and the populace, to get Soldiers and Marines to act against local enemies, to learn about the reaction time and patterns of U.S. operations, and to confuse the COP. This means HUMINT collectors should screen informants and get feedback from analysts and operations personnel on the accuracy of intelligence provided.

878
879
880
881
882
3-127. Insurgents have their own reconnaissance and surveillance networks. Because they usually blend well with the populace, they can execute reconnaissance of targets without easily being identified. They also have an early warning system composed of citizens who inform them of U.S. movements. Identifying the techniques and weaknesses of enemy reconnaissance and surveillance enables commanders to detect signs of insurgent preparations and to surprise insurgents by neutralizing their early warning system.

883
884
885
886
887
888

3-128.   Insurgents may also have a SIGINT capability based on commercially available scanners and ra-dios, wire taps, or captured U.S. equipment. If Soldiers and Marines use unencrypted radios or phones, it is possible for insurgents to collect information from them. From an operations security standpoint, it is there-fore preferable for troops to not use commercial radios or phones. If Soldiers and Marines have to use commercial equipment or unencrypted communications, they should employ brevity codes to reduce insur-gents' ability to collect on them.

889
SECTION V – ANALYSIS

890
891
892
893

3-129.   The purpose of analysis is to fuse raw intelligence into products that support COIN operations. In-telligence analysis in COIN is very challenging. This is due to the need to understand perceptions and cul-ture, the need to track hundreds or thousands of personalities, the local nature of insurgencies, and the ten-dency of insurgencies to change over time.

894
895
896
897

3-130.   Databases are very important for analyzing insurgent activities and personalities. At a minimum, there should be a common theater database of insurgent actions and another containing all intelligence re-porting. These should be accessible by all analysts in theater and include all reporting from all units and organizations in theater. This is to ensure a common operational picture can be developed.

898
899
900
901
902

3-131.   Because all echelons collect and use intelligence, all staffs are heavily involved in analysis. Units are simultaneously intelligence producers and consumers. While this is normal at brigade and above, bat-talion staffs often do not have the personnel to collect patrol debriefs, analyze incoming intelligence from multiple sources, produce finished intelligence products, and disseminate products to the appropriate con-sumers. In many cases brigade intelligence sections may also not be adequate for a COIN environment.

903
904
905
906
907
908
909
910

3-132.   Pushing additional analysts down to battalion and brigade staffs is important to ensure tactical units have the analytical support they require. There are instances when an analyst can be beneficial at the company level. This is the case when a maneuver company must collect large amounts of information on the local populace and insurgents. An analyst can aid the company commander and leaders in the collection and processing of this information and develop an operational picture of the AO. Pushing analysts down to the brigade level and below places analysts closer to collectors, improves the COP, and helps ensure higher echelon staffs receive answers to their PIRs. If no additional analysts are available, commanders may as-sign nonintelligence personnel to work in the intelligence section.

911
912
913
914
915

3-133.   Tactical analysis at the brigade and below is the basis for operational-level intelligence developed at higher echelons. This is due to the bottom-up flow of intelligence in COIN. Battalions and brigades de-velop intelligence for their AOs, while higher echelons fuse it into theaterwide intelligence of the insur-gency. Operational-level intelligence adds national and international politics and their effects on the opera-tional environment.

916
917
918
919
920
921

3-134.   There are two basic kinds of analysis done at all echelons: analysis of enemy actions and network analysis. Analysis of enemy actions is commonly known as current operations. It focuses on what the en-emy is doing now. Network analysis focuses on the people in an AO and develops an understanding of in-terrelationships between them and the ideas and beliefs driving their actions. Generally speaking, current-operations information provides threat warning and metrics of enemy capabilities, while network analysis provides intelligence for targeting and planning.

922
CURRENT OPERATIONS

923
924

3-135.   Current operations intelligence supports a commander's understanding of what the enemy is cur-rently doing. The basic tasks of analysts working in current operations are—

925
926

● Conduct analysis on past and current enemy actions (event-pattern analysis) and look for changes in insurgent tactics or strategy.

927
● Track the effects of friendly operations on the populace and enemy.

928
● Provide intelligence support to ongoing operations.

929
● Disseminate immediate threat warnings to appropriate consumers.

930 3-136. Intelligence for current operations comes from a variety of sources, but operations reports are par-
931 ticularly important. This is because current enemy activities are more often reported by patrols, units con-
932 ducting raids, or observation posts than they are by dedicated intelligence collectors. OSINT will also be
933 important for tracking the effects of information operations. Current operations analysis depends on the in-
934 surgent actions database for determining changes in insurgent tactics and techniques.

NETWORK ANALYSIS

936 3-137. The purpose of network analysis is to provide commanders with an understanding of the insur-
937 gency and what is driving it. Network analysis requires a large investment of time. Analysts may have to
938 spend weeks or months going through large amounts of multiple-source intelligence reporting in order to
939 provide an accurate picture of insurgent groups. The investment is worth it, however. The more accurate
940 and thorough the intelligence on insurgent organizations and the operational environment, the more effec-
941 tive friendly operations become. It is easy for intelligence personnel to be drawn into current operations
942 due to the time-sensitive and high-profile demands of such activity. However, commanders must ensure
943 that network analysis still occurs. The tasks of network analysts are to—

944 • Identify insurgent goals and motivations.
945 • Identify the grievances exploited by insurgents.
946 • Determine how culture, interests, and history inform insurgent and HN decision making.
947 • Understand the links between political, religious, tribal, criminal, and other social networks.
948 • Determine how various networks and groups interact with insurgent networks.
949 • Determine the structure and function of insurgent organizations.
950 • Identify key insurgent activities and leaders.
951 • Understand popular and insurgent perceptions of the host nation, the insurgency, and U.S. mili-
952 tary forces and how these affect the insurgency.

953 3-138. There are many techniques and tools used in network analysis. What makes them all similar is that
954 they examine interactions among individuals, groups, and beliefs within the historic and cultural context of
955 the operational environment. One of the more important products of network analysis is an understanding
956 of how the people in an AO think. This knowledge allows predictive analysis of enemy actions along with
957 the development of effective information operations and civil-military operations. Such cultural analysis is
958 difficult because it requires a Soldier or Marine to understand the mindset of an insurgent or average citi-
959 zen. However, this knowledge is critical to ensuring operations are effective. (Network analysis is covered
960 in more depth in appendix E.)

INTELLIGENCE REACH

962 3-139. Intelligence reach (sometimes called reachback) refers to the ability to exploit resources, capabili-
963 ties, and expertise not physically located in the theater of operations. Deployed or deploying units often use
964 reach capabilities to "outsource" time-intensive aspects of network analysis or socio-cultural factors analy-
965 sis. This is particularly useful when deployments occur with little warning and when organizations used for
966 reach have a great deal of expertise available on a given subject. Analysts may receive reach assistance
967 from the commands above them or from external sources. Many of the resource organizations exist in the
968 continental U.S. and are connected to DOD, while others are not. Most organizations affiliated with DOD
969 regard assisting field commanders as one of their primary missions. Lists of reach organizations and their
970 expertise are available from the Center for Army Lessons Learned at Fort Leavenworth, Kansas <
971 http://call.army.mil/>.

CONTINUITY

973 3-140. The complexity of analyzing an insurgency means it often takes analysts months to fully under-
974 stand the operational environment and the insurgency. For this reason, analysts should maintain situational
975 awareness of the insurgency for as long as possible. This can be accomplished by having intelligence and

976 other staff sections track the fight from their home station. This flattens the learning curve of units rotating
977 in to a theater and increases their effectiveness during the critical first months of deployment.

978 ## SECTION VI – INTELLIGENCE COLLABORATION AND FUSION

979 3-141. The purpose of intelligence collaboration and fusion is to organize the collection and analytical ef-
980 forts of various units and organizations into a coherent and mutually supportive intelligence effort. The
981 common operational picture for an insurgency is complex. Insurgencies don't normally lend themselves to
982 generalizations like "if this leader is removed the insurgency is over" or "this group drives the insurgency."
983 It is important not to oversimplify an insurgency. However, analysts and commanders still require a means
984 of defining and describing the enemy that can be commonly understood. One means of aiding this process
985 is to track and report the insurgency by region, insurgent organization, key personalities, or by insurgent
986 goals and motivations.

987 3-142. As noted earlier, insurgencies are often localized but have national or international aspects to
988 them. This complicates intelligence fusion between adjacent units and various echelons. For instance, if
989 numerous insurgent groups operate in one country, adjacent battalions may face very different threats. The
990 higher-echelon analysts must then understand multiple insurgent organizations and determine what links
991 there are, if any, between them. Usually, battalions focus on the population and insurgents in their AO.
992 Higher-echelon analysts determine links and interactions among the people and insurgents across unit
993 boundaries. Theater-level analysts determine the major linkages within the theater and internationally.
994 Based on these requirements, a common database based on intelligence reporting is a prerequisite for effec-
995 tive intelligence fusion.

996 3-143. Also complicating fusion is the fact that many government agencies and foreign security forces are
997 involved in COIN operations. Analysts must establish good working relationships with various agencies in
998 order to ensure they are able to fuse intelligence.

999 # INTELLIGENCE CELL AND WORKING GROUPS

1000 3-144. Intelligence community assets operating in an AO work in or coordinate with the intelligence cell
1001 in one of the unit's command posts under the staff supervision of the unit intelligence officer. Examples of
1002 intelligence community assets operating in a division AO include the following:

1003 - Defense Intelligence Agency.
1004   - Defense HUMINT case officers.
1005   - Reports Officers.
1006   - Document exploitation teams.
1007 - Central Intelligence Agency.
1008   - Chief of base.
1009   - Case officers.
1010   - Reports officers.
1011 - Air Force Office of Special Investigations special agents.
1012 - U.S. Special Operations Command.
1013   - Special forces operational detachment–alpha teams.
1014   - Special mission units.
1015   - Civil Affairs teams.
1016   - PSYOPS teams.
1017 - Federal Bureau of Investigation (FBI) Agents.
1018 - Immigration and Customs Enforcement/Department of Homeland Security agents.
1019 - State Department political advisor.
1020 - National intelligence support team.
1021 - National Security Agency.

- National Geospatial-Intelligence Agency.
- Treasury Department.

3-145. As necessary, intelligence officers form working groups/boards to synchronize collection, analysis, and targeting efforts. (See FMI 5-0.1, paragraphs 2-30, 2-38, and 2-43.) Cells and working groups conduct regular meetings to establish and maintain joint situational awareness, share collection priorities, deconflict activities and operations, discuss target development, and share results of operations. These meetings build mutual trust and understanding of each member's mission, capabilities, and limitations. They are integrated with meetings of other staff cells and working groups/boards (for example, the targeting board) as part of the command post's battle rhythm established by the chief of staff. (See FMI 5-0.1, paragraphs 2-89 through 2-91.)

3-146. An effective intelligence cell enhances the commander's knowledge of enemy activity, local peculiarities, and friendly forces operating in the unit's AO. Incorporating HN representatives (for example, intelligence services, military forces, and local government officials) and multinational partners into the intelligence cell should also be considered in a COIN environment to foster teamwork, gain insight into local customs and activities, and prepare the host nation to assume the mission when multinational forces depart the area.

PROTECTING SOURCES

3-147. Protection of sources is another important consideration when sharing intelligence. Acting on intelligence can compromise its sources. This is why organizations may sometimes choose to not share intelligence. However, use of the targeting process to synchronize targeting decisions is usually a better means of protecting sources. This is because the targeting board draws on information from the COP rather than specific sources. (See the targeting discussion in chapter 5.)

HOST-NATION INTEGRATION

3-148. COIN operations require U.S. personnel to work closely with the host nation. Sharing of intelligence with HN security forces and government personnel is an important and effective means of supporting their COIN efforts. However, it should be noted that HN intelligence services may not be well developed and their personnel may not be well trained. For this reason, HN intelligence should be seen as another useful source of intelligence, but definitely not as the only one, and usually not as the most important one. It is essential for U.S. personnel to evaluate the HN's intelligence capabilities and offer training, as required.

3-149. In addition, infiltration of the HN security forces by insurgents or foreign intelligence services can create drawbacks to intelligence sharing. Insurgents may learn what is known about them, gain insight into intelligence sources and capabilities, and get early warning of targeting efforts.

3-150. When sharing intelligence with the host nation, it is important to understand the level of infiltration by insurgents or foreign intelligence services. Insofar as possible, intelligence should be tailored so required intelligence still gets to the HN consumers, but it does not give away information about sources and capabilities. In addition, care should be taken when providing targeting information and it should be done in such a way that insurgents do not receive early warning of an upcoming operation. As trust develops between the HN and U.S. personnel, the amount of intelligence shared should grow. This will make the COIN effort more effective.

SECTION VII – SUMMARY

3-151. What makes intelligence analysis for COIN so distinct and so challenging is the amount of cultural information that must be gathered and understood. However, to truly grasp the environment of operations, commanders and their staffs must expend at least as much effort understanding the people they are supporting as the enemy. All this information is essential to get at the root causes of the insurgency and to determine the best ways to combat it. Identifying the real problem and developing solutions is the essence of the operational design process that is discussed in the next chapter.

1069
1070

1 | **Chapter 4**

2 | **Designing Counterinsurgency Operations**

3 | *The first, the supreme, the most far-reaching act of judgment that the statesman and the*
4 | *commander have to make is to establish...the kind of war on which they are embarking;*
5 | *neither mistaking it for, nor trying to make it into something that is alien to its nature.*
6 | *This is the first of all strategic questions and the most comprehensive.*
7 | Carl von Clausewitz

8 | This chapter describes considerations for designing counterinsurgency campaigns
9 | and the associated planning and operations. While campaign design is most often as-
10 | sociated with a joint force command, understanding of design is necessary by all
11 | commanders and staffs.

THE IMPORTANCE OF CAMPAIGN DESIGN

13 | 4-1. In chapter 1, insurgency was defined as an organized, protracted politico-military struggle designed
14 | to weaken government control and legitimacy while increasing insurgent control. Ultimately the long-term
15 | objective for both sides in that struggle remains acceptance by the people of the state or region of the le-
16 | gitimacy of one side's claim to political power. If insurgency is a struggle between an insurgent group and
17 | government authorities over the granting of legitimacy by the populace, then where does the struggle begin
18 | and over what? The reason why an insurgency forms to challenge the existing order is different in each
19 | case. For a U.S. military commander directed to counter an insurgency, knowing "why" an insurgency
20 | movement has gained support and the purpose of American involvement is essential in designing the coun-
21 | terinsurgency (COIN) campaign. Failure to do this can have disastrous consequences, as illustrated by Na-
22 | poleon's experience in Spain.

23 | **The Necessity for Campaign Design in COIN: Napoleon in Spain**
24 |

25 | Counterinsurgents always face a series of intricate and shifting challenges. The best
26 | course of action to solve this complex set of problems is often difficult to recognize
27 | because of complicated interdependencies between them. The apparent resolution
28 | of one issue may reveal or create another, even more complicated problem. The
29 | purpose of COIN campaign design is to achieve a greater understanding of the
30 | situation, create a proposed solution based on that understanding, and maintain the
31 | ability to learn and adapt as the campaign is executed. During Napoleon's occupa-
32 | tion of Spain in 1808, little thought was given to the potential challenges of subduing
33 | the Spanish populace. Conditioned by the decisive victories at Austerlitz and Jena,
34 | Napoleon believed the conquest of Spain would be little more than a "military prome-
35 | nade." Napoleon's campaign included a rapid conventional military victory over
36 | Spanish armies but ignored the immediate requirement to provide a stable and se-
37 | cure environment for the people and the countryside.

38 | The French should have expected ferocious resistance. The Spanish people were
39 | accustomed to hardship, suspicious of foreigners, and constantly involved in skir-
40 | mishes with security forces. The French failed to analyze the history, culture, and
41 | motivations of the Spanish people, or to seriously consider their potential to support
42 | or hinder the achievement of French political objectives. Napoleon's cultural miscal-

43
44
45
46
47
48
49
50

culation resulted in a protracted struggle that lasted nearly six years and ultimately required approximately three-fifths of the French Empire's total armed strength, almost four times the force of 80,000 Napoleon originally had designated for this theater. The Spanish resistance drained the Empire's resources and was the beginning of the end of Napoleon's reign. Despite his reputation for brilliant campaign planning, in this instance Napoleon had failed to grasp the real situation in the theater, and his forces were not capable of learning and adapting for the unexpected demands of counterinsurgency.

51
52
53
54
55
56
57
58
59

4-2. Design and planning are qualitatively different yet interrelated activities essential for solving complex problems. While planning activities receive consistent emphasis in both doctrine and practice, discussion of design remains largely abstract and is rarely practiced. Presented a problem, staffs often rush directly into planning without a clear understanding of the complex environment of the situation, the purpose of military involvement, and the approach required to address the real core issues. This situation is particularly problematic when dealing with insurgencies. Campaign design informs and is informed by planning and operations with an intellectual foundation that aids continuous assessment of operations and the operational environment. The commander should lead this design process and communicate the resulting design framework to other commanders for planning and execution.

60

THE RELATIONSHIP BETWEEN DESIGN AND PLANNING

61
62
63
64
65
66
67

4-3. It is important to understand the distinction between design and planning. (See figure 4-1.) While both activities seek to formulate ways to bring about preferable futures, they are cognitively different. Planning applies established procedures to solve a largely understood problem within an accepted framework. Design inquires into the nature of a problem to conceive a framework that can be used for solving that problem. In general, planning is problem *solving*, while design is problem *setting*, Where planning focuses on generating a plan—a series of executable actions—design focuses on learning about the nature of an unfamiliar problem.

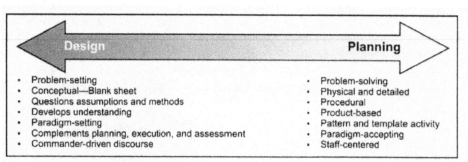

68

Figure 4-1. Design planning continuum

69
70
71
72
73
74
75
76
77
78

4-4. When situations do not conform to established frames of reference—when the hardest part of the problem is figuring out what the problem is—planning alone is inadequate and design becomes essential. In these situations, absent a design process to come to grips with the essential nature of the problem, planners default to doctrinal norms, developing plans based on the familiar rather than an understanding of the real situation. Design provides a means to conceptualize and hypothesize about the underlying causes and dynamics that explain an unfamiliar problem. Design provides a means to gain understanding of a complex problem and insights towards achieving a workable solution. Although design should precede planning, design is continuous throughout the operation, continuously testing and refining to ensure the relevance of military action to the situation. In this sense, design guides and informs planning, preparation, execution, and assessment. However, planning is necessary to translate a design into execution.

79
80
81

4-5. Planning focuses on the physical actions intended to have a direct effect on the enemy or environment. A planner typically is assigned a mission and a set of resources, and required to devise a plan to use those resources to accomplish that mission. The planner starts with a design (whether explicit or implicit)

82 and focuses on generating a plan—a series of executable actions and control measures. Planning generally
83 is analytical and reductionist. It breaks the design into manageable pieces assignable as tasks, which is es-
84 sential to transforming the design into an executable plan. Planning implies a stepwise process in which
85 each step produces an output that is the necessary input for the next step.

THE NATURE OF DESIGN

4-6.   Given the difficult and multifaceted problems an insurgency presents, dialogue among the commander, principal planners, members of the interagency team, and host-nation (HN) representatives provides many benefits to the design process. This involvement of all participants is essential for effective
COIN. The object of this discourse is to achieve a level of situational understanding at which the approach
to the solution of the problem becomes clear. The underlying premise is that when participants achieve a
sufficient level of understanding the situation no longer appears complex, they can exercise logic and intuition effectively. As a result, design focuses on rationalizing the problem rather than explicitly developing
courses of action.

4-7.   COIN campaign design must be iterative because, by their nature, COIN efforts require repeated assessments from different perspectives to see all the various factors and relationships required to gain adequate understanding. Every iteration is an opportunity to learn more about the situation and make incremental improvements to the design. The design team engages in constructing and continuously modifying
two complementary logics, or mental models. The first is the governing logic of the problem. The aim here
is to rationalize the problem situation—to construct a logical explanation, in the form of an abstract model,
of events observed in the real world. The second is the counterlogic—the guiding logic of the campaign
that unravels the problem logic. The essence of this counterlogic is the success mechanism or sequence of
interactions envisioned to achieve a desired end state. This mechanism may not be a military activity—or it
may involve military actions in support of activities in other lines of operations. Once the designers have
constructed these models they are obligated to identify measures of effectiveness and related information
requirements that allow them to test the models over time. This feedback becomes the basis for subsequent
design iterations, which refine or reconstruct the logics.

ELEMENTS OF DESIGN

4-8.   The key elements of the design process include the following:
- Discourse.
- Systems thinking.
- Model making.
- Intuitive decision making.
- Continuous assessment.
- Structured learning.

4-9.   *Discourse* is rigorous and structured critical discussion that provides an opportunity for interactive
learning, deepening shared understanding, and leveraging the collective intelligence and experiences of
many actors to enable design.

4-10. *Systems thinking* involves developing an understanding of the relationships within the insurgency
and the environment, also the relationships of actions within the various lines of operations. This element is
based on the perspective of the system sciences that seeks to understand the interconnectedness, complexity, and wholeness of the elements of systems in relation to one another.

4-11. In *model making*, the model describes an approach to the COIN campaign, initially as a hypothesis.
The model includes operational terms of reference and concepts that shape the language governing the
conduct (planning, preparation, execution, and assessment) of the operation. It addresses questions like
these: Will planning, preparation, execution, and assessment activities use traditional constructs like center
of gravity, decisive points, and lines of operations? Or are other constructs—such as leverage points, fault
lines, or critical variables—more appropriate to the situation?

129
130
131
132
4-12. *Intuitive decision making* involves the act of reaching a conclusion that emphasizes pattern recognition based on knowledge, judgment, experience, education, intelligence, boldness, perception, and character. This approach focuses on assessment of the situation vice comparison of multiple options (Army/Marine Corps). The design emerges intuitively as understanding of the insurgency deepens.

133
134
135
4-13. *Continuous assessment* is essential as an operation unfolds because of the inherent complexity of COIN. Any design or model necessarily has mismatches with reality. The object of continuous assessment is to identify where and how the design is working or failing and to consider opportunities for redesign.

136
137
138
4-14. The objective of *structured learning* is to develop a reasonable initial design and then learn, adapt, and iteratively and continuously improve that design as more about the dynamics of the COIN problem become evident.

CAMPAIGN DESIGN FOR COUNTERINSURGENCY

139

140
141
142
143
144
145
146
147
148
149
4-15. Through design commanders gain an understanding of the problem and the intervention's purpose within the strategic context. Communicating this understanding of the problem, purpose, and context to subordinates allows them to exercise subordinates' initiative. (See FM 6-0, paragraphs 2-83 through 2-92.) It facilitates decentralized execution and iterative assessment of operations at all levels throughout the campaign. While traditional aspects of campaign design as expressed in joint and Service doctrine remain relevant, they are not adequate for a discussion of the broader design construct for a COIN environment. Inherent in this construct is the tension created by understanding that military capabilities provide only one component of an overall approach to a COIN campaign. Design of a COIN campaign must be viewed holistically. Only a comprehensive approach employing all relevant design components, including the other instruments of national power, is likely to reach the desired end state.

150
151
152
153
154
155
156
157
4-16. Campaign design begins with identification of the campaign's purpose, as derived from the policy aim. The purpose of the campaign provides context and logic for operational and tactical decision making. Consequently, strategic goals must be communicated clearly to commanders at every level. While strategy drives campaign design, which in turn drives tactical actions, the reverse is also true. The observations of tactical actions result in learning and greater understanding that may generate modifications to the campaign design, which in turn may have strategic implications. The COIN imperative to "learn and adapt" is essential in making the campaign design process work correctly. Figure 4-2 (below) illustrates the iterative nature of COIN campaign design and the large number of factors involved.

158

THE OPERATIONAL NARRATIVE

159
160
161
162
163
164
165
166
167
168
4-17. Guided by the purpose of the campaign, commanders articulate an operational logic for the campaign that expresses in clear, concise, conceptual language a broad vision of what they plan to accomplish. The operational logic is the commander's assessment of the problem and approach toward solving it. Ideally, the operational logic is expressed in comprehensive terms, such as what the commander envisions achieving with various components or particular lines of operations. Included in this operational logic is a narrative of how the commander sees the campaign achieving the desired end state and a rationale for the design and required actions. This narrative, or "vision of resolution," is expressed as a broad goal. It helps subordinate commanders and planners, as well as members of other agencies and organizations, see the direction the campaign is going. In the same way, the vision of resolution can act as a unifying theme for interagency planning.

169

COMPLEX PROBLEMS DEMAND COMPREHENSIVE SOLUTIONS

170
171
172
173
174
175
176
4-18. As in any other operation, effective problem solving is dependant on a detailed understanding of the environment. This is particularly important in COIN. Although members of the force never achieve perfect situational understanding, their understanding should be aspirational—a continuous learning affair. The COIN environment is more complex and uncertain than that associated with conventional military operations due to the preeminence of a difficult and tangled web of complicated societal issues. Identification of the problem is one of the most challenging aspects of COIN operations, because that task demands an appreciation of how social, cultural, political, economic, and physical conditions have supported or even fo-

177 mented discontent among the population. The problem and potential solutions must therefore be viewed
178 from a comprehensive campaign perspective. As has been noted earlier, the development of a campaign
179 should include discourse among joint, interagency, multinational and HN partners. Given a shared appre-
180 ciation of the root causes of discontent, the respective capabilities of the partners involved in COIN can be
181 applied in a cohesive way. In order to develop a comprehensive campaign design, the following lines of
182 operations can be employed as a means to both assess the problem and articulate the commander's vision
183 of resolution:

184 ● Training and advising HN security forces.
185 ● Essential services.
186 ● Economic development.
187 ● Promotion of governance.
188 ● Information.
189 ● Combat operations (protection of the civil populace).

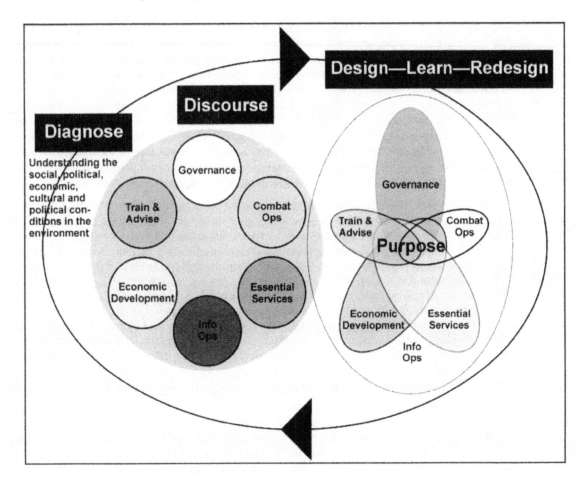

190 **Figure 4-2. Iterative campaign design**

191 4-19. Not intended as a "success template," these lines of operations require judgment in application. The
192 mosaic nature of COIN and the shifting circumstances within each area of operations (AO) requires a dif-
193 ferent emphasis on and interrelationship among the various lines. The situation may also require that mili-
194 tary forces closely support, or temporarily assume responsibility for, tasks normally accomplished by other
195 government agencies and private organizations. By broadly describing how the lines of operations interact
196 to achieve the end state, the commander provides the operational logic to link the various components in a
197 comprehensive campaign framework. This framework guides the initiative of subordinate commanders as

198 they establish local conditions that support the overall campaign design. It also promotes unity of effort
199 among joint, interagency, multinational, and HN partners.

LOCAL RELEVANCE

201 4-20. Informed by that operational logic, subordinate commanders tailor and prioritize their actions within
202 the lines of operations based on the distinct and evolving circumstances within their respective AOs. While
203 military forces are accustomed to unity of command, the interagency and multinational nature of COIN op-
204 erations will likely make such arrangements unlikely. Not all participating organizations share the same at-
205 titudes and goals. General cooperation on matters of mutual concern, established through informal agree-
206 ments, may be the most practicable arrangement. The campaign design must therefore guide and empower
207 subordinate leaders to conduct the coordination, cooperation, and innovation required to achieve the cam-
208 paign purpose in a manner best suited to local conditions.

LEARNING IN EXECUTION

210 4-21. Before commanders deploy their units, they make every effort to mentally prepare their Soldiers or
211 Marines for the anticipated challenges, with a particular focus on environmental understanding of the an-
212 ticipated AO. This environmental understanding is only an initial appreciation, but it is a chance for com-
213 manders to establish a common frame of reference. Design begins based on this initial understanding. Ele-
214 ments of the problem and the means of solving them do not remain static. Conditions are seldom consistent
215 throughout any given operational environment and continue to change based on stimulus from friendly, en-
216 emy, neutral, and other involved parties. Rather than being uniform in character, such environments are
217 likely to consist of a complex and shifting mosaic of conditions. To be effective, commanders—and indeed
218 all personnel—must continually develop and enhance their understanding of the mosaic peculiar to their
219 AO. Observing tactical actions and the resulting changing conditions deepens understanding of the envi-
220 ronment and enables commanders to relearn and refine their design and implementation actions.

221 4-22. Initially, understanding of the environment will probably be relatively low and the campaign design
222 will, by necessity, require a number of assumptions, especially with respect to the populace and the force's
223 ability to positively influence their perception of events. The campaign design can be viewed as an experi-
224 ment that tests the operational logic, with the expectation of a less than perfect solution. As the experiment
225 unfolds, interaction with the population and the insurgents reveals the validity of those assumptions, identi-
226 fying the strengths and weaknesses of the design. Effective assessment, the ability to recognize changing
227 conditions and determine their meaning, is crucial to successful adaptation and innovation by commanders
228 within their respective AOs. A continuous discourse among commanders at all echelons provides the feed-
229 back necessary to refine the campaign design. The discourse is supported by formal assessment techniques
230 and red teaming to ensure commanders are fully cognizant of the causal relationships between their actions
231 and the adversary's adaptations. Accordingly, assessment is a learning activity and a critical aspect of de-
232 sign. This learning leads to redesign. Therefore, design can be viewed as a perpetual design-learn-redesign
233 activity, with the campaign purpose and commander's vision of resolution providing the unifying themes.

234 4-23. The critical role of assessment necessitates establishing measures of effectiveness during the design
235 of COIN operations. Commanders should choose these carefully so that they always align with the cam-
236 paign's purpose and reflect the emphasis on and interrelationship among the lines of operations. Sound as-
237 sessment blends qualitative and quantitative analysis with the judgment and intuition of all leaders. Great
238 care must be applied here, as COIN often involves complex societal issues that may not lend themselves to
239 quantifiable measures of effectiveness. Moreover, bad assumptions and false data can undermine the valid-
240 ity of both the measures of effectiveness and the conclusions drawn from them. Data and metrics can in-
241 form a commander's assessment. However they must not be allowed to dominate it in uncertain situations.
242 Subjective and intuitive assessment must not be replaced by a focus on data or metrics. Commanders must
243 exercise their professional judgment in determining the proper balance.

GOALS IN INTERVENTION

245 4-24. In an ideal world, the commander of military forces engaged in COIN would enjoy clear and well-
246 defined goals for the campaign from the very beginning. However, the reality is that many goals emerge

247 only as the campaign develops. For this reason, COIN forces usually have a combination of defined and
248 emerging goals on which to work. Likewise, the complex problems encountered during COIN operations
249 can be so difficult to understand that a clear design cannot be developed at the outset. Often, the best
250 choice is to create iterative solutions to better understand the problem. In this case, these iterative solutions
251 allow the initiation of intelligent interaction with the environment. The experiences of the 1st Marine Divi-
252 sion during Operation Iraqi Freedom II illustrate this situation.

253
254

Iterative Design During Operation Iraqi Freedom II

255 During Operation Iraqi Freedom (OIF) II, the 1st Marine Division employed an opera-
256 tional design similar to that used during the Philippine Insurrection. (See figure 4-3,
257 below.) The commanding general, Major General James N. Mattis, began with an
258 assessment of the people that the Marines, Soldiers, and Sailors would encounter
259 within his division's AO. The division's AO during OIF II was in western Iraq's Al An-
260 bar Province, with a considerably different demographic than the *imam*-led Shia ar-
261 eas that dominated OIF I operations.

262 Major General Mattis classified provincial constituents into three basic groups: the
263 tribes, the former regime elements, and the foreign fighters. The tribes constituted
264 the primary identity group in Al Anbar. They had various internal tribal affiliations and
265 looked to a diverse array of sheiks and elders for leadership. The former regime ele-
266 ments were a minority that included individuals with personal, political, business, and
267 professional ties to the Ba'ath Party. These included the civil servants and career
268 military personnel with the skills to run government institutions. Initially, they saw little
269 gain from a democratic Iraq. The foreign fighters were a small but dangerous minority
270 of transnational Islamic *jihadists*. To be successful, U.S. forces had to apply a differ-
271 ent approach to each of these groups within the framework of an overarching plan.
272 As in any society, some portion of each of these groups was composed of a criminal
273 element, further complicating planning and interaction. Major General Mattis's "vision
274 of resolution" was composed of two major elements encompassed in an overarching
275 "bodyguard" of information operations.

276 The first element, and the main effort, was reducing support for insurgency. Guided
277 by the maxims of "first do no harm" and "no better friend, no worse enemy," the ob-
278 jective was to establish a secure local environment for the indigenous population so
279 people could pursue their economic, social, cultural, and political well-being, and
280 achieve some degree of local normalcy. Establishing a secure environment involved
281 both offensive and defensive operations, with a heavy emphasis on training and ad-
282 vising the security forces of the fledgling Iraqi government. It also included putting the
283 population to work. Simply put, an Iraqi with a job was less likely to succumb to ideo-
284 logical or economic pressure to support the insurgency. Other tasks included the de-
285 livery of essential services, economic development, and the promotion of govern-
286 ance, all geared towards increasing employment opportunities and furthering the
287 establishment of local normalcy. Essentially, diminishing support for insurgency was
288 about gaining and maintaining the support of the tribes, as well as converting as
289 many of the former regime members as possible. "Fence-sitters" were considered a
290 winnable constituency and addressed as such.

291 The second element involved neutralizing the bad actors, a combination of irrecon-
292 cilable former regime elements and foreign fighters. Offensive combat operations
293 were conducted to defeat recalcitrant former regime members. The task was to make
294 those who were not killed outright see the futility of resistance and give up the fight.
295 With respect to the hard-core extremists, who would never give up, the task was
296 more straightforward: their complete and utter destruction. Neutralizing the bad ac-
297 tors supported the main effort by improving the local security environment. Neutrali-

298
299
zation had to be accomplished in a discrete and discriminate manner, however, to avoid unintentionally increasing support for insurgency.

300
301
302
303
304
305
306
307
308
Both elements described above were wrapped in an overarching "bodyguard" of information operations. Information operations, both proactive and responsive, were aggressively employed to favorably influence the populace's perception of all coalition actions while simultaneously discrediting the insurgents. These tasks were incredibly difficult for a number of reasons. Corruption had historically been prevalent among Iraqi officials, generating cynicism toward government. Additionally, decades of Arab media mischaracterization of U.S. actions had instilled distrust of American motives. The magnitude of that cynicism and distrust highlighted the critical importance of using information operations to influence every situation.

309
310
311
312
313
314
315
316
317
In pursuing this "vision of resolution" the 1st Marine Division faced an adaptive adversary. Persistent American presence and interaction with the populace threatened the insurgents and caused the adversary to employ more open violence in selected areas of Al Anbar Province. This response resulted in learning and adaptation within the 1st Marine Division. The design enabled 1st Marine Division to adjust the blend of "diminishing support for insurgents" and "neutralizing bad actors" to meet the local challenges. Throughout the operation, 1st Marine Division continued learning and adapting with the espoused vision providing a constant guide to direct and unify the effort.

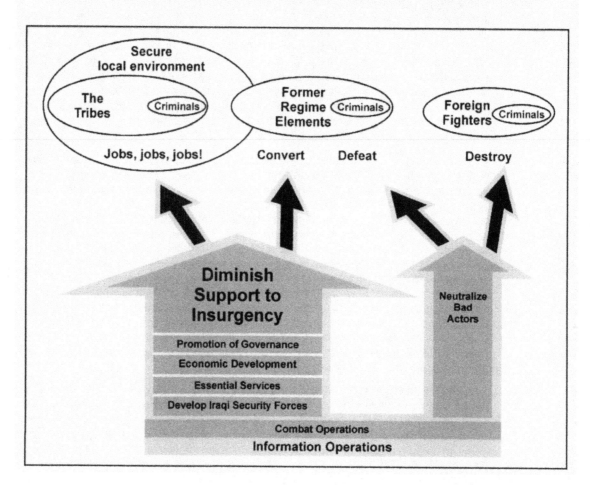

318
Figure 4-3: 1st Marine Division's operational design for Operation Iraq Freedom II

319 **SUMMARY**

320 4-25. Campaign design may very well be the most important aspect in countering an insurgency. It is cer-
321 tainly the area in which the commander and staff can have the most influence. Design is not a function to
322 be accomplished, but rather a living process. Design should reflect ongoing learning and adaptation and the
323 growing appreciation COIN forces share for the environment and all actors within it, especially the insur-
324 gents, populace, and HN government. Though design precedes planning, it does not stop when aspects are
325 turned over for planning, preparation, and execution. Rather, it is dynamic, even as the environment and
326 the counterinsurgents' understanding of the environment is dynamic. The resulting growth in understand-
327 ing requires integrated assessment and a rich dialogue among leaders at various levels and throughout the
328 entire cross-section of the COIN force. Design should reflect a comprehensive approach that works across
329 all relevant lines of operations in a manner applicable for the stage of a campaign in which it exists. There
330 should only be one campaign and therefore one campaign design. This single campaign should bring in all
331 players, with particular attention placed on the HN participants. Campaign design and operations are inte-
332 gral to the COIN imperative to "learn and adapt," enabling a continuous cycle of design-learn-redesign to
333 achieve the campaign's purpose.

334
335
336
337

Chapter 5

Executing Counterinsurgency Operations

It is a persistently methodical approach and steady pressure which will gradually wear the insurgent down. The government must not allow itself to be diverted either by countermoves on the part of the insurgent or by the critics on its own side who will be seeking a simpler and quicker solution. There are no short-cuts and no gimmicks.

Sir Robert Thompson,
Defeating Communist Insurgency: The Lessons of Malaya and Vietnam

This chapter addresses principles and tactics for executing counterinsurgency (COIN) operations. It begins by describing the different stages of a COIN operation and logical lines of operations that commanders can use to design one. It continues with discussions of three COIN approaches and how to continuously assess a COIN operation. The chapter concludes by describing lethal and nonlethal targeting in a COIN environment.

THE NATURE OF COUNTERINSURGENCY OPERATIONS

5-1. Counterinsurgency (COIN) is a violent political struggle waged with military means. The political issues at stake for the contestants defy nonviolent solutions because they are often rooted in culture, ideology, societal tensions, and injustice. Military forces can compel and secure but cannot, by themselves, achieve the necessary political compromise among the protagonists. A successful COIN force—which includes civilian agencies, U.S. military forces, and multinational forces—purposefully attacks the insurgency rather than just its fighters and addresses the host nation's core problems in a comprehensive fashion. The host-nation (HN) leaders must be purposefully engaged in this comprehensive approach.

5-2. There are five overarching considerations in COIN operations:
- The commander and HN government together must select the logical lines of operations for attacking the insurgents' strategy and focusing their effort to establish government legitimacy.
- The HN and COIN forces must establish control of one or more areas from which to operate. The HN forces must secure the people continuously within these areas.
- Operations should be initiated from the HN government's areas of strength against areas under the insurgents' control. The HN must regain control of the major population centers to achieve stability.
- Regaining control of insurgent areas requires the HN government to expand operations to secure and support the population. If the insurgency has established firm control of a region, its military apparatus there must be eliminated and its politico-administrative apparatus rooted out.
- Information operations must be aggressively employed to accomplish the following:
  - Favorably influence perceptions of HN legitimacy and capabilities.
  - Garner local, regional, and international support for COIN operations.
  - Publicize insurgent violence.
  - Discredit insurgent propaganda.

5-3. COIN operations combine offensive, defensive, and stability operations to achieve a stable environment for the growth of government legitimacy. COIN operations are conducted by an integrated joint, interagency, and multinational team working with a HN government. The weight of effort in COIN opera-

43 tions generally progresses through three indistinct stages of development that can be envisioned as a medi-
44 cal analogy. Understanding this evolution and recognizing the relative maturity of the operational envi-
45 ronment are important for operational design, planning, and execution. With this knowledge, commanders
46 can ensure that their activities are appropriate to the current situation.

47 ### INITIAL STAGE: "STOP THE BLEEDING"

48 5-4. Initially, COIN operations are characterized by emergency first aid for the patient. The intent is to
49 protect the population from further injury, set the insurgency back (break their initiative and momentum),
50 and set the conditions for further engagement. Limited offensive operations may be undertaken, but not at
51 the expense of stability operations focused on civil security. During this stage, civil security, force protec-
52 tion, the common operational picture, intelligence collection, and initial assessments and estimates are be-
53 ing developed. COIN forces also begin shaping the information environment, including the expectations of
54 the local populace.

55 ### MIDDLE STAGE: "IN-PATIENT CARE—RECOVERY"

56 5-5. This stage is characterized by efforts aimed at assisting the patient through long-term recovery or
57 restoration of health—which in this case is stability. The COIN force is most active here, working aggres-
58 sively along all lines of operations. The desire in this stage is to develop and build resident capability and
59 capacity in the HN government and security forces. As civil security is assured, focus expands to include
60 civil control, provision of essential services, and the stimulation of economic development. Relationships
61 with HN counterparts and the local population are developed and strengthened. With the establishment of
62 these relationships with the local government agencies, the public, and the HN security forces comes an in-
63 crease in the flow of human intelligence and other types of intelligence. These facilitate measured offen-
64 sive operations against the insurgency in conjunction with the HN security forces. The host nation in-
65 creases its legitimacy through the provision of security for its people, the expansion of effective HN
66 governance and provision of essential services, and achieving incremental success in meeting the public's
67 expectations.

68 ### LATE STAGE: "OUT-PATIENT CARE—MOVEMENT TO SELF-SUFFICIENCY"

69 5-6. Stage three is characterized by the expansion of stability operations across contested regions, ideally
70 using HN forces. The main goal for this stage is on transition to HN leadership and execution of opera-
71 tions. In this mature stage, the multinational force works with the host nation in an increasingly supporting
72 role, turning over responsibility wherever and whenever appropriate. Reaction forces and fire support ca-
73 pabilities may still be needed in some areas, but more functions along all lines of operations will be per-
74 formed by HN forces with the low-key assistance of multinational advisors. As the security, governing, and
75 economic capacity of the host nation increases, the need for foreign assistance is reduced. At this stage, the
76 host nation has established (or re-established) the requisite systems needed to provide effective and stable
77 government that supports and sustains the rule of law. The government secures its citizens continuously,
78 sustains and builds legitimacy through effective government, effectively isolates the insurgency, and is able
79 to manage and incrementally meet the expectations of the nation's entire population.

80 # LOGICAL LINES OF OPERATIONS IN COUNTERINSURGENCY

81 5-7. The technique of logical lines of operations (LLOs) emerged to address situations when positional
82 reference to enemy or adversary forces has little relevance. Using LLOs is an appropriate technique for
83 synchronizing operations against an enemy that hides among the population. A plan based on LLOs unifies
84 the efforts of joint, interagency, multinational, and HN forces toward a common purpose. Each LLO repre-
85 sents a conceptual category along which the HN government and commander of the COIN force intend to
86 attack the insurgent strategy and establish government legitimacy. The LLOs are closely related and de-
87 pend upon closely coordinated action among all elements of the COIN force.

88 5-8. Success in one LLO complements activities in other LLOs. Progress in each LLO contributes toward
89 the attainment of a stable and secure environment for the host nation. Stability is reinforced by popular

90 recognition of the legitimacy of the HN government, improved governance, and progressive, substantive
91 reduction of the root causes of the insurgency. LLOs combine the following:

92 ● Information operations.

93 ● Offensive, defensive, and stability operations.

94 ● Training and employment of HN security forces.

95 ● Establishment or restoration of essential services.

96 ● Better governance.

97 ● Support for economic development.

98 LLOs attack the insurgency's subversive strategy on multiple fronts to weaken the insurgency's ability and
99 will to continue the struggle.

100 5-9. LLOs are not exclusive to any level of conflict. In the same way that a comprehensive campaign is
101 designed at the theater-strategic or operational level by a combatant commander, commanders and their
102 staffs at lower levels may also use LLOs. Their operations, also comprehensive in nature, are nested with
103 the higher echelon's operational design, but should be specifically formulated to each unit's specific envi-
104 ronment.

105 5-10. The commander's vision of resolution and LLOs serve as expressions of the design for a COIN op-
106 eration where conventional objectives like terrain features or enemy positions have little relevance. The
107 commander and staff orchestrate and integrate the LLOs to gain unity of effort. This approach ensures the
108 LLOs converge on a well-defined, commonly understood end state.

109 5-11. LLOs are directly related to one another and internally connect actions within the operation that,
110 when achieved or accomplished, support the overall purpose. Operations designed to use LLOs typically
111 employ an extended, event-driven timeline with short-, mid-, and long-term goals combining the comple-
112 mentary, long-term effects of civil-military operations as well as cyclic, short-term events like combat op-
113 erations and provision of essential services. (See Figure 5-1.)

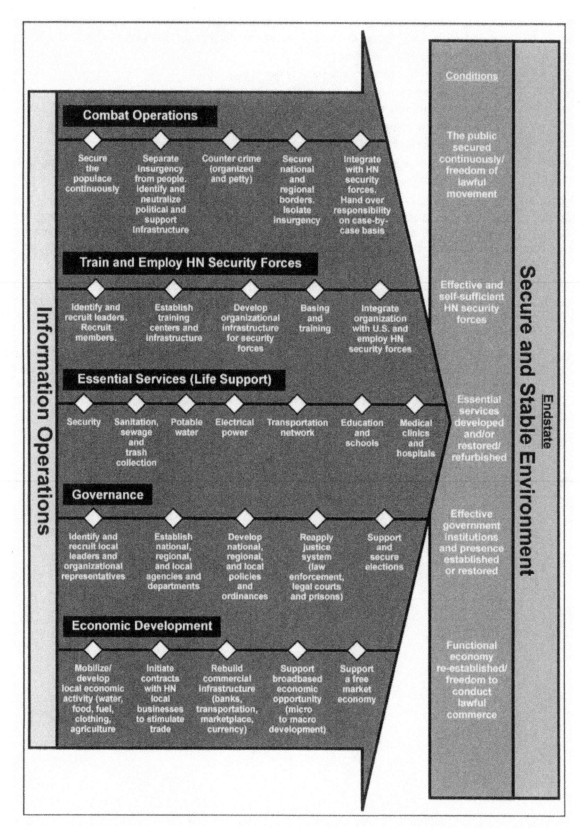

114   **Figure 5-1. Common counterinsurgency operational-level logical lines of operations**

115 5-12. Commanders determine which LLOs are applicable to their operational environment and how they
116 connect with and support one another. For example, establishing continuous security for the public as part
117 of stability operations behind a shield of offensive and defensive operations creates a safe environment in
118 which activities for life support and reconstruction operations can take place. People may perceive that the
119 environment is safe enough to leave their family at home while they seek employment or conduct public
120 economic activity. This perception facilitates the further provision of essential services and development of
121 greater economic activity, and can facilitate the attraction of outside capital for further development. Ne-
122 glecting one LLO will inevitably create vulnerable conditions in another LLO that the insurgency can ex-
123 ploit. Achieving the desired end state requires linked successes on all LLOs. At the operational level,
124 proper execution along the LLOs should lead to increased popular support for a legitimate government.
125 (See figure 5-2.)

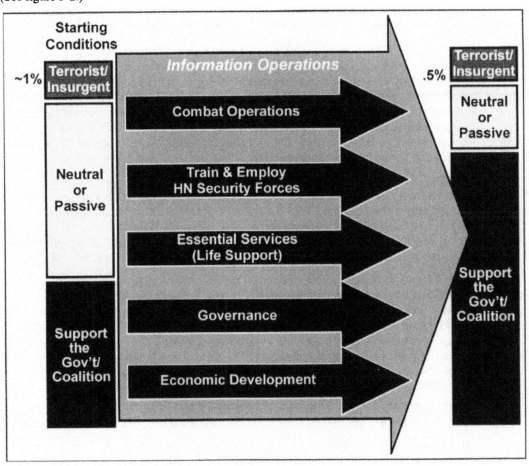

126
127 **Figure 5-2. The effect of proper application of LLOs in counterinsurgency**

128 5-13. The LLOs are analogous to a rope. Each LLO is a separate string, which by itself may be relatively
129 weak and incapable of addressing the weight of issues in a COIN operation. However, a strong rope is cre-
130 ated when strands are woven together. The overall COIN effort is further strengthened through information
131 operations, which support and enhance the operation by highlighting the successes in each LLO. (See Fig-
132 ure 5-3.)

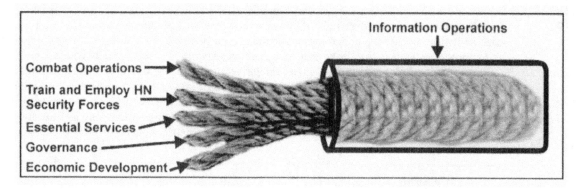

133   **Figure 5-3. The strengthening effect of interrelated LLOs**

134   5-14. Using LLOs, tactical commanders develop missions and tasks, allocate resources, and assess the ef-
135   fectiveness of operations. The commander ought to specify the LLO that is the decisive operation and
136   those that have a shaping role. This prioritization usually changes as insurgency vulnerabilities are created,
137   the insurgency reacts or adjusts its activities, or the environment evolves. In this sense, commanders adapt
138   their operations not only to the state of the insurgency, but also to the condition of the overall environment,
139   which ideally should be moving toward greater stability.

140   5-15. LLOs are mutually supportive between echelons as well as adjacent organizations. For example,
141   similar LLOs among brigade combat teams produce complementary effects, while the collective set of bri-
142   gade accomplishments reinforce the goals at the division level. LLOs are normally used at brigade and
143   higher levels, where the leadership, staff, and unit resources are sufficiently robust to plan and employ
144   them; however, battalions can use the LLOs in a similar fashion to that of their higher headquarters. Com-
145   manders at various levels may expect their subordinate commanders to converse with them in these terms.

146   5-16. LLOs are formulated and coordinated during the course of action development step of the military
147   decision-making process. Beginning with the broadly defined end state, conditions for success, and other
148   aspects of the commander's intent, the commander and staff perform a mission analysis of the situation and
149   environment to determine the objectives necessary to set those conditions and achieve the intent. Related
150   objectives are grouped based on their purpose to ensure unity of effort. If decisive actions can be identified,
151   these are expressed within the context of an LLO, and a logical connection is made between these actions
152   deemed decisive and the commander's intent. The operational design, including the LLOs, is translated
153   into missions, objectives, and tasks, and then assigned to subordinate commanders.

154   ## The Importance of Multiple Lines of operations in COIN
155

156   In China during 1945, Chiang Kai-Sheik adopted a strategy to secure and defend the
157   coastal financial and industrial centers from the communist insurgency led by Mao
158   Zedong. These areas had been the Chinese government's prewar core support ar-
159   eas. Although a logical plan, this strategy suffered in implementation. Republic of
160   China administration and military forces were often corrupt and did not provide good
161   governance and security or facilitate the provision of essential services to the people.
162   Furthermore, the government, because of an insufficient number of soldiers, was
163   forced to rely on warlord forces, which lacked quality and discipline. Their actions
164   undermined the legitimacy of and popular support for the government within the very
165   core areas vital to government rejuvenation. Likewise, when government forces at-
166   tempted to re-establish their presence in the Chinese countryside, their undisciplined
167   conduct towards the rural citizenry further delegitimized the Chinese government.
168   Because of these actions, Chiang's forces were unable to secure or expand their
169   support base.

170
171
172
173
174
175
176

As a result there was increasing lack of material and political support for the government, whose legitimacy was undercut. The government's inability to enforce ethical adherence to the rule of law by its officials and forces, combined with widespread corruption and economic collapse, served to move millions from being supporters into the undecided middle. When economic chaos eliminated any ability of the government to fund even those efforts that were proper and justified, an insurgent victory led by the Chinese Communist Party became inevitable.

177
178
179
180
181
182
183
184
185
186
187
188

As government defeat followed defeat, a collapse of morale magnified the impact of material shortages. Chiang's defeat in Manchuria, in particular, created a psychological loss of support within China, caused economic dislocation due to substantial price inflation of foodstuffs, sowed discord and dissension among government allies, and generated increased though fragmented international support for the Chinese Communist Party. As the regime lost moral authority, it also faced a decreasing ability to govern. All these factors served to create a mythical yet very powerful psychological impression that the success of the Chinese Communist Party was historically inevitable. The failure of the leaders of the Republic of China to address the requirements of lines of operations like good governance, economic development, and essential services magnified their military shortcomings and forced their abandonment of the Chinese mainland.

189
190

COMMON LOGICAL LINES OF OPERATIONS IN COUNTERINSURGENCY

191
192
193

5-17. Commanders at all levels should select the LLOs that relate best to what they desire to achieve in the manner they envision achieving it. This following list is not all inclusive, but gives commanders a place to start:

194
195
196
197
198
199

- Information.
- Combat operations.
- Development of HN security forces.
- Essential services.
- Governance.
- Economic development.

200
201
202
203
204
205

These lines can be customized, renamed, changed altogether, or simply not used. LLOs should be used to isolate the insurgency from the population, address and correct the root causes of the insurgency, and create or reinforce the societal systems required to sustain the legitimacy of the HN government. The sections that follow discuss six common LLOs that appear in most operations. Though information operations (IO) is a separate strand in the coil of LLOs, it is probably the most important, since it is interwoven throughout all the others and surrounds them.

206

INFORMATION

207
208
209

5-18. The information LLO is decisive in that it assists in setting conditions for success of all other LLOs. By publicizing government policies, the actual situation, and counterinsurgent accomplishments, IO can neutralize insurgent propaganda and false claims. The major task categories follow:

210
211

- Identify all the audiences (local, regional, and international), the various news cycles, and how to reach them with the HN government's message.

212

- Manage local population expectations regarding what COIN forces can achieve.

213
214

- Develop common, multiechelon themes based on and consistent with HN government policies and the operation's objectives, and sustain unity of the message.

215
216

- Coordinate and provide a comprehensive information preparation of the operational environment using all communications means and incorporating the activities of all other LLOs.

217
218
219

- Remember actions always speak louder than words—every Soldier and Marine is an integral part of IO communications. IO are executed every day through the actions of firm, fair, professional, and alert Soldiers and Marines on the streets with the people.

220 **Information Operations**

221
222
223
224

5-19. Commanders and staffs orchestrate IO in harmony and cooperation with the activities in the other LLOs. The information LLO should address and manage the public's expectations by explaining what the HN government and COIN forces are doing and why. The information LLO can effectively address the subject of root causes the insurgency uses to gain support.

225
226
227
228

5-20. Information should be localized to address the concerns of the inhabitants of each AO. IO should educate the public regarding successfully completed projects and improvements within all LLOs, including accomplishments in security, infrastructure, essential services, and economic development. This information facilitates popular acceptance of the legitimacy of the HN government.

229
230
231
232
233
234

5-21. The information LLO should use consistent themes based on policy, facts, and deeds—not claims or future plans, because these can be thwarted. Themes must be reinforced by actions along all LLOs. Making unsubstantiated claims can undermine the credibility and legitimacy of the HN government in the long term. COIN forces should never knowingly make a promise or obligate themselves to an action that cannot be completed. However, when a promise cannot be fulfilled, COIN forces should take the initiative and publicly address the reasons why this occurred in order to reduce the negative impact.

235
236
237
238

5-22. Information themes are based on policy and should be distributed simultaneously or as soon as possible using all available means, such as radio, television, newspapers, flyers, billboards, and the Internet. Polling and analysis should be conducted to determine which media allow the widest dissemination of the themes to the desired audiences at the local, regional, national, and international levels.

239
240
241
242
243
244

5-23. Insurgencies are not constrained by truth and create propaganda that serves their purpose. Insurgent propaganda may include lying, deception, and creating false causes to gain support for the insurgency. Historically, insurgencies change their message to address whatever issue gains them support in the changing environment. The information LLO should publicly identify the insurgency's propaganda and lies to the local populace. Doing so creates doubt in the mind of the uncommitted public and supporters of the insurgency regarding the viability of the insurgents' short- and long-term intentions.

245
246
247

5-24. Impartiality is a common theme for information activities when there are political, social, and sectarian divisions in the host nation. COIN forces should avoid taking sides, if possible, as perceived favoritism can exacerbate civil strife and make the COIN force a more desirable target for sectarian violence.

248 **Media and the Battle for Perceptions**

249
250
251
252
253
254
255
256

5-25. Effective commanders directly engage in a dialogue with the media and communicate IO themes. With the proliferation of sophisticated communication technologies throughout the global information environment, the nature of media coverage has a significant impact on COIN operations at all echelons. Civilian and military media coverage influences the perceptions of the political leaders and public in the host nation, the United States, and throughout the international community. The media directly influence the support of key audiences for COIN forces, the execution of their operations, and the opposing insurgency. Recognition of this influence creates a war of perceptions between insurgents and COIN forces that is conducted continuously through the communications media.

257
258
259
260
261
262
263
264

5-26. The media are a permanent part of the information environment and effective media/public affairs operations are critical to successful military operations. Every aspect of a military operation is subject to immediate scrutiny. Well planned, properly coordinated, and clearly expressed IO significantly clear the fog of war and improve the effectiveness and morale of COIN forces, the will of the U.S. public, and the support of the HN people for their government. The right messages can reduce misinformation, distractions, confusion, uncertainty, and other factors that cause public distress and undermine the COIN effort. Constructive and transparent information enhances understanding and support for continuing operations against the insurgency.

265   5-27. There are numerous methods available to commanders for working with the media to facilitate accu-
266   rate and timely information flow. These include the following:
267   ● Embedded media.
268   ● Press conferences.
269   ● Applying resources.
270   ● Network with media outlets.

271   5-28. Embedded media representatives get to know the Soldiers' and Marines' perspectives in the context
272   of the COIN environment. Embedding for days rather than weeks runs the risk of media representatives not
273   gaining any real understanding of the context of operations and may lead to unintended misinformation.
274   Media representatives embedding for weeks become better prepared to present an informed report. The
275   media should be given access to the young men and women in the field. These young people nearly always
276   do a fantastic job articulating the important issues for a broad audience and, given a chance, can share their
277   courage and sense of purpose with the American people and the world.

278   5-29. Weekly press conferences might be held, particularly with HN media, to explain operations and pro-
279   vide transparency to the people most affected by COIN efforts. Such venues also provide an opportunity to
280   highlight the accomplishments of COIN forces and the HN government.

281   5-30. Commanders should apply time, effort, and money to establish the proper combination of media out-
282   lets and communications to transmit the repetitive themes of HN government accomplishments and insur-
283   gent violence against the populace. This might require COIN forces to be proactive, alerting media to news
284   opportunities and perhaps providing transportation or other services to ensure proper coverage. Assisting in
285   the establishment of effective HN media is another important COIN requirement. A word of caution here:
286   there can be no perception by the populace or the HN media that the media is being manipulated by COIN
287   or HN forces. Even the slightest appearance of impropriety here can undermine the credibility of the COIN
288   force and the host nation.

289   5-31. Establishing relationships between COIN leaders and responsible members of the U.S. media is in
290   the interests of the Nation. Without a clear understanding of the COIN efforts, U.S. media representatives
291   relay a message to the American public as best they can. Through these professional relationships, the mili-
292   tary can ensure U.S. citizens better understand what their military is doing in support of the Nation's inter-
293   ests. Operations security must always be maintained, but that should not be used as an excuse to create a
294   media blackout that can lead to misinformation. Similar relationships can be established with international
295   media sources.

296   5-32. The media are ever present and influence perceptions of the COIN environment. Therefore, success-
297   ful leaders engage the media, create positive relationships, and help the media tell the story. Otherwise the
298   media develops a story of its own that may not be as accurate and may not include the COIN force perspec-
299   tive. (See JP 3-61, FM 46-1, FM 3-61.1.)

300   5-33. In developing the information LLO consider the following:
301   ● Consider word choices carefully. Words are important—they have specific meanings and de-
302   scribe policy. For example, is the COIN force a liberator or an occupier? Occupiers generate a
303   "resistance," whereas liberators may be welcomed for a time. Soldiers and Marines can be influ-
304   enced likewise. In a conflict among the people, terms like "battlefield" influence perceptions
305   and confuse the critical nature of a synchronized approach. Refrain from referring to and con-
306   sidering the area of operations (AO) as a "battlefield" or it may continue to be one.
307   ● Publicize insurgent violence and use of terror to discredit the insurgency.
308   ● Admit mistakes (or actions perceived as mistakes) quickly, and explain these mistakes and ac-
309   tions—including mistakes committed by COIN forces. However, do not attempt to explain ac-
310   tions by the HN government. Instead encourage HN officials to handle such information them-
311   selves. They know better the cultural implications of their actions, and honesty should help to
312   build legitimacy.
313   ● Highlight HN government and COIN force successes promptly. Positive results speak loudly
314   and resonate with people. Do not delay communications waiting for all results. Initiate commu-

315
316
nications immediately to let people know what the COIN force is doing and why. Delaying announcements creates "old news" and misses news cycles.

317
318
319
320
- Respond quickly to insurgent propaganda. As stated above, delaying a response can let the insurgent story dominate many news cycles, allowing their version of events to become widespread and accepted. This consideration may require giving increased information assets and responsibilities to lower-level leaders, since they are also at the "point of the spear" for IO.

321
322
323
- Shape expectations of the populace. Generally people expect too much too soon, and when the government or COIN force is slow to deliver, people become easily and perhaps unfairly disgruntled.

324
325
326
327
328
- Give the people some means to voice their opinions and grievances, even if that activity appears at first to cause short-term friction with ongoing efforts. This applies not just to the formal political process, but even more to the informal, local issues (where government actually touches the people). Develop a feedback loop from populace to local government to ensure needs are identified and perceptions aligned.

329
330
331
- Keep troops engaged with the people. Presence patrols facilitate Soldiers and Marines mingling with the populace. The communication flow works both ways, as the people and COIN forces learn to know each other better.

332
333
334
- Conduct ongoing audience analysis and seek to identify key personnel that influence the people at the local, regional, and national levels. Seek to determine with great specificity the relevant lines of loyalty of a population.

335
336
- Take a census as soon as is practicable. Help the local government do this. This information can be helpful for learning about the people and meeting their needs.

337
338
339
- Assist the government in the production and distribution of identification cards. This is an effort to register all citizens—or, at least those nearing a predetermined, adult age. Identification cards may help to track people's movements, which are useful in identifying illicit activity.

340
341
342
343
- Treat detainees professionally and publicize their treatment. Arrange for local HN leaders to visit and tour your detention facility. Consider allowing them to speak to detainees. If news media visit your detention facility, allow them as much access as is prudent. (Provide a guided tour and explain your procedures.)

344
345
346
347
348
349
- Consider initiating a dialogue with the opposition. This does not equate to "negotiating with terrorists," and is an attempt to open the door to mutual understanding. There may be no common ground and the enmity may be such that nothing specifically or directly comes of the dialogue. However, if COIN forces are talking, they are using a positive approach—and may learn something. Do not rely solely on the host nation to do this. Consider adopting a "We understand why you fight" mentality—and maybe stating this to the insurgents.

350
351
- Work to convince insurgent leadership that the time for resistance has ended and that there are other ways to accomplish what they desire.

352
353
354
- Take the insurgents' demands and turn them on the insurgents. Examine the disputed issues objectively, and then work with the host nation to resolve them where possible. Then communicate any success as a sign of responsiveness and improvement.

355
356
- Communicate the message that the COIN force is robust and persistent, and will assist the population through their present difficulty.

357
358
359
- Learn the insurgent's messages or narratives and develop countermessages and counternarratives to attack the insurgents' ideology. An understanding of indigenous culture is required for this activity, and HN personnel can play a key role.

360
COMBAT OPERATIONS

361
362
363
5-34. This line of operations is the most familiar to military forces. Care must be taken not to apply too many resources to this LLO at the expense of other LLOs that facilitate the development or reinforcement of the legitimacy of the HN government.

5-35. An inherent trait of an insurgency is the use of unlawful violence to weaken a HN government, intimidate people into passive or active support, and murder those who oppose the insurgency. There is always some element of measured combat operations required to address insurgents who cannot be co-opted into operating inside the rule of law. However, COIN is "war amongst the people." Combat operations must therefore be executed with an appropriate level of restraint on the use of force to minimize or avoid killing or injuring innocent people who are not involved with the conflict. Not only is there a moral basis for the use of restraint or measured force; there are practical reasons as well. The COIN force does not want to turn the will of the people against the COIN effort by harming innocents. Discriminating use of fires and calculated, disciplined response should characterize performance in COIN, where kindness and compassion can be as important as killing and capturing.

5-36. In general, battalion-sized and smaller unit operations are often the most effective for countering insurgent activities. The COIN force needs to get as close as possible to the people to secure them, glean the maximum amount of quality information, and thereby gain a fluidity of action equal or superior to that of the enemy. This does not mean larger unit operations are not required. The brigade is usually a synchronizing headquarters while the division shapes the environment to set conditions and facilitate brigade and battalion success. The sooner the COIN force can execute small unit operations effectively, the better.

5-37. In developing this LLO, consider the following:

- Develop cultural intelligence, which assumes a prominent role. Make every effort to learn as much about the environment as possible. Human dynamics tend to matter the most.
- Ensure that rules of engagement adequately guide Soldiers and Marines engaged in combat while encouraging the prudent use of force commensurate with mission accomplishment and self-defense.
- Consider how the populace might react when planning tactical situations, even for something as simple as a traffic control point. Understand how people will respond to the operation.
- Identify and focus the COIN force on legitimate tasks for the supporting COIN force. These are tasks the HN government and population generally perceive to be productive and appropriate for an outside force.
- Win over, exhaust, divide, capture, or eliminate the senior- and mid-level insurgent leaders as well as network links.
- Frustrate insurgency recruitment.
- Disrupt base areas and sanctuaries.
- Organize local security and support forces.
- Deny outside patronage (external support). Make every effort to stop the insurgency from importing material support across international and territorial borders.
- When executing any form of unit sweep operation, clear only what the unit intends to hold—otherwise, reconsider the operation and its goals.
- When Soldiers and Marines interact with the population, encourage them to treat people with respect to avoid alienating anyone and assisting the insurgency.

DEVELOPMENT OF HOST-NATION SECURITY FORCES

5-38. Security is the key facilitating mechanism for most societal and government functions. Although U.S. and multinational forces can provide direct assistance to establish and maintain security, this situation is at best a provisional solution. Ultimately, the host nation must secure its own people.

5-39. The U.S. military, working with other government agencies and multinational partners, is able to help the host nation develop the security forces required to establish and sustain stability within its borders. This assistance can take many forms, to include the development, equipping, training, and employment of HN security forces. The assistance may extend to operations in which multinational military units fight alongside the newly formed, expanded, or reformed HN forces.

5-40. Though this LLO is discussed in detail in chapter 6, some ideas are worth highlighting here. In developing this LLO, consider the following:

413
414
- Understand the security problem. The function, capabilities, and capacities required for HN security (military and police) forces should align with the theater strategy and the threat they face.

415
416
417
418
419
420
- Start at the earliest planning stage with a comprehensive approach to local needs in consultation with local representatives. This process should be undertaken in partnership with the HN government and military authorities, and in consultation with multinational partners and those international organizations that may be involved. While the commitment of U.S. military power and money may entail taking a leadership role, that role should always be partnered with local authorities to achieve legitimacy in the eyes of the populace.

421
422
- Avoid mirror-imaging—trying to make HN forces look like the U.S. military. That solution fits few cultures or situations.

423
424
- Establish separate training academies for military and police forces. Staff them with multinational personnel. (Tap into the talents of as many nations as possible for this).

425
- Establish mobile training teams.

426
427
428
- Train the HN cadres first, and focus on identifying leaders. When trained, these key personnel can establish new units, staff the training academies, and in some cases, staff mobile training teams.

429
430
431
432
433
- Create general-purpose forces and special-purpose forces. These special-purpose forces should be based on need and ideally should be limited. For police, this consideration could entail the development of a special intelligence branch like the British have to prevent espionage and subversion as well as to provide security for ports and public figures For the HN military it could mean riverine operations forces, explosive ordinance disposal, or other specialized units.

434
- Put HN personnel in charge of as much as possible as soon as possible.

435
436
437
438
- Conduct "joint" operations with HN forces, and show them that you respect their partnership. All plans should be prepared in partnership with HN forces once those forces are ready to work with the COIN force. Encourage their ownership of plans and operations as they move toward self-sufficiency.

439
440
441
442
- In public, show appropriate respect to leaders of HN security forces. The people should know that their security forces have earned the respect of the COIN force. But abuses cannot be tolerated, so that respect should depend upon generally upright comportment on the part of the HN security forces.

443
444
- Put liaison officers with HN units. Additionally, provide unit advisors for HN units under development (noting that liaison officers and military advisors serve distinctly different purposes).

445
446
447
- Establish competent military and police administrative structures early. Troops need to be provisioned and paid in a timely manner. Pay should come from the HN organization—not through the COIN force.

448
449
450
- Identify insurgents who might seek to join the security forces under false pretext. However, encourage insurgents to change sides—welcome them in with an "open-arms" policy. Vetting "turncoats" is a task for the HN government in partnership with the country team.

451
ESSENTIAL SERVICES

452
453
454
455
456
457
5-41. The essential services LLO addresses the life support needs of HN population. The U.S. military's primary involvement is normally to provide a safe and secure environment that facilitates planning and actions to develop the supporting services or infrastructure needed to provide or restore essential services. Due to security problems, the military may initially have the leading role. Other agencies may not be initially present to assume this role or might not have sufficient capability or capacity. Therefore, the U.S. military must be prepared to conduct these tasks for an extended period.

458
459
460
461
462
5-42. The COIN force should work closely with the HN in establishing achievable goals. If lofty goals are set and not achieved, both the COIN force and the HN government can lose the respect of the population. The long-term objective for this LLO is to transition to complete HN responsibility and accountability. Establishing activities that the HN government will be unable to sustain on their own may be counterproductive. IO must be nested within this LLO to manage expectations and ensure that the public has a coherent

463
464
understanding of the problems requiring redress, to include infrastructure sabotage by insurgency forces. The major categories follow:

465   • Transportation and public works.

466   • Public utilities.

467   • Telecommunications and public communications.

468   • Public health and sanitation.

469   • Life-support and humanitarian relief.

470   • Education.

471
472
473
5-43. Figure 5-4 (below) illustrates how a unit might apply the essential services LLO through the acronym SWEAT-MS. Each category includes a set of missions that set the conditions for success. Those conditions, in turn, contribute to achieving the higher commander's desired end state.

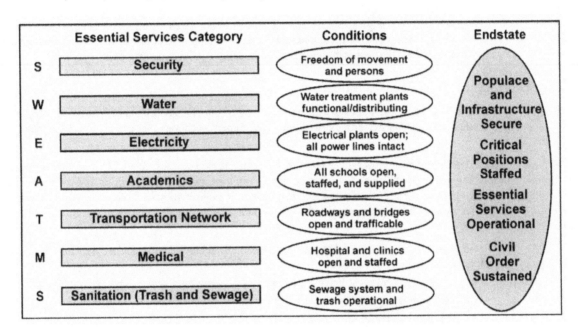

474   **Figure 5-4. Unit application of the essential services logical lines of operations**

475   5-44. In developing this LLO, consider the following:

476
477
• Make this effort a genuine partnership between the COIN force and HN authorities. Use as much local leadership, talent, and labor as soon as possible.

478
479
480
481
482
• Plan for a macro and a micro assessment effort. Acknowledge up front what is known and not known about the environment—and begin an honest appraisal of what needs to be accomplished. The macro assessment concerns operational design functions and is long term in focus. The micro assessment effort, by necessity, focuses on the local level and determines, with regional sensitivity, what the actual and specific needs are in the immediate future.

483
484
485
486
487
• Appreciate local desires. A needs assessment must reflect a great deal of cultural sensitivity; otherwise great attention (time and expense) could be wasted on something the people do not consider to be of real value. Ask, How do I know this effort is important from a local perspective? If there is no answer to that question, it may not be. HN authorities are a good place to start with this question, and local polling of the populace may be useful.

488
489
• Make a point of establishing realistic, measurable goals, and put in place methods of assessment towards the achievement of those goals.

490
491
• Form interagency planning teams to discuss design, assessment, and redesign. Learn early to understand and use an interagency language and recognize and understand other agencies' insti-

492
493
tutional cultures. COIN activities are interagency activities—whether agencies beyond the military are initially present or not.

494
495
496
497
498
● Meet with representatives from organizations beyond the government team. Nongovernmental organizations seldom want to give the appearance of being too closely aligned with the COIN effort. Encourage their participation in planning, even if it means holding meetings in neutral areas. In your meetings with these organizations, help them understand that there are mutual interests in achieving the COIN force objectives of local security, stability, and relief.

499
500
● Be as transparent as possible with the local people. Do your best to help people understand what the COIN force is doing and the basis for decisions that affect them.

501
502
● Consider the role women play in the society and how this cultural factor may influence activities in this LLO.

503
504
505
506
507
● Consider that, in some situations, the local people may form an impression that the COIN force (especially the military side of it) has arrived to "save the day," or in contrast, that their arrival will only cause greater problems. Such reactions are especially likely concerning essential services. Understanding these phenomena and working to keep expectations manageable avoids the frustrations that inevitably result from unrealized local expectations.

508
GOVERNANCE

509
510
511
5-45. This LLO relates to the capability and capacity of the HN government to secure its people, establish and maintain order, and perform all necessary government activities that pertain to a legitimate, sovereign nation.

512
513
514
515
516
5-46. In more rudimentary cases, where no government exists at all, this LLO may involve creating and organizing a HN capability and capacity to govern. In the long run, the activities in this LLO will probably affect the lives of the people the most and may lead to the elimination of the root causes for the insurgency. Activities in this LLO are among the most important of all in establishing lasting stability for a region or nation.

517
518
5-47. Good governance is normally a key requirement to achieve legitimacy for the HN government. Necessary activities include the following:

519
● Security activities (military and police isolating the insurgency as part of public safety).

520
● Establishing and enforcing the rule of law.

521
● Public administration.

522
● Justice (judiciary system, prosecutor/defense representation, and corrections).

523
● Property records and control.

524
● Public finance.

525
● Civil information.

526
● Historical, cultural, and recreational services.

527
● Electoral process for representative government.

528
● Disaster preparedness and response.

529
5-48. In developing this LLO, consider the following:

530
531
532
533
● Encourage local leaders to step forward and participate. If no local council exists, encourage the populace to create such a body. Teachers, businessmen, and others who enjoy the respect of the community should be strongly encouraged to come together and form a temporary council to serve in such capacity until a more permanent organization can be elected.

534
535
536
● Assist (or encourage) the HN government to remove or reduce genuine grievances, expose imaginary ones, and resolve contradictions where possible immediately. Note that this consideration may be very difficult to achieve because—

537
■ Genuine grievances may be hard to ascertain.

538
539
■ Solutions might involve the HN giving up power or control in a fashion they are unable or unwilling to accommodate.

- Make only such promises as can be fulfilled in the foreseeable future.
- Help the HN develop and empower competent and responsive leaders and strengthen their civil service and security forces. This is traditionally difficult to do, and backing an incompetent (or worse) HN leader can backfire. Do not be afraid to step in and make a bold change where necessary. A corrupt official, such as a chief of police who is working at the behest of both sides, can be doing more harm than good. You may be forced to replace him. If so, move decisively. Arrange the removal of all officials necessary such that, although the pain of the affair may be acute, it is brief and final. Wherever possible, have other HN authorities conduct the actual removal.
- Provide accessibility to ensure two-way communications with the populace. Establish rapport with the local public for HN government and COIN forces.
- Encourage the host nation to grant local demands and meet acceptable aspirations. Some of these might be driving the insurgency.
- Emphasize the national perspective in all HN government activities, downplaying sectarian divides.
- Provide liaison officers to HN government ministries or agencies. Even better, do this in an interagency fashion, using a team approach. Obviously these proposed teams differ in composition, depending on function.
- Once the legal system is established or reestablished, send someone to observe firsthand a person or persons moving through the legal system (arrest by police, trial, and punishment by confinement to a correctional facility). Ask to see the docket of the judges at the provincial courthouse. If there is no one on the docket or if it is full and there are no proceedings, there may be a problem.
- Create a system for citizens to pursue legal redress for perceived wrongs by authorities. Rule of law must include a citizen's right and ability to petition the government for redress of wrongs committed by that government—or to petition the COIN force for redress of wrongs perpetrated by that force (intentionally or otherwise).
- Provide adequate security for the populace to resume their lives and livelihoods. This is an essential requirement for effective governance.
- Whenever and wherever possible, build on extant capabilities. The host nation often has some capability and supporting COIN forces may just need to help it develop greater capacity.

ECONOMIC DEVELOPMENT

5-49. The economic development LLO includes both a short-term and a long-term aspect. The short-term aspect deals with immediate problems, such as large-scale unemployment and means to jump-start an economy from the local level through the national level. The long-term aspects of the LLOs should work to stimulate an indigenous capability and capacity that produces robust and broad economic activity for the HN population. Any stability a nation enjoys may be directly related to its population's economic situation and adherence to the rule of law. The basic economic health of a nation is also dependent on a government's ability to continuously secure its population.

5-50. Formulating a plan for economic development requires planners to understand the society, culture, and the operational environment. For example, in a rural society, land ownership and the availability of agricultural equipment, seed, and fertilizer may be the chief components of any economic development plan. In a more urban and diversified society, the availability of jobs and the infrastructure to support commercial activities may be more important for constructive economic development. With the exception of completely socialist economies, governments do not create jobs except in the public bureaucracy. However, the micro-economy can be positively stimulated through the advancement of small businesses. To jump-start small businesses, there is a requirement for microfinance in the form of some sort of banking activities. So then, work in this LLO requires attention to both the macroeconomy and the microeconomy.

5-51. Without a viable economy and employment opportunities, the public is likely to pursue false promises offered by an insurgency, which may be fostering the very conditions keeping the economy stagnant. Insurgencies attempt to exploit a lack of employment/job opportunities to gain active and passive support

591   for their cause and ultimately delegitimize a government. The major categories of economic activity in-
592   clude the following:

- Fossil fuels, mining, and related refining infrastructure.
- Generation and transmission of power and energy.
- Transportation and movement networks.
- Stock and commodities exchange, and the banking industry.
- Manufacturing and warehousing.
- Building trades and services.
- Agriculture, food processing, fisheries, and stockyard processing.
- Labor relations.
- Education and training.

5-52. In developing this LLO, consider the following:

- To draw the most out of the local population, work with the host nation to strengthen the economy and quality of life. In the long run, success in COIN is about supporting the livelihoods of
  the populace.
- Create an environment where business can thrive. In every state (except perhaps a completely
  socialist one), business drives the economy. To strengthen the economy, find ways of encouraging and supporting legitimate business activities. Even the provision of security is part of this
  positive business environment.
- Work with the HN government to reduce unemployment to a manageable level.
- Seek to understand the impact of military operations on business activities and vice versa in the
  AO. Understand the impact of outsourcing and military support on the local economy and the
  level of employment. Focus on HN capability and capacity.
- Use economic leverage for entry into new areas and to reach new people. Remember that in
  many societies, monies are distributed though the tribal or clan networks. For instance, making
  sure an important clan leader gets a large contracting job may ensure that many local men are
  employed—and therefore not as available to the insurgency. It may be necessary to pay more
  than seems fair for a job, but this form of economic leverage is cheap if it keeps people from
  supporting the insurgency.
- Ensure that noncompliance with government policies has an economic price. Likewise, show
  early that compliance with those policies is profitable. In the broadest sense, the operational design should reflect that "peace pays."
- Program funds for commanders to use from the beginning of any intervention for economic projects in their AO. No one has a better appreciation of the specific situation than those "on the
  ground." Creating these funds may require congressional action. (A more detailed description of
  relevant funding sources is in appendix D.)

COUNTERINSURGENCY APPROACHES

5-53. There are many approaches to using LLOs to achieve success in a COIN effort. Each approach consists of components that are not mutually exclusive to any particular strategy and are often shared among
multiple strategies. The strategies described below are not the only choices available, are neither discrete
nor exclusive, and may be combined depending on the environment and resources available. But the following methods and their components are offered as examples that have proven effective in the past. Making any of these methods effective in the future requires adapting them to the demands of the local environment. The three examples of approaches discussed are—

- Clear-hold-build.
- Combined action.
- Limited support.

CLEAR-HOLD-BUILD

5-54. COIN efforts should begin by controlling key areas. Security and influence then spread out from these secured areas like an oil spot. The pattern of this approach is to clear, hold, and build one village, area, or city, and then to expand to other areas, reinforcing success. This approach aims to develop a long-term, effective HN government framework and presence that secures the people and facilitates the meeting of their basic needs, thereby reinforcing the legitimacy of the HN government. The primary tasks to accomplish clear-hold-build are—

- Provide continuous security for the local populace.
- Eliminate insurgent presence.
- Reinforce political primacy.
- Enforce the rule of law.

To create success that can spread, a clear-hold-build operation should not begin by assaulting the main insurgent stronghold or by focusing on areas where the population overtly supports the insurgents.

5-55. Clear-hold-build objectives require a considerable expenditure of resources and time. U.S. commanders and their HN counterparts should prepare for a long-term effort. All operations require a focused unity of effort by civil authorities, intelligence agencies, and security forces. Coherent IO are also needed. A clear-hold-build operation is executed in a specific, high-priority area experiencing overt insurgency and has the following objectives:

- Creating a secure physical and psychological environment.
- Establishing firm government control of the population and the area.
- Gaining the support of the population as manifested by their participation in the government programs to counter the insurgency.

5-56. Clear-hold-build operations should expand outward from a secure base, such as an urban industrial complex whose population supports the government effort and where security forces are in firm control. No area or its population that has been subjected to the intensive organizational efforts of a subversive insurgency organization can be won back until certain conditions are created:

- The commander responsible for the clear-hold-build operation has security forces clearly superior to the insurgent force known or suspected to be operating in the area or immediately available in an adjacent area.
- Sufficient nonmilitary resources are allocated to effectively carry out all essential improvements needed to provide basic services and control the population.
- The insurgency is cleared from the area.
- The insurgency organizational infrastructure and its support have been neutralized or eliminated.
- A government presence is established to replace that of the insurgency, and the local population willingly supports this presence.

5-57. The following discussion describes some examples of activities involved in a clear-hold-build approach. It execution involves activities across all LLOs. There can be overlap between required steps, especially between hold and build, where relevant activities are often conducted simultaneously.

Clearing the Area

5-58. *Clear* is a tactical mission task that requires the commander to remove all enemy forces and eliminate organized resistance in an assigned area (FM 3-90). Clearing operations lead to the destruction or retreat of insurgent combatants. This task is most effectively initiated by a clear-in-zone or cordon-and-search operation to disperse insurgent forces or force reaction by major insurgent elements in the area. Once this action has been initiated, units employ a combination of offensive small unit operations, such as area saturation patrolling and interdiction ambushes.

5-59. These combat operations are only the beginning, not the end state. Removal of the visible insurgent forces does not remove the entrenched insurgent infrastructure. While their infrastructure exists, insurgents continue to recruit among the population, attempt to undermine the HN government, and try to coerce the

686
687
688
populace through intimidation and violence. After insurgent combatant forces have been eliminated, removing the infrastructure begins. This should be done in a way to minimize the impact on the local population and is essentially a police action that relies heavily on HN courts and legal processes.

689
690
691
692
693
694
695
5-60. If insurgent combat forces are not eliminated but instead are expelled or have broken into smaller groups, they must be prevented from re-entering the area in force or re-establishing an organizational structure inside the area. Once COIN forces have established their support bases, security elements cannot remain static. They should be mobile and establish a constant presence throughout the area. Use of special funds should be readily available for all units to pay compensation for damages that occur while clearing the area of insurgent forces. Combat actions continue to maintain gains and set the stage for future activities by—

696
● Isolating the area to cut off external support and to catch insurgents trying to escape.

697
● Conducting periodic sweeps of the area to identify, disrupt, eliminate, and/or expel insurgents.

698
699
● Employing security forces and government representatives throughout the area to secure the people and facilitate follow-on stages of development.

700
701
702
703
704
705
706
5-61. Clearing operations are supplemented by IO focused particularly on two key audiences: the local population and the insurgents. The purpose of the message to the population is to gain and maintain their overt support for HN security forces with promises to protect the people from insurgent forces and activities. Conversely, the population should understand that if they actively support the insurgency, they are prolonging combat operations and creating a risk to themselves and their neighbors. The purpose of the IO message to the insurgents is to convince them that they cannot win against the HN efforts, and that the most constructive course of action is to surrender or cease their activities.

707
Holding with Security Forces

708
709
710
711
712
713
5-62. Ideally this part of the clear-hold-build approach is handled by HN elements. Establishment of HN security forces in bases among the population facilitates the continued disruption, identification, and elimination of the local insurgent leadership and infrastructure. The success or failure of the effort depends on effectively and continuously securing the populace, and then on the effectiveness of re-establishing a HN government presence (and operational systems) at the local level. Measured combat operations will continue against insurgent combatants as opportunities arise, but the main effort is focused on the population.

714
715
716
717
5-63. Key infrastructure must be secured. Since resources are always limited, those elements of the infrastructure that are both vital for stability and vulnerable to attack should be protected. Ask in planning, What elements of infrastructure are in jeopardy? For instance, a glass-making factory may be important for economic recovery, but it may not be at risk of insurgent attack and therefore may not require security.

718
5-64. There are three key IO audiences during the hold stage:

719
● Populace.

720
● Insurgents.

721
● COIN force.

722
723
5-65. The IO message to the population should affirm that security forces supporting the HN government are in the area to accomplish the following:

724
● Protect the population from so-called "revolutionary justice."

725
● Eliminate insurgent leaders and infrastructure.

726
● Improve basic life support where possible.

727
● Reinstate HN government presence.

728
729
730
IO should also state that the security forces will remain for several years and will not leave. This message of a persistent presence can be reinforced by making long-term contracts with the local populace for various supply or construction requirements.

731
732
5-66. The IO message to the insurgents calls on them to surrender or leave the area, emphasizing the permanent nature of the government victory and presence. The HN government might try to exploit success by

offering a local amnesty. Insurgent forces will most likely not surrender, but may cease hostile actions against the HN government in that area.

5-67. The commander's message to the COIN force should explain the changes in missions and responsibilities that are associated with creating or reinforcing the legitimacy of the HN government. The importance of securing the people to protect them, gaining the people's support by assisting them, and using measured force when fighting the insurgents should be reinforced and understood.

5-68. Operations during this stage are designed to—

- Continuously secure the population to separate them from the insurgency.
- Establish a firm government presence and control over the area and population.
- Recruit, organize, equip, and train local people to provide area security.
- Establish a government political apparatus to replace the insurgency apparatus.
- Develop a dependable network of informants.

5-69. Major actions occurring during this stage include—

- Designating and allocating area-oriented security forces to continue offensive operations in the area. Other security forces that participated in clearing actions are released or are assigned to other tasks.
- Thorough population screening to identify and eliminate remaining insurgents and to identify any lingering insurgent support structures.
- Conducting area surveys to determine resources and specific needs of the people and area. Local leaders should be involved.
- Constructively influencing the people's motivation through environmental improvements designed to convince the population to support and participate in the active security of their area and in the area reconstruction effort.
- Training of local paramilitary security forces, including arming and integrating them in successful operations against the insurgents.
- Establishing a communications system integrating the area into the larger secure communications grid and system.

Building Support and Protecting the Population

5-70. The population in the AO must be protected from the insurgents in order to make any progress in building support for the HN government. If the people are not secure or they perceive they are not, then COIN efforts are greatly hindered. The HN security forces must conduct patrols and use measured force against insurgent targets of opportunity on a continuous basis to ensure the protection of the people in the AO.

5-71. This contact with the population is critical to the success of the local COIN effort. Actions designed to completely eliminate the remaining covert insurgent political infrastructure must be continued, as its presence will continue to threaten and influence people.

5-72. Special funds (or other available resources) should be available as wages to reimburse the population for accomplishing applicable required tasks. Tasks that provide an overt and direct benefit for the population and for which reimbursement is appropriate are key, initial priorities. These tasks can begin the process of securing and reconciling the population to the HN government. Some sample tasks are—

- Collecting and clearing trash from the streets.
- Removing or painting-over insurgency symbols or colors.
- Building and improving roads.
- Digging wells.
- Preparing and building indigenous local security force infrastructure.
- Securing, moving, and distributing supplies.
- Providing guides, sentries, and translators.

Monitoring the Population

5-73. Population control measures include determining who lives in the area and what they do. This process requires research into family, clan, tribe, interpersonal, and professional relationships. Establishing control normally begins with a census and the issuing of identification cards. A census must be advertised and executed systematically. Census tasks include establishing who resides in which building and the responsible family head for each household. Those responsible for the household are held accountable to report any future changes to the appropriate agencies. The collected census records provide intelligence regarding real property ownership, relationships, and business associations.

5-74. Insurgents may try to force the people to destroy their identification cards. The incentive for the people to retain their identification cards must be sufficient. Insurgents will participate in the census to gain valid identification cards themselves. Requiring applicants to bring two men from outside their family to swear to their identity is a method to reduce this probability. Those who affirm the status of the identification card applicant are accountable for their official statements made on behalf of the applicant. Identification cards should have a code that indicates where the holders live.

5-75. Some other control measures are—
- Establishing and enforcing a curfew.
- Setting up a pass system administered by security forces or civil authorities and establishing limits on the length of time people can travel.
- Establishing limits on the number of visitors from outside the area.

5-76. Once control measures are in place, the HN government should have an established system of punishments for various offenses. These should be announced to the population and enforced. The establishment of this system should be initiated by the host nation to ensure the same system is enforced everywhere, and there is consistency with the rule of law throughout its territory. The HN government must have the means to impose fines and other punishments for civil infractions of this nature.

Increasing Popular Support

5-77. COIN forces should use all available resources and capabilities to capitalize on every opportunity to help the population and meet its needs and expectations. Projects to improve their economic, social, cultural, and medical needs can begin immediately. Actions speak louder than words. Once the insurgent political infrastructure has been destroyed and local leaders begin to establish themselves, any necessary political reforms can be implemented. Other important tasks include—
- Establishing agencies of the HN government to carry out routine administrative functions and to begin improvement programs.
- Providing HN government support to those willing to participate in reconstruction. Selection for participation should be based on determinations of need and ability to help. People should also be willing to secure what they create.
- Begin progressive efforts—such as participation of the population in local elections of leaders, community-sponsored environmental improvement, formation of youth clubs, and other projects—to develop regional and national consciousness, and rapport between the population and its government.

5-78. There are key IO audiences and messages in this stage, as well.

5-79. The IO message to the population has three facets.
- First, it should aim to gain understanding or approval for the actions of security forces that are affecting the population, such as the control methods or a census. This is relatively easy to achieve by telling the people what is to be done and for what purpose.
- The second aim is to establish human intelligence sources that lead to identification and destruction of any remaining insurgent infrastructure in the area.
- The third aim should be to win over those people who are still passive or neutral by demonstrating how the HN government is going to make their life better.

828
829
830
5-80. The IO message to the insurgents should aim to create divisions between the leadership and the mass base by emphasizing failures of the insurgency and successes of the government. Success is indicated when elements of the insurgency abandon it and return to work with the government.

831
832
833
5-81. The commander should emphasize to COIN forces that they must remain friendly towards the populace while staying vigilant against insurgent actions. Commanders must ensure Soldiers and Marines understand their rules of engagement, which will become even more restrictive as peace and stability return.

834
835
836
837
838
839
5-82. The most important activities during the build stage are conducted by nonmilitary agencies. HN government representatives reestablish political offices and normal administrative procedures. National and international development agencies rebuild infrastructure and key facilities. Local leaders are developed and given authority. Life for the area's inhabitants begins to return to normal. Activities along the combat and security force LLOs become secondary to those involved in essential services, good governance, and essential services LLOs.

840
841
Clear-Hold-Build in Tal Afar, 2005-2006

842
843
844
845
846
847
848
849
In early 2005, the northern Iraqi city of Tal Afar had become a focal point for Iraqi insurgent efforts. The insurgents tried to assert control over the population. They used violence and intimidation to inflame ethnic and sectarian tensions. They took control of all schools and mosques, while destroying police stations. There were frequent abductions and executions. The insurgents achieved at least some success as the population divided into communities defined by sectarian boundaries. Additionally, Tal Afar became a support base and sanctuary that was used to launch attacks in the major regional city of Mosul and throughout Nineveh province.

850
851
852
During the summer of 2005, the 3d Armored Calvary Regiment (3d ACR), assumed the lead for military efforts in and around Tal Afar. In the months that followed, the 3d ACR applied a clear-hold-build approach to reclaim Tal Afar from the insurgents.

853
Destruction or Expulsion of Insurgency Forces (Clear)

854
855
856
857
858
859
860
861
862
863
864
865
866
867
868
869
870
In August 2005, the 3d ACR and Iraqi forces began the process of destroying the insurgency in Tal Afar. Their first step was to conduct reconnaissance to understand the enemy situation, understand the ethnic, tribal, and sectarian dynamics, and set the conditions for effective operations. Iraqi security forces and U.S. Soldiers isolated the insurgents from external support by controlling nearby border areas and creating an eight-foot-high berm around the city. The berm's purpose was to deny the enemy freedom of movement or safe haven in outlying communities. The berm prevented free movement of fighters and weapons and forced all traffic to go through security checkpoints that were manned by U.S. and Iraqi forces. Multinational checkpoints frequently included informants who could identify insurgents. Multinational forces supervised the movement of civilians out of contentious areas. The forces conducted house-to-house searches. When significant violent resistance led to battle, combat included the use of precision fires from artillery and aviation. Targets were chosen through area reconnaissance operations, interaction with the local population, and information from U.S. and Iraqi sources. Hundreds of insurgents were killed or captured during the encirclement and clearing of the city. Carefully controlled application of violence limited the cost to residents.

871
Deployment of Security Forces (Hold)

872
873
874
875
Following the defeat of enemy fighters, U.S. and Iraqi forces established security inside Tal Afar. The security forces immediately enhanced personnel screening at checkpoints based on information from the local population. To enhance police legitimacy in the eyes of the populace, the multinational forces began recruiting a more

876
877
878
879
880
881
882
883
884
885
886

diverse and representative mix of city residents and residents of surrounding communities for the Iraqi police. The police recruits received extensive training in a police academy program. U.S. forces and the Iraqi Army also trained Iraqi police in military skills. Concurrent with this effort, the local and provincial government fired or prosecuted Iraqi police who were involved in offenses against the population and assigned new police leaders to the city from Mosul and other locations. U.S. forces assisted to ensure Iraqi Army, police, and their own forces shared common boundaries and were positioned to provide mutual support to one another. At the same time, U.S. forces continued to equip and train a border defense brigade that increased the capability to interdict external support. Among its successes, the multinational force defeated an insurgent network that included a chain of safe houses between Syria and Tal Afar.

887
Improving Living Conditions and Restoring Normalcy (Build)

888
889
890
891
The local population generally accepted guidance and projects to re-establish control by the Iraqi government. The 3d ACR commander noted, "The people of Tal Afar understood that this was an operation for them—an operation to bring back security to the city."

892
893
894
895
896
897
With the assistance of the U.S. Agency for International Development's Office of Transition Initiatives and the State Department, efforts to re-establish municipal and economic systems began in earnest. Among such initiatives were the provision of essential services (water, electricity, sewage, and trash collection), education projects, police stations, parks, reconstruction efforts, as well as a legal claims process and compensation program to address local grievances for damages.

898
899
900
901
902
As security and living conditions in Tal Afar improved, citizens began providing intelligence that facilitated the elimination of the insurgent infrastructure. In addition to information received on the streets, multinational forces established joint coordination centers in Tal Afar and nearby communities that became multinational command posts and intelligence-sharing facilities with the Iraqi Army and the Iraqi police.

903
904
Unity of effort by local Iraqi leaders, Iraqi security forces, and U.S. forces was critical to success. Many families who had fled the area returned to the secured city.

905
COMBINED ACTION

906-914
5-83. Combined action is a technique that involves joining United States and HN troops in a single organization, usually a platoon or company, to conduct counterinsurgency operations. This technique is appropriate in environments where large insurgent forces do not exist or they lack resources and freedom of maneuver. Combined action normally involves joining a U.S. rifle squad or platoon with a HN platoon or company, respectively. The purpose of this approach is to hold and build while providing a persistent COIN force presence among the populace. This approach attempts to first achieve security and stability in a local area, followed by offensive operations nearby enemy forces now denied access or support. Not designed for offensive operations themselves, combined action units rely on more robust combat units for this task. Combined action units can also establish mutual support among villages to secure a wider area.

915
916
Combined Action Program

917
918
919
920
Building on their COIN experiences in Haiti and Nicaragua, the Marine Corps implemented an innovative program in South Vietnam in 1965 called the Combined Action Program (CAP). This program paired small teams of about 15 Marines led by a noncommissioned officer with approximately 20 indigenous security personnel. These

921
922
923
924
925
926
927
928
929
930

combined action platoons operated in the hamlets and villages in the northern two provinces of South Vietnam adjacent to the demilitarized zone. These CAP Marines earned the trust of the villagers by living among them while helping them defend themselves. Marines trained and led the local people's defense forces, learned the customs and language of the villagers, and were very successful in denying access to areas under their control to the Viet Cong insurgency. CAP became a model for local success in countering insurgencies. Many of the lessons learned from it were copied and used in various peace enforcement and humanitarian assistance operations Marines conducted during the 1990s, such as Operations Provide Comfort in Northern Iraq (1991) and Restore Hope in Somalia (1992-93).

931
932
933

5-84. Combined action programs can only be instituted in areas with limited insurgent activity and should not be used to isolate or expel a well established and supported insurgent force. Combined action is most effective after an area has been cleared of armed insurgents.

934

5-85. The following geographic and demographic factors can also influence the probability of success:

935
- Towns relatively isolated from other population centers are simpler to continuously secure.

936
937
- Towns and villages with a limited number of roads passing through are easier to secure than towns with many routes in and out, because all these approaches must be guarded.

938
939
- Existing avenues of approach into a town should be observable from the town to facilitate the interdiction of an insurgent force and the control of population movements.

940
941
942
943
- The town's population should be small and constant, where people know one another and can easily identify outsiders. In towns or small cities where a large number of inhabitants are not acquainted with each other, conducting a census is the most effective tool to establish initial accountability for everyone.

944
945
946
947
- The town and combined action/local defense force must establish mutual support with other adjacent or nearby towns where combined action forces are operating. Larger reaction or reserve forces as well as close air support, attack aviation, and air assault support should be quickly available to assist each combined action unit.

948
949
950
951
952
953
954

5-86. Combined action unit members must develop and build positive relationships with their associated HN security forces and with the town leadership. By living among the people, combined action units serve an important purpose. They demonstrate the commitment and competence of COIN forces while sharing experiences and relationships with the local people. These working relationships build trust and enhance the legitimacy of the HN government. To further build trust, U.S. members should request that the HN security forces provide training on local customs, key terrain, possible insurgent hideouts, and relevant cultural dynamics, as well as provide a description of recent local events.

955
956
957
958
959

5-87. While combined action units are integrated into a regional scheme of mutually supporting security and influence, they should remain an organic asset of the parent unit from which they are drawn. Positioning reinforced squad-sized units (13 to 15 Soldiers or Marines) among HN citizens creates a dispersal risk the parent unit can mitigate with on-call reserve and reaction forces along with mutual support from adjacent villages and towns.

960
961
962
963
964
965

5-88. The thorough integration of U.S. and HN combined action personnel supports the effective teamwork that is critical to the success of each team and the overall program. U.S. members should be drawn from some of the parent unit's best personnel. Commanders should designate potential members prior to deployment to facilitate specific training and team-building to enable the success of the combined action unit once it arrives in theater. Preferably, team members should have had prior experience in the host nation. Other desirable characteristics for combined action members include—

966
- Ability to operate effectively as part of a team.
967
- Strong leadership qualities, among them—
968
    - Communicates clearly.
969
    - Shows maturity.
970
    - Leads by example.

971   ■   Makes good decisions.

972   ●   Ability to apply the commander's intent in the absence of orders.

973   ●   Possession of cultural awareness and understanding of the HN environment.

974   ●   Absence of any obvious prejudices.

975   ●   Mutual respect when operating with HN personnel.

976   ●   Experience with the HN language or the ability to learn languages.

977   ●   Patience and tolerance when dealing with language and translation barriers.

978
979
980
981
982
5-89. Combined action units should be used in cleared and semicleared areas to work with the local people, help provide continuous security in towns, and train and conduct operations with HN security forces. Secondary missions include conducting civic action along established LLOs in support of their parent unit, gathering information for COIN forces, and providing other appropriate assistance to the HN local government to further develop its legitimacy.

983   5-90. Appropriate tasks include but are not limited to the following:

984
985
     ●   Assisting HN security forces in maintaining entry control points and providing reaction force capabilities through the parent unit.

986   ●   Conducting multinational, coordinated day and night patrols to secure the town and area.

987
988
989
     ●   Facilitating local contacts to gather information in conjunction with local HN security force representatives. (Ensure information gathered is made available promptly and on a regular basis to the parent unit for timely fusion and action.)

990
991
     ●   Conducting training in leadership and general military subjects so HN security forces can secure the town/area on their own.

992   ●   Conducting operations with other multinational forces and HN units, if required.

993   ●   Operating as a team with HN security forces to instill pride, leadership, and patriotism.

994
995
     ●   Assisting HN government representatives with civic action programs to establish an environment where the people have a stake in the future of their town and nation.

996
997
     ●   Protecting and assisting HN judicial and government representatives in establishing the rule of law.

LIMITED SUPPORT

998

999
1000
1001
1002
1003
5-91. Not all COIN efforts require deployments of large combat formations. In many cases, U.S. support is limited, focused on missions like security forces advisory duty, fire support, or logistic sustainment. The purpose of the limited support approach is to focus on building HN capacity and capability. HN security forces are expected to conduct combat operations, including any clearing and holding that has to be accomplished.

ASSESSMENT OF COUNTERINSURGENCY OPERATIONS

1004

1005
1006
1007
1008
1009
1010
The two best guides, which can not be readily reduced to statistics or processed through a computer, are an improvement in intelligence voluntarily given by the population and a decrease in the insurgents' recruiting rate. Much can be learnt merely from the faces of the population in villages that are subject to clear-and-hold operations, if these are visited at regular intervals. Faces which are at first resigned and apathetic, or even sullen, six months later are full of cheerful welcoming smiles. The people know who is winning.

1011
1012
Sir Robert Thompson
Defeating Communist Insurgency: The Lessons of Malaya and Vietnam

1013
1014
1015
1016
5-92. *Assessment* is the continuous monitoring and evaluation of the current situation and progress of an operation (FMI 5-0.1). Commanders, assisted by the staff, continuously assess the environment, situation, and the progress of the operation, and compare it with the commander's visualization and intent. Assessment precedes and is integrated into every operations-process activity. (See FM 3-0, chapter 6, and FMI 5-

1017 0.1, chapter 3.) Assessment involves comparing projected outcomes to actual events, using conceptual
1018 tools to determine progress toward success. It entails two distinct tasks:

1019 ● Continuously monitoring the current situation (including the environment) and progress of the
1020   operation.

1021 ● Evaluating the operation against established criteria using measurement tools and criteria.

1022 Based on their assessments, commanders adjust the operation and associated activities to better achieve the
1023 desired end state. (See FMI 5-0.1, paragraphs 4-12 through 4-20, and FM 6-0, paragraphs 6-90 through 6-
1024 92 and 6-110 through 6-121.)

1025 ## DEVELOPING MEASUREMENT CRITERIA

1026 5-93. Commanders and staffs want to know why and when they are achieving success within the LLOs.
1027 Traditionally, they look for discrete quantitative and qualitative measurements to indicate success. How-
1028 ever, the complex nature of COIN makes progress difficult to measure. Subjective assessment at all levels
1029 is essential to understand the diverse and complex nature of COIN problems and to measure local success
1030 or failure against the overall operation's purpose. Additionally, since commanders want to know how ef-
1031 fectively the LLO are working together, planners must evaluate not only progress along each LLO, but also
1032 how they are affecting activities along other LLOs.

1033 ## ASSESSMENT TOOLS

1034 5-94. Assessment tools help commanders and staffs determine—

1035 ● Completion of tasks and their impact.
1036 ● Level of achievement of objectives.
1037 ● Whether a condition of success has been developed or established.
1038 ● Whether the operation's end state has been attained.
1039 ● Whether the commander's intent was achieved.

1040 For example, planning for transition of responsibility to the host nation is an integral part of COIN opera-
1041 tional design and planning because of the fundamental requirement to support development of the HN gov-
1042 ernment's legitimacy. Assessment tools may be used to assess the geographic and administrative transfer of
1043 control and responsibility to the HN government as it develops its capabilities. Assessments differ for
1044 every mission, task, and LLO, and for different phases of an operation. Leaders adjust assessment methods
1045 as the insurgents adapt to COIN force tactics and the environment changes.

1046 5-95. The two most common assessment measures are measures of effectiveness (MOEs) and measures of
1047 performance (MOPs).

1048 5-96. A *measure of effectiveness* is a criterion used to assess changes in system behavior, capability, or op-
1049 erational environment that is tied to measuring the attainment of an end state, achievement of an objective,
1050 or creation of an effect (FMI 5-0.1). MOEs focus on the results or consequences of actions. MOEs answer
1051 the question, Are we achieving results that move us towards our desired end state, or are additional or al-
1052 ternative actions required?

1053 5-97. A *measure of performance* is a criterion to assess friendly actions that is tied to measuring task ac-
1054 complishment (FMI 5-0.1). An MOP answers the question, Was the task or action performed as the com-
1055 mander intended?

1056 5-98. Leaders may use observable, quantifiable, objective data as well as subjective indicators to assess
1057 progress measured against expectations. A combination of both types of indicators is recommended to re-
1058 duce the chance of misconstruing trends.

1059 5-99. All MOEs and MOPs for assessing COIN operations should be designed with the same characteris-
1060 tics. These four characteristics are—

1061 ● **Measurable**. MOEs and MOPs must have quantitative or qualitative standards against which
1062   they can be measured. The most effective measurement would be a combination of quantitative
1063   and qualitative measures to guard against an inaccurate view of results.

- **Discrete**. Each MOE and MOP must measure a separate and distinct aspect of the task, purpose, or condition.
- **Relevant**. MOEs and MOPs must be relevant to the measured task, outcome, and condition. HN local, regional, and national leaders and nongovernmental organization personnel may provide practical, astute, and professional ideas and feedback to craft relevant MOPs and MOEs.
- **Responsive**. Assessment tools must detect environmental/situational changes quickly and accurately enough to facilitate the commander's development of an effective response or counter.

BROAD INDICATORS OF PROGRESS

5-100.   Numerical and statistical indicators have limits when measuring social environments. For example, in South Vietnam U.S. forces used the body count as an indicator of success or failure within what this chapter describes as the combat LLO. Yet, the body count indicator only communicated a small part of the information commanders needed to assess their operations and was therefore misleading. Body count can only be a partial and effective indicator when adversaries and their identities can be verified through a uniform or possession of an insurgent political identification card. Additionally, COIN forces require the exact number of insurgent armed fighters initially present to gain an accurate appreciation of what the number of insurgent casualties might indicate in respect to enemy strength or capability. This discrete indicator fails to inform as to which side the local population blames for collateral damage, whether this fighting and resultant casualties damaged the insurgent infrastructure and affected the insurgency strategy in that area, and where the families of the dead insurgents reside and how they might react. For another example, within the essential services LLO the number of schools built or renovated does not equate to the effective operation of an educational system.

5-101.   Planners should start with broad measures of social and economic health or weakness when assessing environmental conditions. Here are some possible examples of useful indicators in COIN:

- **Refugees**. Refugees are a product of the deliberate violence associated with insurgencies and their counteraction. People and families exiled from or fleeing their homes and property and people returning to them are measurable and revealing. The number, population, and demographics of refugee camps or the lack thereof are a resultant indicator.
- **Human movement and religious attendance**. In societies where the culture is dominated by religion, activities related to the predominant faith may indicate the ease of movement and confidence in security, people's use of free will and volition, and the presence of freedom of religion. Possible indicators include the following:
  - Flow of religious pilgrims or lack thereof.
  - Development and active use of places of worship.
  - Number of temples and churches closed by a government.
- **Value of the regional and/or national currency**. If the currency is traded on international markets, the international perception of its stability, and by inference that of the economy as a whole, can be evaluated by whether the currency is rising, falling, or stationary against other currencies.
- **Presence and activity of small- and medium-sized businesses**. When danger or insecure conditions exist, these businesses close.
- **Level of agricultural activity**. Is a region/nation self-sustaining, or must life-support-type foodstuffs be imported? Has the annual need increased or decreased?
- **Presence or absence of associations**. Formation and presence of multiple political parties, independent, professional associations, and trade unions.
- **Participation in elections**, especially when an insurgency publicly threatens violence against those who participate.
- **Government services available**, such as—
  - Police stations operational and police officers present throughout the area.
  - Clinics and hospitals in full operation, and new whether facilities sponsored by the private sector are open and operational.

1114   ■   Virtually all schools and universities are open and functioning.

1115   ●   **Industry exports**.

1116   ●   **Employment/unemployment rate**.

TARGETING

1117

1118   5-102. The targeting process focuses operations and the use of limited assets and time. Commanders and
1119 staffs use the targeting process to achieve effects that support the lines of operations in a COIN campaign
1120 plan. It is important to understand that targeting is done for all operations, not just attacks against insur-
1121 gents. The targeting process can support IO, civil-military operations, and even meetings between com-
1122 manders and HN leaders, based on the commander's desires. The targeting process occurs in the targeting
1123 cell of the appropriate command post.

1124   5-103. Targeting in a COIN environment requires the creation of a targeting board or working group at
1125 all echelons. (See FMI 5-0.1, paragraphs 2-31, 2-38, and 3-58.) The intelligence cell provides representa-
1126 tives to the targeting board/working group to synchronize targeting with intelligence sharing and ISR op-
1127 erations. The goal is to prioritize targets and determine the best means of engaging them in order to have
1128 effects on the people, host nation, and insurgency that support the commander's intent and operation plan.

1129   5-104. The focus for targeting is on people, both insurgents and noncombatants. There are two major
1130 types of targets: key personalities and areas. Key personalities include insurgent personnel, community or
1131 political elites, and other individuals with power over the populace. Areas include segments of the popula-
1132 tion and regions under insurgent control.

1133   5-105. Personality and area targets can be further broken down into lethal targets and nonlethal targets.
1134 Lethal targets are best addressed via operations to capture or kill, while nonlethal targets are best dealt with
1135 via civil-military operations, IO, negotiation, political programs, economic programs, social programs and
1136 other noncombat methods. Nonlethal targets are usually more important than lethal targets and will never
1137 be less important.

1138   5-106. Personality targets include the following:

1139   ●   **Lethal**. Insurgent leaders to be captured or killed.

1140   ●   **Nonlethal**.

1141   ■   Personnel such as community leaders and insurgents, who should be engaged through out-
1142 reach, negotiation, meetings, and other interaction.

1143   ■   Corrupt HN leaders who may have be replaced.

1144   5-107. Area targets include the following:

1145   ●   **Lethal**.

1146   ■   Insurgent bases and insurgent logistic depots or caches.

1147   ■   Smuggling routes that can be interdicted.

1148   ●   **Lethal and nonlethal mix**.

1149   ■   Populated areas where insurgents commonly operate.

1150   ■   Populated areas controlled by insurgents where the presence of U.S. or HN personnel pro-
1151 viding security could undermine support to insurgents.

1152   ●   **Nonlethal**. Populations potentially receptive to civil-military operations or IO.

1153   5-108. The basic targeting process activities are similar to those described in FM 3-09.31:

1154   ●   Decide what targets must be addressed.

1155   ●   Detect the targets.

1156   ●   Deliver (conduct the operation).

1157   ●   Assess the effects of the operation.

1158   5-109. The first two activities are often switched in sequence, however. Rather than *decide-detect-deliver,*
1159 *assess*, the process may become *detect-decide-deliver-assess*. This is because the targets are individuals or

1160   groups of people rather than a fixed enemy order of battle. It is impossible to decide who/what to target
1161   without first knowing who/what the targets are.

1162   ## DETECT

1163   5-110.   The detect activity is performed on a continuous basis. It requires a great deal of analytical work
1164   by intelligence personnel. They analyze large quantities of all-source intelligence reporting to determine
1165   the validity of threats, how important different potential targets actually are, what means should be used to
1166   engage the targets, and the expected effects of engaging the targets. This requires a detailed understanding
1167   of social networks, insurgent networks, insurgent actions, and community atmospherics. (See appendix E.)

1168   5-111.   As noted earlier, attention must be paid to intelligence regarding the perceptions and interests of
1169   the populace. This intelligence is crucial to IO and civil-military operations targeting. It is also important
1170   for developing political, social, and economic programs. (See FM 3-13, appendix E, for IO targeting tech-
1171   niques.)

1172   ## DECIDE

1173   5-112.   The decide activity is the culmination of previous intelligence work. Intelligence personnel, with
1174   the commander and other staff members, must decide when analysis is sufficiently developed to warrant
1175   engagement. Continuous staff integration and regular meetings of the intelligence cell and targeting board
1176   smooth and enable this process. Staff personnel take finished intelligence products, fuse them with their
1177   own understanding of the AO, and help the commander and operations officer make decisions about
1178   who/what to target and how. Intelligence personnel provide information on the relative importance of dif-
1179   ferent target personalities and areas, and the projected effects of lethal or nonlethal engagement.

1180   5-113.   At the end of the decide activity, the targeting board has produced a prioritized list of targets and a
1181   recommended course of action associated with each. Fragmentary orders by operations personnel often
1182   come about based on the decide activity. Each of these orders is an operation that should be nested with the
1183   higher headquarters' plan and the commander's intent. The intelligence synchronization plan may have to
1184   change, based on targeting decisions.

1185   ## DELIVER

1186   5-114.   The deliver activity is simply executing the missions decided upon by the commander.

1187   ## ASSESS

1188   5-115.   The assess activity occurs continuously throughout an operation. During assessment, it is critical
1189   that collectors and analysts review the results of the operation and change intelligence synchronization
1190   plans and analyses appropriately. In addition to assessing changes to their own operations, intelligence per-
1191   sonnel look for reporting that indicates the effects on all aspects of the operational environment, to include
1192   insurgents and civilians. Relevant reporting can come from any of the intelligence disciplines, open
1193   sources, or operational reporting. Depending on the effects, commanders may choose to expand the opera-
1194   tion, maintain it as is, halt it, initiate a branch or sequel, or take steps to correct the damage of a mistake.
1195   Therefore, an accurate after-action assessment is very important. Metrics often include the following:
1196   - Changes in local attitudes (friendliness towards U.S. and HN personnel).
1197   - Changes in public perceptions.
1198   - Changes in the quality or quantity of intelligence provided by individuals or groups.
1199   - Changes in the economic or political situation of an area.
1200   - Changes in insurgent patterns.
1201   - Captured/killed insurgent personnel.
1202   - Captured equipment and documents.

1203   5-116.   As indicated in chapter 3, detainees, captured documents, and captured equipment may provide a
204   great deal of additional information. Its exploitation and processing into intelligence often adds to the over-

1205 all understanding of the enemy and precipitates additional targeting decisions. In addition, the assessment
1206 of the operation should be fed back to collectors. This allows them to see if their intelligence sources are
1207 credible. In addition, effective operations often cause locals to provide more intelligence, which drives fu-
1208 ture operations.

LEARNING AND ADAPTATION
1209

1210 5-117. Once the operation plan has been initiated, a commander may conduct operations to develop the
1211 situation to gain a more thorough understanding of specific elements of the environment. This increased
1212 environmental understanding represents a form of operational learning and applies across all LLOs. The
1213 commander and staff should expect to adjust the operation's design and plan based on what they learn. The
1214 result is an ongoing design-learn-redesign cycle.

1215 5-118. COIN operations involve complex, changing relations among all the direct and peripheral partici-
1216 pants, and these participants adapt, respond, and react to each other throughout an operation. A cycle of
1217 adaptation usually develops between insurgents and the counterinsurgents in which both sides continually
1218 adapt to neutralize existing adversary advantages and develop a new (albeit usually short-lived) advantage
1219 of their own. Victory is gained through a tempo or rhythm of adaptation that is beyond the other side's
1220 ability to achieve or sustain. Therefore, the COIN force should seek to gain and sustain advantages over in-
1221 surgents through emphasis on the learning and adaptation that is emphasized throughout this manual.

1222 5-119. This process of learning and adapting in COIN is made even more difficult because of the com-
1223 plexity of the problems to be solved. Generally, there is not a monolithic adversary that can be singularly
1224 classified as the enemy. Many insurgencies consist of multiple competing insurgency groups (more than 40
1225 groups in some historical cases such as Sri Lanka). The HN government and COIN force must adapt based
1226 on understanding this very intricate environment. But the key to effective COIN operational design and
1227 execution remains the ability to adjust better and faster than the insurgents.

SUMMARY
1228

1229 5-120. The challenge of COIN is complex, demanding, and tedious in execution. There are no simple or
1230 quick solutions. Success may often seem elusive. However, contributing to the complexity of the problem
1231 is the manner in which the environment is viewed by the COIN force and how success is defined. The spe-
1232 cific architecture of the COIN operation and the manner in which it is executed must be based on a holistic
1233 treatment of the environment and remain focused on the commander's vision of resolution. Success will be
1234 gained from unity of effort across all LLOs to achieve goals that relate directly to the purpose for the COIN
1235 operation—to establish legitimacy for the HN government and gain popular support. Operational design
1236 and execution cannot really be separated. They are both part of the same whole.
1237
1238
1239

Chapter 6

Developing Host-nation Security Forces

Helping others to help themselves is critical to winning the long war.

2006 Quadrennial Defense Review Report

This chapter addresses aspects of developing host-nation security forces. It begins with a discussion of the challenges involved and the resources required. It provides a framework for organizing the development effort. It concludes with a discussion of the role of police in counterinsurgency operations.

OVERVIEW

6-1.   Success in COIN requires the establishment of a legitimate government which has the support of the people and is able to address the fundamental causes that the insurgents use to gain popular support. Achieving these goals requires the host nation to defeat or render irrelevant the insurgent forces, uphold the rule of law, and provide a basic level of essential services and security for the population. Key to all these tasks is the development of an effective host-nation (HN) security force. In some cases, U.S. armed forces might be actively engaged in fighting insurgents while simultaneously helping the host nation build its own security forces.

6-2.   Just as an insurgency and counterinsurgency (COIN) are defined by a complex array of factors, the job of training the HN security forces is also affected by a variety of determinants. These include whether sovereignty in the host nation is being exercised by an indigenous government or by a U.S. or multinational element. The second gives the COIN force more freedom of maneuver, but the first is important for legitimate governance, a key goal of any COIN effort. If the host nation is sovereign, the quality of the governance it provides also has an impact. The scale of the effort is another factor; what works in a small country might not work in a large one. Terrain and civil considerations are also important. A nation that is compartmentalized by mountains, rivers, or ethnicity presents different challenges for the COIN effort. A large "occupying" force or international COIN effort is a factor that can facilitate success in training HN security forces, but also clearly complicates the situation. There are certainly many more factors: what type of security forces previously existed, whether the effort requires "building from scratch" or "simply" changing what already exists, sectarian challenges within the forces, resources available, popular support, and so fourth. This chapter describes some doctrinal foundations for training HN security forces. However, these foundations must be adapted to the situation on the ground and the particular insurgency at hand.

6-3.   The term "security forces" in this chapter includes all forces under HN control with the mission of protecting against internal and external threats. Elements of the security forces include, but are not limited to, military forces, police, corrections personnel, and border guards (to include the coast guard) at the local through national levels. Only in unusual cases can the United States expect a situation where the host nation has no security force. With this in mind, this chapter addresses methods to develop security forces, realizing that the range of assistance varies depending on the situation.

6-4.   Foreign internal defense (FID) doctrine is available in Joint Publication 3-07.1. JP 3-07.1 addresses the legal and fiscal regulations and responsibilities concerning the planning, development, and administration of FID programs. It also discusses command and supervisory relationships of the U.S. diplomatic mission, the geographic combatant command, and joint task forces in the application of military aid, support, and advisory missions. The tenets presented in this chapter are intended to reinforce and supplement JP 3-07.1.

43 # CHALLENGES, RESOURCES, AND END STATE

44 6-5. Each instance of developing security forces will be as unique as each insurgency. In Vietnam, the
45 United States committed thousands of advisers for South Vietnamese units and hundreds of thousands of
46 combat troops and failed in its COIN efforts. In El Salvador, a relative handful of American advisers were
47 sufficient for success. Many factors influence the amount and type of aid required. These are discussed in
48 more detail later, but include the following:

49 • Existing capabilities of the HN security force.

50 • Character of the insurgency.

51 • Terrain, the culture.

52 • Level of commitment and sovereignty of the host nation.

53 • Level of commitment from the United States and other nations.

54 6-6. While U.S. and multinational forces may be required to assist the host nation in improving security,
55 the insurgency will use the presence of foreign forces as another reason to question the legitimacy of the
56 HN government. A government reliant on foreign forces for internal security runs the risk of not being rec-
57 ognized as legitimate. While combat operations with significant U.S./multinational participation may be
58 necessary, U.S. combat operations are secondary to enabling the host nation's ability to provide for its own
59 security.

60 ## CHALLENGES TO THE DEVELOPMENT OF EFFECTIVE SECURITY FORCES

61 6-7. There have been a number of common problems and issues in the training missions undertaken by
62 U.S. forces since the end of World War II These problems generally fall under differing national perspec-
63 tives in one of four broad categories:

64 • Resources.

65 • Leadership.

66 • Exercising power.

67 • Organizational structures.

68 6-8. Governments must properly balance national resources to meet the expectations of the people. Fund-
69 ing for services, education, and health care can limit resources available for security forces. The result of
70 the host nation's spending priorities may be a security force capable of protecting only the capital and key
71 government facilities, leaving the rest of the country unsecured. Undeveloped countries generally lack the
72 resources to maintain logistic units, resulting in chronic sustainment problems. Conducting effective COIN
73 operations requires an allocation of resources that ensures integration of efforts to develop all aspects of
74 the security force. Recognizing the interrelationship of security and governance, the central government
75 must also ensure adequate resources are devoted to meeting such basic needs as health care, clean water,
76 and electricity.

77 6-9. The existing HN approach to leadership may need to be adjusted. HN leaders may be appointed and
78 promoted on the basis of family ties or on the basis of membership in a party or faction, rather than on
79 demonstrated competence or performance. Leaders may not seek to develop subordinates. The need to en-
80 sure the welfare of the troops may not be a commonly shared trait. In some cases, leaders enforce the obe-
81 dience of their subordinates by pure fear and use their position to exploit them. Positions of power can also
82 lead to corruption, which can also be affected by local culture.

83 6-10. The behavior of HN security force personnel is often a primary cause of public dissatisfaction. The
84 corrupting influences of power must be guarded against. Cultural and ethnic differences within a popula-
85 tion may lead to significant discrimination within the security forces and by the security forces against mi-
86 nority groups. In more ideological struggles, the prejudicial treatment may be manifested against personnel
87 of other political parties, whether in a minority cultural group or not. Security forces that abuse civilians do
88 not win the trust and confidence of the population. A program of comprehensive security force develop-
89 ment requires identifying and addressing biases along with improper or corrupt practices.

90
91
92
93
94
6-11. Perhaps the biggest hurdle for U.S. forces to overcome is accepting that the host nation can ensure security using practices that work but are different from U.S. practices. The typical American bias of, "The American way is best," must be recognized as unhelpful and continuously addressed. While the American relationship between police, customs, and the military works for the United States, it usually does not work for other nations that have developed differently.

RESOURCES

96
97
98
99
100
101
102
103
6-12. For the U.S. military, the mission of developing HN security forces is more than just a task assigned to a few specialists. While FID was traditionally the primary responsibility of the special operations forces, given the scope and scale of training programs today and the scale of programs likely to be required in the future, training foreign forces is now a core competency of regular and reserve units of all Services. Multi-national partners are often willing to help a nation fight insurgency by assisting with training HN forces. Partner nations may develop joint training teams or assign teams to a specific element of the security force or a particular specialty. Training resources may be received from the following organizations and pro-grams:

Special Operations Forces

105
106
107
108
109
110
111
6-13. With their focus on specific regions of the world and study of languages and cultures, special opera-tions forces (SOF) have long been the lead organization in training and advising foreign armed forces. (Army special forces training programs and tactics, techniques, and procedures are outlined in FM 31-20-3.) While SOF personnel may be ideal for some training and advisory roles, their limited numbers restrict their ability to carry out large-scale missions to develop HN security forces. In a low-level COIN, SOF per-sonnel may be the only forces assigned to the training mission; at the higher end of the spectrum, SOF may be allocated to training only their counterparts in the HN force.

Ground Forces

113
114
115
116
117
118
119
120
121
6-14. Large-scale training and advisory missions require the employment of large numbers of land forces (primarily Army and Marine) that may not have SOF language training or regional expertise. However, such conventional forces may have some advantages in training HN counterparts with similar missions. Linguistic augmentation and additional cultural training are required for both SOF and conventional ground forces. (See appendix C.) Assignment of the best-qualified Soldiers and Marines to training and advisory missions is necessary. Those personnel normally come from the active-duty force, but large-scale efforts require assigning Reserve Component personnel to the training and advisory mission as well. Ade-quate prior training, both predeployment and in-theater, is required for all land forces assigned this high-priority mission.

Joint Forces

123
124
125
126
127
128
6-15. Although other Services normally play smaller roles, they can still make significant contributions because of their considerable experience in training their counterparts. The Navy and Air Force should be utilized in training their HN counterparts. The Coast Guard may also be of value, as its coastal patrol, fish-eries oversight, and port security missions are in keeping with the responsibilities of navies in developing countries. To minimize the burden on the land forces, specialists—such as lawyers and medical person-nel—from other Services should be used for HN training wherever possible.

Interagency/Intergovernmental Resources

130
131
132
133
134
135
136
6-16. Interagency resources can be applied in numerous ways to support training HN security forces. Per-haps most important is training nonmilitary security forces. The Justice and State Departments have the ca-pability of sending law enforcement specialists overseas to train and advise HN police forces, which are best trained by other police. The quick-reaction capability of these agencies is limited, although they can attain necessary levels when given adequate time. Such forces are also expensive. During intensive coun-terinsurgencies, the effectiveness of civilian police advisers and trainers may be limited by the high-threat nature of the environment. These forces are most effective when operating in a benign environment or

137
138
139

when security is separately provided. There are many legal restrictions about training nonmilitary forces, and the State Department normally has the lead in such efforts However, there have been occasions when military forces were assigned that mission by the President.

140 **Multinational Resources**

141
142
143
144
145
146

6-17. Although their support frequently plays more of a legitimizing role, multinational partners also assist materially in training HN security forces. Some nations may be more willing to train HN forces than to provide troops for combat operations. Some multinational forces come with significant employment restrictions. Each international contribution is assessed on its own merits, but decisions to decline such assistance are rarely made. Good faith efforts to integrate multinational partners and achieve optimum effectiveness are required.

147 **International Military Education and Training Program**

148
149
150
151
152

6-18. For more than 50 years, the U.S. military has run the International Military Education and Training (IMET) program to provide opportunities for foreign personnel to attend U.S. military schools and courses. Most of these commissioned officers and noncommissioned officers (NCOs) are provided English language training before attending the U.S. courses. In the case of Latin American armed forces, the U.S. operates courses in Spanish.

153 **Contract Support**

154
155
156

6-19. In some cases, additional training support is contracted to enable uniformed forces to be more efficiently utilized. Contractor support can be used to provide HN training and education to include the following:

157
158
159

- Institutional training.
- Development of security ministries and headquarters.
- Establishment of administrative and logistic systems.

160
161

Contracted police development capabilities through the State Department's Bureau of International Narcotics and Law Enforcement Affairs can provide expertise that is not resident in the uniformed military.

162 **Organizing U.S. Forces to Develop Host Nation Security Forces**

163
164
165
166
167
168

6-20. Developing HN security forces is a complex and challenging mission. The United States and multinational partners can only succeed if that mission is approached with the same deliberate planning and preparation, energetic execution, and appropriate resourcing as the combat aspects of the COIN operation. Accordingly, the joint force commander and staff need to consider the task of developing HN security forces as part of their initial mission analysis. That task must then be an integral part of all assessments, planning, coordination, and preparation.

169
170
171

6-21. As planning unfolds, mission requirements should drive the initial organization for the unit charged with developing security forces. To achieve unity of effort, a single organization should be given this responsibility.

172
173
174

6-22. For small-scale efforts this could be a SOF mission. SOF organizations at the company, battalion, and group level are ideally suited for developing security forces through the FID portion of their doctrinal missions.

175
176
177

6-23. If only a single component (ground, naval, air) or special forces is being developed, a Service-specific organization could be designated. For example, a Naval task force could be given the mission of developing HN naval and marine forces to guard oil distribution platforms.

178
179
180

6-24. In an area in which COIN operations are already being conducted, the development of security forces could be assigned to a specific unit, such as a brigade combat team, division, or Marine air-ground task force.

6-25. For large, multi-Service, long-duration missions, a separate organization with the sole responsibility of developing security forces and subordinate to the joint force commander may be required. Such an organization may be multi-Service, multinational, and interagency.

6-26. The internal structure of the organization charged with developing security forces must reflect the desired end state of those security forces. For example, if army, police, air, naval, and special forces are being developed, the organization in charge of those programs requires teams charged specifically with each of those tasks. If civilian security components, such as a ministry of defense or interior, are being developed, then ministerial teams are also required. Because the quality of the developing security forces in terms of professionalism and ethics is so important, a separate training element focused on those values may be needed.

6-27. The U.S./multinational force responsible for these programs also requires a headquarters and staff task-organized for all the functions required to support the full all aspects of the development of security forces. (See paragraph 6-32.) In addition to traditional staff functions, some or all of the following may require a more robust capability than normal:

- Comptroller, for managing the significant monetary resources required for training, equipping, and building security forces. A separate internal auditor may be required as a check to ensure the host nation's resources are safeguarded and effectively managed.
- Staff judge advocate, with specific specialties and a robust capability for contract law, military justice, and law of land warfare.
- Construction engineer management, to oversee and manage the construction of security forces infrastructure, such as bases, ranges and training areas, depots and logistics facilities, and police stations.
- Political-military advisers, to ensure integration of the development of security forces with the development of civilian ministries and capabilities.
- Public affairs, with a focused capability to build confidence of the population in their security forces and to develop the public affairs capability of the HN forces.
- Force protection and focused intelligence staff, to address the challenge of and threats to the relatively small teams that may be embedded with HN security forces and not co-located with robust U.S. or partner forces or bases.
- Materiel management, until such a capability is developed in the HN forces. The equipping and supplying of new security forces is critical to their development and employment. It may not be able to wait until that capability is developed by the host nation.
- Health affairs, since most developing countries have poor health care systems. HN personnel will be more likely to stay in new units and fight if they are confident that they will be properly treated if wounded. Additionally, disease is a significant threat that must be addressed with preventive medicine and robust care.
- Security assistance, specifically IMET, to manage the external training efforts and foreign military sales, and employ well-developed procedures for purchasing weapons, equipment, goods, and services. In a COIN these functions are likely embedded in the staff elements of the higher headquarters, rather than a stand-alone office as in a traditional office of military cooperation.
- Staff officers with a civilian law enforcement background or actual civilian law enforcement personnel can play a vitally important role in advising the commander. Traditionally, officers from the Reserve Components have played this role quite well.

6-28. An effective security force development organization is flexible and adaptive. The requirements for developing the type, character, composition, and quantity of security forces will change with the growth of security forces and the maturity of the COIN operation. The organization must be able to anticipate such changes, since joint manning document procedures and requests for forces have limited responsiveness. Temporary duty and contract personnel support may be used to provide niche capabilities or to fill gaps until more permanent individuals or units can be added to the organization.

230 **DESIRED END STATE**

231 6-29. Training HN security forces is a slow and painstaking process, and does not lend itself to any "quick
232 fixes." Real success does not appear as a single decisive victory. To ensure long-term success, the end state
233 of training programs needs to be clearly stated at the start. The end state is usually a set of military charac-
234 teristics common to all militaries. Those characteristics have nuances in different states, but a well-trained
235 HN security forces should—

236 ● Be able to provide reasonable levels of security from external threats while not threatening re-
237 gional security.

238 ● Be able to provide reasonable levels of internal security without infringing upon the citizens'
239 civil liberties or posing a coup threat.

240 ● Be founded upon the rule of law.

241 ● Be sustainable by the host nation after U.S. and multinational forces depart.

242 6-30. Because of the demands of dealing with an insurgency, the HN military and police forces may be
243 performing functions that are not normally considered conventional. The military may be filling an internal
244 security role usually reserved for the police, and the police may have forces that are so heavily armed that
245 they would normally be part of the military. In the near term, the HN security forces should—

246 ● Be focused on COIN operations, integrating military capabilities with those of national, re-
247 gional, and local police.

248 ● Maintain the flexibility to transition to more conventional roles of external and internal defense,
249 based on long-term requirements.

250 6-31. To meet both near- and long-term objectives, trainers remember the cumulative effects of training.
251 Effective training programs have impacts in the short, mid, and long-terms.

252 6-32. To achieve this end state and intermediate objectives, the host nation should develop a plan—with
253 multinational assistance, when necessary—that holistically addresses all aspects of force development.
254 U.S. doctrine divides force development into the following domains: doctrine, organization, training, mate-
255 riel, leadership and education, personnel and facilities (DOTMLPF). Doctrine is listed first, but these ele-
256 ments are tightly linked, simultaneously pursued, and difficult to prioritize. The commander monitors pro-
257 gress in all domains. There is always a temptation for Soldiers and Marines involved in such programs to
258 impose their own doctrine and judgment on the host nation. The first U.S. advisers and trainers working
259 with the South Vietnamese Army aimed to create a conventional force to fight another Korean War. They
260 did not appreciate the capabilities of their allies or the real nature of the threat. The organization and doc-
261 trine adopted did not suit the South Vietnamese situation and proved very vulnerable to North Vietnamese
262 guerrilla tactics. HN security force doctrine, like the remaining DOTMLPF domains discussed throughout
263 this chapter, must be appropriate to HN capabilities and requirements.

264 6-33. The objective of development programs is to create security forces with the following characteris-
265 tics:

266 ● Flexible, a force that is capable of accomplishing the broad missions required by the nation—not
267 just to defeat insurgents or defend against outside aggression, but to increase security in all ar-
268 eas. This requires an effective command and organizational structure that makes sense for the
269 host nation.

270 ● Proficient.

271 ■ Security forces that are capable of working effectively in close coordination with each other
272 to suppress lawlessness and insurgency.

273 ■ Military units that are tactically and technically proficient, capable of ensuring their aspect
274 of national security, and capable of integrating with allies.

275 ■ Nonmilitary security forces that are competent in maintaining civil order, enforcing laws,
276 controlling borders, and detaining criminal suspects.

277 ■ Nonmilitary security forces that are thoroughly trained in modern police ethos and proce-
278 dures, and understand the basics of investigation, evidence collection, and proper court and legal
279 procedures.

280	● Self-sustained, forces capable of managing their own equipment throughout its lifecycle (pro-
281	curement to disposal) and conducting administrative support.
282	● Well led, leaders at all levels who possess sound professional standards and appropriate military
283	values, and are selected and promoted based on competence and merit.
284	● Professional.
285	■ Security forces that are honest, impartial, and committed to protecting and serving the entire
286	population, operating under the rule of law, and respecting human rights.
287	■ Security forces that are loyal to the central government and serve national interests, recog-
288	nizing their role as servants of the people and not their masters.
289	● Integrated into society, a force that is representative of the nation's major ethnic groups and is
290	not seen as the instrument of just one faction. Cultural sensitivities toward the incorporation of
291	women must be observed, but efforts should also be made to include women in police and mili-
292	tary organizations.

FRAMEWORK FOR DEVELOPMENT

6-34. The mission to develop HN security forces can be organized around these processes:
- Assess.
- Organize.
- Build.
- Train.
- Equip.
- Advise.

These incorporate all DOTMLPF requirements. Although described sequentially, these processes will normally be conducted concurrently. For example, training and equipping operations must be integrated and, as the operation progresses, assessments will lead to changes. If U.S. forces are directly involved in operations against the insurgency, the U.S. program also includes a transition period during which major COIN operations are handed over to HN security forces.

ASSESS

6-35. As with every major military operation, the first step is to assess the situation. The assessment should be one part of the comprehensive program of analyzing the insurgency and normally includes a social and economic analysis. The analysis is conducted in close collaboration with the U.S. country team, the host nation, and multinational partners. Assessment of the security situation and its influence on other lines of operation must be continuous. From the assessment, planners develop short-, mid-, and long-range goals and programs, but those goals and programs must remain responsive to changing circumstances. A raging insurgency might require the employment of HN forces at various stages of development. Some existing security forces might be discovered to be so dysfunctional or corrupt that they have to be disbanded rather than rehabilitated. In some cases, leaders will need to be replaced in order for the unit to become functional.

6-36. While every situation is different, the following factors should be assessed throughout planning, preparation, and execution of the operation:
- Social structure, organization, demographics, interrelationships, and education level of the elements of the security force.
- Methods, successes, and failures of HN COIN efforts.
- State of training at all levels and the specialties and education of leaders.
- Equipment and priority placed on maintenance.
- Logistics and support structure and its ability to meet the requirements of the force.
- Level of sovereignty of the HN government.
- Extent of acceptance of ethnic and religious minorities.

327 • Laws and regulations governing the security forces and their relationship to national leaders.

328 6-37. The mission analysis should provide a basis for determining the scope of effort required. The HN se-
329 curity forces may require complete re-establishment, or they may only require assistance to increase capac-
330 ity. They may be completely devoid of a capability (for example, internal affairs, federal investigative de-
331 partment, corrections, logistics for military forces, formal schools for leaders), or they may only require
332 temporary reinforcement. As with other military operations, efforts to assist security forces should rein-
333 force success. For example, instead of building new police stations in every town, improve the good sta-
334 tions and use them as a model for weaker organizations.

335 6-38. Decisions are needed on what shortfalls to address first. The extent of the insurgency combined with
336 resource limitations inevitably forces the commander to set priorities. Follow-on assessments should start
337 by reviewing areas to which resources have been restricted, determining where resources should be com-
338 mitted or redirected, and deciding where additional resources should be requested. If the U.S. or some
339 other nation or international entity is exercising sovereignty, such as during an occupation or regime
340 change, decisions about security force actions can be imposed on a host nation; however, it is always better
341 to take efforts to legitimize the HN leaders by including them in decisions.

342 6-39. The process of developing a strategic analysis and outlining a strategic plan for training the forces of
343 a country facing an insurgency is not necessarily a long one. In fact, the exigencies of a situation featuring
344 a security vacuum or very active insurgency often require the initiation of necessary programs as soon as
345 possible. Assessment is a continuous, and a quick initial assessment and the programs it inspires can be ad-
346 justed as more experience and information is gained. In 1981, when El Salvador faced a major insurgency,
347 a team of ten American officers visited El Salvador and consulted with the HN command and the U.S. mili-
348 tary assistance advisory group (referred to as the MILGRP) for ten days. At the end of that time the U.S.
349 team outlined a five-year comprehensive plan to rebuild, train, and re-equip the Salvadoran armed forces to
350 counter the insurgency. The U.S. plan effectively addressed the critical issues of reorganizing, training, and
351 equipping the Salvadoran armed forces and became part of the foundation of a successful national COIN
352 strategy. A team with a similar mission today should also include specialists from the Departments of State,
353 Justice, and Homeland Defense to assess the security force requirements.

354 ## ORGANIZE

355 6-40. The best organization for HN forces depends on the social and economic conditions of the country as
356 well as cultural and historical factors and the security threat the nation faces. The aim is to develop an ef-
357 fective an efficient organization with a command, intelligence, logistic, and operational structure that
358 makes sense for the host nation. General-purpose forces with limited special purpose teams (such as explo-
359 sive ordnance disposal and special weapons and tactics [SWAT]) are preferred, as elite units divert a large
360 share of the best leadership and remove critical talent from the regular forces. Doctrine should be standard
361 across the force, as should unit structures. The organization must facilitate the collection, processing, and
362 dissemination of intelligence across and throughout all security forces.

363 ### General Organizational Considerations

364 6-41. To the maximum extent possible, decisions on the structure of security force organization should be
365 made by the host nation. The host nation may be amenable to proposals from U.S./multinational forces, but
366 should at least approve all organizational designs. As the HN government gets stronger, U.S. leaders and
367 trainers should expect increasingly independent organizational decisions. These may include changing the
368 numbers of forces, types of units, and internal organizational designs. Culture and conditions might result
369 in security forces with what U.S. experience considers nontraditional roles and missions. Police may be
370 more paramilitary than the United States' own model, and the military may have a role in internal security.
371 Eventually, the role of police should be to counter crime and the role of the military should be to address
372 greater threats, though the exact nature of these missions depends on the situation in that nation. Police and
373 military roles should be clearly delineated.

374 6-42. Units should be organized to include all appropriate warfighting functions (formerly, the battlefield
375 operating systems) or some adaptation for police forces. (FMI 5-0.1, paragraphs 1-23 through 1-30, dis-

376 cusses the warfighting functions.) Some systems may be initially excluded for a variety of reasons (for ex-
377 ample, artillery is expensive, takes a large investment in training and is not easily applied to a counterin-
378 surgency), but organizational plans should foresee eventual establishment of all appropriate capabilities.

379 6-43. Organization should address all elements of the security forces, from the ministerial level to the pa-
380 trolling police officer and soldier. The complex matrix of simultaneous development programs is illustrated
381 in figure 6-1 (below). Building a competent HN civilian infrastructure—including civilian command and
382 control systems—is critical for success in COIN. The joint force commander should work with the HN
383 ministries responsible for national and internal security, including the ministry of national defense, the inte-
384 rior ministry, and the justice ministry. The strengths and weaknesses of the ministerial organization should
385 be assessed, as well as the requirements for training of civilian defense officials and employees. The U.S.
386 and multinational advisory team at the ministry level is responsible for assisting the host nation in develop-
387 ing a procurement and management system that effectively meets its requirements.

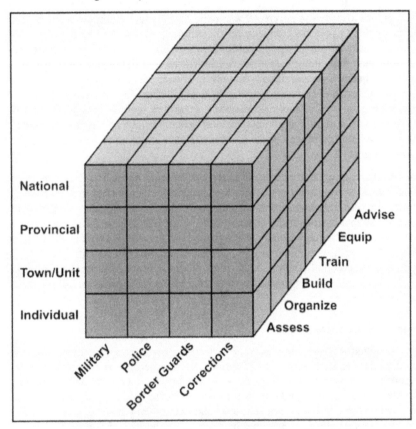

388 **Figure 6-1. Matrix of security force development**

389 6-44. A thorough review of HN military and police doctrine is a necessary first step in setting up a training
390 program. Advisers should review security force regulations to ensure they provide clear and complete in-
391 structions for discipline, acquisitions, and support activities. Doctrine (including tactics, techniques, and
392 procedures) should be reviewed and refined to address COIN operations. Regulations should be appropri-
393 ate for the level of education and sophistication of the security force personnel. The treatment of prisoners,
394 detainees and suspected persons should be spelled out clearly and be consistent with the norms of interna-
395 tional and military law.

396 **Organizational Personnel Issues**

397 6-45. Organizing a security force requires resolving issues related to the following areas:
398   • Recruitment.

399 • Promotion screening/selection.

400 • Pay and benefits.

401 • Leader recruitment and selection.

402 • Personnel accountability.

403 • Demobilization of security force personnel.

404 *Recruitment*

405 6-46. Recruitment is critical to the establishment of security forces. The recruitment program should be
406 crafted by the host nation and take local culture into account, using themes that resonate with the local
407 population. It should ensure that all the major demographic groups are properly represented in the security
408 forces. U.S. and multinational partners should encourage and support HN efforts to recruit from among the
409 minority populations. A mobile recruiting capability should be established in order to target specific areas,
410 ethnic groups, or tribes to ensure demographic distribution within the security forces. Moderate groups and
411 factions within hostile or potentially hostile ethnic groups should be contacted and the members of minor-
412 ity factions encouraged to support recruitment of their group members into the security forces. Recruitment
413 of disaffected ethnic groups into the security forces will likely become a major issue of contention and be
414 resisted by most HN governments. However, even moderate success in recruiting from disaffected ethnic
415 groups provides an enormous payoff in terms of building the legitimacy of the security forces and in quiet-
416 ing the often legitimate fears of such ethnic groups regarding their relationship with the government. An
417 effective disarmament, demobilization, and reintegration strategy for former insurgent individuals and
418 groups must be a component of the overall COIN operation and must therefore be accounted for in the re-
419 cruitment effort.

420 6-47. A clear set of appropriate mental, physical, and moral standards needs to be established and en-
421 forced. Ideally, recruits are centrally screened and inducted. Recruitment centers need to be in areas that
422 are safe and secure from insurgent attacks, as these centers are attractive targets for insurgents. All recruits
423 should undergo a basic security check and be vetted against lists of suspected insurgents. As much as pos-
424 sible, this process should be conducted by HN agencies and personnel. Membership in illegal organizations
425 needs to be carefully monitored. Past membership need not preclude joining the security forces, but any
426 ongoing relationship of a recruit with an illegal unit needs constant monitoring, and care is required to en-
427 sure that no single military or police unit contains too many prior members of an illegal unit, tribal militia,
428 or other militant faction.

429 *Promotion Screening/Selection*

430 6-48. Selection for promotion based on proven performance and aptitude for increased responsibility is es-
431 sential. Objective evaluations ensure promotion is by merit and not through influence or family ties. Two
432 methods may be worth considering for selecting leaders. One is to identify the most competent performers,
433 train them, and recommend them for promotion. The second is to identify those with social or professional
434 status within the training group, then train and recommend them for promotion. The first method may lead
435 to more competent leaders but could be resisted for cultural reasons. The second method ensures the new
436 leader will be accepted culturally, but may sacrifice competence. The most effective solution is often a
437 combination of the two methods.

438 *Pay and Benefits*

439 6-49. Appropriate compensation levels help prevent a culture of corruption in the security forces. It is
440 cheaper to spend the money for adequate wages and produce effective security forces than to pay less and
441 end up with corrupt and abusive forces that alienate the population. This is especially important for the po-
442 lice, who have the greatest opportunity for corruption in the nature of their duties and contact with the ci-
443 vilian community. Some important considerations concerning pay are the following:

444 • Pay for commissioned officers, NCOs, and technical specialists should be competitive with
445   other professions in the host nation. Police officers need to be paid a sufficient wage so that they
446   do not have to supplement their income with part-time jobs, or resort to illegal methods to sup-
447   plement their salary.

448 ● Pay should be disbursed through HN government channels, not U.S. channels.

449
450 ● Cultural norms should be addressed to ensure that any questionable practices such as the "tax-ing" of subordinates are minimized.

451
452 ● Good pay and attractive benefits must be combined with a strict code of conduct that allows for the immediate dismissal of corrupt security personnel.

453
454 ● Pensions should be available to compensate the families of security force members in the event of a service-related death.

455
456
457 ● Pay for military forces should come from central government budgets and specifically not from kickbacks or locally procured revenue. If this happens, the military's loyalty will always be in question and corruption will be likely.

458
459
460 6-50. Effective security forces can help improve the social and economic development of the nation through the benefits each member receives. Every recruit should be provided a basic level of literacy, job training, and morals/values inculcation.

461 *Leader Recruitment and Selection*

462
463
464 6-51. Officer candidate standards should be high. Candidates should be in good health and pass an academic test set to a higher standard than enlisted recruits. Officer candidates should be carefully vetted to ensure that they do not have close ties to any radical or insurgent organization.

465
466
467
468 6-52. NCOs should be selected from the best of the enlisted security force members. Objective standards, including proficiency tests, should be established and enforced to ensure that promotion to the NCO ranks is by merit and not through influence or family ties. Many armies do not have a professional NCO corps, and establishing one for a host nation may be a difficult process.

469 *Personnel Accountability*

470
471
472
473 6-53. Accountability of personnel in the security force units must be tracked carefully. Proper personnel accountability reduces corruption, particularly in manual banking systems where pay is provided in cash. In addition, the number of personnel failing to report for duty can be an indicator of possible attacks, unit morale, or insurgent and militia influences upon the security forces.

474 *Demobilization of Security Force Personnel*

475
476
477
478
479
480
481
482 6-54. Programs should be developed to prevent the formation of a class of impoverished and disgruntled former officers and soldiers who have lost their livelihood. During the conflict it will be necessary to remove officers and NCOs from the security forces for poor performance or for failure to meet the new, higher standards of the force. Some form of government-provided education grants or low-interest business loans will enable them to earn a living outside the military. Officers and NCOs who have served for several years and who are then removed should be given a lump sum payment or a small pension to ease their transition to civilian life. These programs should not apply to those who are guilty of human rights abuses or major corruption.

483
484
485
486
487
488 6-55. As a conflict winds down, some security forces may need to be disarmed, demobilized, and reintegrated into civil society. To avoid producing a pool of recruits for the insurgency, programs should be established to ensure that large numbers of demobilized security force members do not become immediately unemployed. Civil service departments should provide a strong hiring preference to people who have completed an honorable term of service. Government-financed education programs for the demobilized members are another possibility.

489 ## BUILD/REBUILD FACILITIES

490
491
492 6-56. Requirements for sustaining HN security forces include the infrastructure necessary to support the force: barracks, ranges, motor pools, and other military facilities. Because of the long lead times required for construction, early investment in such facilities is essential if they are to be available when needed.

493 Force protection must be considered in any infrastructure design, including headquarters facilities, as they
494 will be attractive targets for insurgents.

495 6-57. During an insurgency, the HN military and police forces are likely to be operating from a variety of
496 local bases. A long-term force-basing plan needs to be established for building training centers and unit
497 garrisons. If possible, garrisons should include housing for the commissioned officers, NCOs, and families;
498 government-provided medical care for the families; and other amenities that make national service attrac-
499 tive.

500 6-58. Extensive investment of time and resources might be required to restore or create the nationwide in-
501 frastructure necessary to effectively command and control HN security forces. Communications facilities
502 are especially important. Besides building local bases and police stations, functional regional and national
503 headquarters and ministries are required.

504 ## TRAIN

505 6-59. U.S. and multinational training assistance should address shortfalls at every level with the purpose of
506 establishing training systems that are self sustaining.

507 ### Training the U.S. Trainers

508 6-60. Soldiers and Marines assigned training missions should receive a course of preparation to deal with
509 the specific requirements of developing HN forces. The course should emphasize the cultural background
510 of the host nation while introducing its language and providing cultural tips for developing a good rapport
511 with HN personnel. The course should also include force protection training for troops working with the
512 HN forces. U.S. trainees must become familiar with the HN organization and equipment, especially weap-
513 ons not found in the U.S. inventory. Key points to be emphasized by this training include the following:

514 • Training must be sustained and include reinforcement of individual and team skills.
515 • Use the smallest possible student-to-instructor ratio.
516 • Develop HN trainers.
517 • Standards—not time—should drive training.
518 • Provide immediate feedback; use after action reviews.
519 • Respect HN culture but be able to tell the difference between cultural practices and excuses.

520 6-61. U.S. personnel should show respect for local religions and traditions, and willingly accept many as-
521 pects of the local/national culture, including the local food (if sanitation standards permit). U.S. personnel
522 need to make it clear that they are not in the host nation to undermine or change the local religion or tradi-
523 tions. On the other hand, U.S. personnel do have a mission to reduce the effects of dysfunctional social
524 practices that affect the ability to conduct effective security operations. U.S. trainers and advisers must
525 have enough awareness to identify inappropriate behavior and see that it is stopped, or at least reported to
526 both the multinational and HN chains of command.

527 ### Establishing Training Standards

528 6-62. Although the nature of insurgencies and their corresponding responses from targeted governments,
529 vary widely, clear metrics for individual, leader, and unit training for security forces can still be estab-
530 lished. These sets of standards include many of the same skills required for individuals and units to be ef-
531 fective in conventional military operations as well as additional requirements for COIN. Due to the small
532 unit nature of most COIN operations, effective COIN forces require a high level of junior leadership. Val-
533 ues training should be integrated into all levels of training for all components. Metrics to evaluate units
534 should include subjective measures, such as loyalty to the HN government, as well as competence in purely
535 military tasks. Soldiers and Marines are familiar with the evaluation of military training. However, the
536 more difficult task of gauging the acceptance of values, such as ethnic equality or the rejection of corrup-
537 tion, could very well provide more important measures of the training effectiveness in some COIN situa-
538 tions.

6-63. Effective training programs require the establishment of clear individual, leader, and unit perform-ance standards in detail, taking into account cultural factors that directly affect the ability of the individual or unit to operate. For example, training a unit to conduct effective operations requires more time in a country where the average soldier is illiterate. Similarly, staff training is more difficult in a country with a low educational level. Building a security force from scratch takes far more time than when there is a trained cadre of HN personnel already available. With this in mind, it is usually advantageous to take ad-vantage of existing military personnel to form units and cadres for units, rather than starting from the be-ginning. As previously mentioned, a vetting process may be required, but this is still usually better than the alternative.

6-64. Poorly trained leaders and units are far more prone to human rights violations than well-trained, well-led units. Leaders and units unprepared for the pressure of active operations tend to employ indis-criminate force, target civilians, and abuse prisoners—all actions that can threaten the popular support and government legitimacy so essential for success in COIN. Badly disciplined and poorly led security forces have served very effectively as recruiters and propagandists for the insurgents.

6-65. Setting realistic objective and subjective metrics for HN security forces and following through on training plans are time consuming. The pressure will be strong to find training shortcuts, employ "quick fixes," or to train personnel on the job. Such approaches should be resisted for, in the long term, they create more problems than they solve. However, trainers should also avoid the temptation to create long, complex training programs based on unrealistic standards. The program must take into account the host nation's cul-ture, resources, and short-term security needs. No firm rules exist on how long particular training programs should take, but existing U.S. or multinational training programs can be considered as starting points for planning. To a certain extent, the insurgent threat dictates how long training can take. As security im-proves, training programs can be expanded to facilitate achievement of the long-term end state.

Training Methods

6-66. Training programs should be designed to prepare HN personnel to eventually train themselves. In-digenous trainers are the best trainers and should be used to the maximum extent possible. There are a number of possible training methods that have proven successful, many of which also enhance the devel-opment of HN training capability:

- Formal schools run by U.S. forces with graduates selected to return as instructors. This includes entry-level individual training.
- Mobile training teams to reinforce individual or collective training on an as-needed basis.
- Partnership training, with U.S. combat units tasked to train and advise HN units with whom they are partnered. The U.S. unit provides support to the HN unit. As training progresses, HN squads, platoons, and companies may work with their U.S. partners in security or combat operations. In this manner, the whole U.S. unit mentors their partners. Habitual training relationships should be maintained between partners until HN units meet established standards for full capability.
- Advisor teams detailed to assist HN units with minimal segregation between U.S. and HN per-sonnel. Advisor teams are a particularly good method for training senior ministry personnel.
- U.S. personnel can be embedded in key billets in HN units. This may be required where HN se-curity forces are needed but leader training is still in its early stages. This approach has the dis-advantage of increasing dependency on U.S. forces and should only be used in extreme circum-stances. As HN capabilities improve, their personnel should be moved back into those key positions.
- Contractors can also be used to assist with training, though care must be taken to ensure the training is closely supervised and meets standards.

Training Soldiers

6-67. Security force members must be developed through a systematic training program that first builds their basic skills, then teaches them to work together as a team, and finally allows them to function as a unit. Basic military training for forces involved in COIN should focus first on COIN-related skills, such as first aid, marksmanship, and fire discipline. Leaders should be educated on tactics, including patrolling and

589 urban operations. Everyone must master rules of engagement and the law of armed conflict. HN units
590 should train to standard for conducting the major COIN missions they will face. Requirements include the
591 following:

592 • Manage their own security.
593 • Provide effective personnel management.
594 • Conduct logistic (planning, maintenance, sustainment, and movement) operations.
595 • Conduct basic intelligence functions.
596 • Coordinate indirect fires.
597 • Provide for effective medical support

598 ## Training Leaders

599 6-68. The effectiveness of the security forces is directly related to the quality of their leadership. Building
600 an effective leadership cadre requires a comprehensive program of officer, staff, and specialized training.
601 The ultimate success of any U.S. involvement in COIN depends on the ability to create viable HN leader-
602 ship capable of carrying on the fight at all levels and building their nation on their own.

603 ### *Leader Training Standards*

604 6-69. The training methodology adopted must reinforce the different levels of authority within the HN se-
605 curity force. The roles and responsibilities of each commissioned officer and NCO rank must be firmly es-
606 tablished so recruits can understand what is expected of them. Their subordinate relationship to civilian au-
607 thorities must also be reinforced to ensure civilian control. In addition, training should be conducted in a
608 way that establishes team dynamics. Military forces in some cultures may require training to understand the
609 vital role played by those who are not necessarily in primary leadership positions.

610 6-70. In addition to tactical skills, commissioned officers should be trained in accountability, decision
611 making, delegation of authority, values, and ethics. Special requirements for COIN should be the primary
612 focus of the initial curriculum. These subjects include the following:

613 • Intelligence collection.
614 • Day and night patrolling.
615 • Point security.
616 • Cordon and search operations.
617 • Operations with police.
618 • Treatment of detainees and prisoners.
619 • Psychological operations.
620 • Civic action.
621 As the insurgency declines, the curriculum can be adjusted for a long-term focus.

622 6-71. In addition, training of leaders should be conducted in a way that shows—
623 • How to work as a team.
624 • How to develop and take advantage of the skills of subordinates.
625 • How to train subordinates.
626 • How to maintain discipline and assume responsibility for one's actions and the actions of subor-
627 dinates.
628 • How to understand and enforce the rules of engagement.

629 ### *Basic Officer Education*

630 6-72. Various models for basic officer education exist. These include Sandhurst, West Point, Officer Can-
631 didate School (OCS), and Reserve Officers Training Corps (ROTC). Time available may be the key indica-
632 tor as to which model should be used as the primary commissioning source. If the state of the insurgency
633 allows, the four-year programs at West Point and ROTC may be the best choice; if not, the OCS and Sand-
634 hurst models may be better. Theoretically, it may be better to have a few high-quality officers than many

635
636

adequate ones, but the insurgents may not allow this luxury. Citizens being attacked would much rather have an adequate officer and unit now than to wait years for the better leader and organization.

637
638
639
640

6-73. The Sandhurst model is used in the British army. It is an intensive, one-year course that includes a rigorous program of basic training and a thorough grounding in the British Army's history and culture. It also emphasizes building each future officer as a leader. At the end of the training, each new officer attends a shorter specialty branch course. The end product is a competent junior leader.

641
642
643

6-74. The West Point or ROTC four-year model provides perhaps the best overall education, while simultaneously preparing officers for work at the tactical and operational levels. The longer program also is best for inculcating values. However, it requires significant time and resources.

644
645
646

6-75. Under an OCS-style program, outstanding individuals are taken from the enlisted ranks or from society and given the opportunity to go to an officer training course. An OCS-style course should be followed by additional specialized branch training.

647

Intermediate and Advanced Commissioned Officer Education

648
649
650
651
652
653

6-76. Military units only become effective when their commanders and staffs can effectively plan, prepare, execute, and assess operations. Initial intermediate-level commissioned officer training should focus on building effective commanders and staffs for small units, then progressively move to higher echelons. Thus, initial intermediate-level officer training should be focused at the company- and battalion-levels (or police station-level). Later courses address higher echelons, depending on the size of the overall force to be developed.

654
655
656
657
658
659
660
661
662
663
664

6-77. A cadre of carefully selected low- and mid-level commissioned officers can be provided advanced-level education at existing formal schools in the United States or other partner nations through IMET-like programs. The object of this type of program is to build a qualified leadership cadre. This cadre can, upon their return home, assume leadership positions and become faculty for HN schools. These officers should have increased credibility when they return to their own country. Officer students usually make and maintain strong personal connections with their foreign hosts during and after their stay abroad. Since one of the key goals for developing HN security forces is to professionalize them, the first-hand experience of the officers training in foreign military schools, living abroad, and seeing military professional standards practiced, is invaluable. As with officer commissioning programs, time is a key consideration. IMET-like programs are expensive and time-consuming. The best officers—those normally selected for such training—may be needed more in the country's combat forces fighting the insurgency today.

665

Operational Employment of Newly Trained Forces

666
667
668
669
670
671
672
673
674
675
676
677
678
679

6-78. Building up the morale and confidence of security forces should be a primary strategic objective. Committing poorly trained and badly led forces results in high casualties and invites tactical defeats. While defeat in a small operation may have little strategic consequence in a conventional war, even a small tactical defeat of HN forces can have serious strategic consequences in COIN. Insurgent warfare is largely about perceptions, and minor insurgent wins can be quickly turned into major propaganda victories. Defeat of one government force can quickly degrade the morale of the others. Failure by HN forces can also result in a loss of confidence of the public in the government's ability to protect them. A string of relatively minor insurgent victories can cause widespread loss of morale in the HN forces and encourage the "neutral majority" to side with the insurgents. In short, the HN security forces must be prepared for operations so that they have every possible advantage. The decision to commit units into their first actions and their employment method requires careful consideration. As much as possible, they should begin with easy missions and, as their confidence and competence grows, assume more difficult assignments. Partnering with U.S. or multinational units can be an effective way to allow new HN units to become gradually accustomed to the stresses of combat.

680
681
682
683

6-79. Newly trained units should enter their first combat supporting more experienced HN, U.S., or multinational forces. Operational performance of such inexperienced organizations should be carefully monitored and assessed so that weaknesses can be quickly corrected. The employment plan for HN security units should allow considerable time after each operation for additional training. By introducing units into

684   combat gradually, the poor leaders can be weeded out while the most competent leaders are identified and
685   given greater authority and responsibility.

686   ## Training Civilians in the Defense Ministry

687   6-80. U.S. forces tasked with training HN personnel must also ensure that the military and security forces
688   have capable management in the top ranks. The combatant commander should consider placing experi-
689   enced U.S. officers and Department of Defense personnel inside the HN defense and interior ministries as
690   trainers and advisers for their managers and leaders. U.S. forces should also develop a training program for
691   civilian personnel of the ministry of defense. Personnel training should address the following:

692   • Equipment acquisition.
693   • Departmental administration.
694   • Personnel management.
695   • Financial and resource management.

696   Selected ministry of defense personnel may receive specialized training in defense management through
697   U.S. or partner nation schools (for example, the National Defense University) or in civilian institutions that
698   specialize in graduate programs for security studies.

699   ## EQUIP

700   6-81. The strategic plan for security force development should outline requirements for appropriate equip-
701   ment for the host nation. Equipment should meet the host nation's requirements. Equipment meets the stan-
702   dard when it is affordable and suitable against the threat. The host nation must also be able to train on the
703   equipment. Interoperability may be a desired goal in some cases. A central consideration must be the host
704   nation's long-term ability to support and maintain the equipment provided.

705   6-82. The initial development plan should include phases with goals to be met by the HN forces over a pe-
706   riod of three to four years. Due to the highly adaptive nature of insurgents and the often rapidly changing
707   situation on the ground, the direction and progress of developing HN security forces must be continually
708   assessed.

709   6-83. The requirement to provide equipment may be as simple as assisting with maintenance of existing
710   formations or as extensive as providing everything from shoes and clothing to vehicles, communications,
711   investigation kits, and so forth. If the insurgents are using heavy machine guns and rocket-propelled gre-
712   nades, HN security forces need comparable or better equipment. This is especially important for police
713   forces, which are generally lightly armed and vulnerable against well-armed insurgents.

714   6-84. Maintainability, ease of operation and long-term sustainment costs should be primary considerations,
715   as few developing nations have the capability to support highly complex equipment. In COIN operations it
716   may be better to have a large number of versatile vehicles that are easy to maintain and operate than a few
717   highly capable armored vehicles or combat systems that require extensive maintenance to keep operational.
718   Developing an effective maintenance system for the host nation often begins with major maintenance con-
719   ducted by contracted firms. The program then progresses to partnership arrangements with U.S. forces as
720   HN personnel are trained to carry out the support mission.

721   6-85. Sources for HN materiel include U.S. foreign military sales, multinational or third-nation resale of
722   property, HN contracts with internal suppliers, or HN purchases on the international market. Appendix D
723   provides amplification on relevant legal considerations. The host nation should have the flexibility neces-
724   sary to obtain equipment that meets the indigenous force needs for quality, timeliness and cost

725   ## ADVISE

726   6-86. Advisers are the most prominent group of U.S. personnel that serve with HN units. Advisers live,
727   work and fight with their HN units and segregation is kept at an absolute minimum. The relationship be-
728   tween advisers and HN forces is vital. U.S. commanders must be aware that advisers are not just liaison of-
729   ficers, nor do they command HN units.

6-87. Effective advisers are an enormous force enhancer. The importance of the job means that the most capable individuals should be picked to fill these positions. Advisers should be Soldiers and Marines known to take the initiative and who set the standards for others. FM 30-20-3 provides additional information and guidelines for advisers.

6-88. More than anything else, professional knowledge and competence win the respect of HN troops. Effective advisers develop a healthy rapport with HN personnel but avoid the temptation to adopt HN positions contrary to U.S. or multinational values or policy.

6-89. Advisers who understand the HN military culture understand that local politics have national effects. It is important to recognize and employ the cultural factors that support HN commitment and teamwork. Part of the art of the good advisor is to employ the positive aspects of the local culture to get the best performance out of each security force member and leader.

6-90. Important guidelines to be followed by advisers include the following:

- Try to learn enough of the language for simple conversation.
- Be patient. Be subtle. In guiding your counterparts, explain the benefits of an action and convince them to accept the idea as their own. Respect the rank and positions of your counterparts.
- Be diplomatic in correcting the HN forces. Praise each success and work to instill pride in the unit.
- Understand that the U.S. advisory team is not the unit commander but an enabler. The HN commander must make decisions and command the unit. You are there to help with this task.
- Keep your counterpart informed; try not to hide your own agendas.
- Work to continually train and improve the unit, even in the combat zone. Help the commander develop unit standing operating procedures.
- Be prepared to act as a liaison to multinational assets, especially air support and logistics. Maintain liaison with civil affairs and humanitarian teams in the area of operations.
- Be ready to advise on the maintenance of equipment and supplies.
- Have a thorough knowledge of light infantry tactics and unit security procedures.
- Use "confidence" missions to validate training.
- Stay integrated with your unit. Eat their food. Don't isolate yourself from them.
- Be aware of the operations around you in order to prevent fratricide.
- Insist on HN adherence to the recognized human rights standards concerning treatment of civilians, detainees, and captured insurgents. Violations that are observed must be reported to the chain of command.
- Be objective in their reports on HN unit and leader proficiency and report gross corruption or incompetence.
- Train HN units to standard and fight alongside them. Consider HN limitations and adjust. Flexibility is key. It is impossible to plan completely for everything in this type of operation. Therefore constantly look forward to the next issue and be ready to develop solutions to problems that cannot be answered with a doctrinal solution.
- Remember that most actions have long-term strategic implications.
- Maintain a proper military bearing and professional manner.

770
771

Multinational Security Transition Command–Iraq

772
773
774

The experience of Multinational Security Transition Command–Iraq (MNSTC-I) demonstrates challenges of developing and assisting HN security forces engaged in an intensive insurgency. MNSTC-I programs were built upon three pillars:

775

- Training and equipping the Iraqi Security Forces to standard.

776

- Use of transition teams to guide the development of leaders and staffs.

777
778

- Partnerships between U.S./multinational forces on the ground and the developing Iraqi forces.

779
780
781
782
783
784
785
786
787
788
789

Initial plans called for only a small army (to deal with external threats) supplemented by conventional police forces (to maintain internal law and order). As the insurgency matured, the decision was made to develop a larger Iraqi Army and to focus it on the internal threat. A more robust police force was also developed. Training regimens also matured, becoming generally longer and more focused on counterinsurgency tasks. Decisions by the Iraqi government also necessitated training and organizational changes. Training programs were adjusted, based on the experience of recruits, and eventually lengthened and changed in response to the increasing lethality of the insurgency and lessons learned. Advisers assigned to Iraqi units were termed "transition teams" and instructed to focus on the development of leaders and staffs at battalion level and above.

790
791
792
793
794
795
796
797

Once it became clear that the civilian infrastructure also required considerable reform and resources under the auspices of the overall security development program, MNSTC-I assumed the additional mission of creating ministerial-level organizations, finding resources, and changing culture. Another significant challenge was the selection and training of competent indigenous leadership, NCOs as well as commissioned officers. This became even more challenging as the host nation regained sovereignty and asserted its independent authority, which led to a greater focus on developing staffs by transition teams.

798
799
800
801
802
803
804
805
806
807

The MNSTC-I experience shows the advantage of doing a prompt initial assessment and then adjusting as conditions change and lessons are learned. Development of indigenous security forces from scratch in an active insurgency environment is often more about overcoming fog and friction than about perfect planning. MNSTC-I also demanded robust interagency and multinational participation in the training effort, and fought for the funding necessary to make it effective. MNSTC-I also found that measures of effectiveness that are easily measured, such as soldiers equipped or battalions fielded, were not as useful as more complicated and more subjective metrics, such as the training level of fielded units and their loyalty to the national government.

808

POLICE IN COUNTERINSURGENCY

809
810
811
812
813
814

6-91. The primary front line COIN force is often the police—not the military. Few military units can match a good police unit in developing an accurate human intelligence picture of their area of operations. Because of their frequent contact with populace, police are often the most appropriate force to counter small insurgent bands that receive support from the civilian population. In COIN operations, special police strike units may be moved to different areas of operations, while patrol police remain in the local area on a daily basis and build a detailed intelligence picture of the insurgent strength, organization, and support.

ORGANIZING THE POLICE

6-92. The police are normally organized into several independent but mutually supporting forces. These may include the following:

- Criminal and traffic police.
- Border police.
- Transport police for security of rail lines and public transport.
- Specialized paramilitary strike forces.

In addition, a country may establish a variety of reserve police units or home guards to provide local security. The force may include paramilitary or gendarmerie-like units. Police might be organized on a national or local basis. Whatever police organization is established, it is important for U.S. personnel to understand the system and help the host nation organize and employ its police forces effectively. This often means dealing with a variety of police organizations and developing appropriate plans for training and advising each one.

6-93. While military forces might have to perform police duties at the start of an insurgency, in the long term it is best to establish police forces to take over these duties as soon as possible. As quickly as possible, the U.S., multinational partners, and the host nation should institute a comprehensive program of police training. Moreover, plans for police training need to envision a several-year program to systematically build institutions and leadership.

6-94. Although the roles of the police and military in COIN may be blurred, there are important distinctions between the two forces. If insurgents are considered criminals, the police may retain the primary responsibility for their arrest, detention and prosecution

6-95. Combating insurgency requires a police force that is visible day and night. The host nation will not gain legitimacy if the populace believes that insurgents and criminal bands control the city and village streets. Well-sited and protected police stations are one means to establish a presence in communities as long as the police do not hide in those stations. Police presence provides security to communities and builds support for the legitimacy of the government. The police are also in daily contact with the civilian community, thus serving as a key information collectors for COIN forces.

6-96. Good pay and attractive benefits must be combined with a strict code of conduct that follows the rule of law and allows for the immediate dismissal of police personnel for corruption. Planning must ensure that police pay, housing, benefits, and work conditions are good enough to attract a high quality of police recruit, and to serve as a shield against the temptation of the petty corruption that undermines the public's confidence in the police and government. One of the most important steps in organizing a police force is to set up an independent review board, with the authority to investigate charges of police abuse and corruption, oversee the complaints process, and dismiss and fine police found guilty of misconduct. This board should be composed of experts, government officials or nongovernmental organization members. It should not be under the direct command of the police force or responsible government ministry.

TRAINING THE POLICE IN COUNTERINSURGENCY

6-97. Police training is best conducted as an interagency and multinational operation. In a multinational effort, a separate multinational police training and advisory command could work in tandem with the military training command. Ideally the leaders for police training are civilian police personnel from the Departments of Justice and State along with senior police officers from multinational partners. Civilian police have personnel with extensive experience in large city operations. Justice Department and multinational police organizations also have extensive experience operating against organized crime groups. Experience countering organized crime is especially relevant to COIN, as most insurgent groups are more similar to organized crime in their organizational structure and relations with the population than they are to military units. U.S. military police units serve best when operating as a support force for the professional civilian police trainers. However, there may be times when military forces are given the primary responsibility for police training, and they must be prepared to assume that role if required. (More details on police training responsibilities are discussed in appendix D.)

864
865

6-98. If mandated to do so, military police are capable of providing much of the initial police training and are especially suited to teach the HN police forces the following skills:

866
- Weapons-handling.

867
- Small unit tactics.

868
- Special weapons employment.

869
- Convoy escort.

870
- Riot control.

871
- Traffic control.

872
- Prisoner/detainee handling and processing.

873
874

Higher-level police—skills such as civilian criminal investigation procedures, anti-organized crime operations, and police intelligence operations—are best taught by civilian experts.

875
876
877
878

6-99. Military police or corrections personnel can also provide training for detention and corrections operations. HN personnel should be trained to handle and interrogate detainees and prisoners in accordance with internationally recognized human rights norms. Prisoner and detainee management procedures should provide for the security and the fair and efficient processing of those detained.

879
880
881

6-100. Police forces, just like military forces, need effective support personnel to be effective. This requires training teams to ensure that training in support functions is established. Specially trained personnel required by police forces include the following:

882
- Armorers.

883
- Supply specialists.

884
- Communications personnel.

885
- Administrative personnel.

886
- Vehicle mechanics.

887
888
889

6-101. Effective policing also requires an effective justice system with trained judges, prosecutors, defense counsel, prison officials, and court personnel who can process arrests, detentions, warrants, and other judicial records. These elements are important components for establishing the rule of law.

890
891
892

6-102. Advisers should assist the host nation in establishing and enforcing the roles and authority of the police. The authority to detain and interrogate, procedures for detention facilities, and human rights standards are important considerations during this process.

893
894

Developing a Police Force in Malaya

895
896
897
898
899
900

In 1948, the Malayan Communist Party began an insurgency against the British colonial government. The first British response was to dramatically expand the security forces. Between 1948 and 1950, the Malaya Police was expanded fivefold to 50,000 men, while the British army garrison was expanded to 40,000 men. However, there was not enough time to provide anything more than a few weeks of rudimentary training to the new police officers before throwing them into operations.

901
902
903
904
905
906
907
908
909
910

Police with little training and little competent leadership were not only ineffective in conducting operations but also routinely abused the civilian population and fell into corrupt practices. The population largely regarded the police as a hostile force and was reluctant to provide information on the insurgents. In response, the British began reforming and retraining the entire Malayan Police force. The officers who had proven to be most competent in operations were pulled from the field and made instructors in new police schools. Police NCOs and officers were sent to advanced courses of three-to-four months duration. A police intelligence school was established, where police officers were taught the latest methods of criminal investigation as well as intelligence collection and analysis. As part of the process of winning the

911
912

ethnic Chinese away from the insurgents, the British worked closely with ethnic Chinese organizations to recruit Chinese for the Malaya Police.

913
914
915
916
917
918
919
920
921
922
923
924

Efforts to bring Chinese into the army and police were fairly successful, and both forces began to be more representative of the Malayan population as a whole. With thorough police and soldier training and with fully trained officers, the discipline of the Malayan security forces improved dramatically. With better discipline in the police, the Malayan civilian population provided more information about the insurgent organization and forces. Thanks to the intelligence training program, intelligence analysis quickly improved. By 1953, the government had gained the initiative and the insurgent military forces and support structure were in a state of rapid decline. By late 1953, the British were able to begin withdrawing their troops and turning the war over to the Malayans, who by then were fully prepared to conduct the COIN mission. The British were extremely successful in developing and utilizing HN security forces to defeat the Malayan insurgency.

925
926

Chapter 7

Leadership and Ethics for Counterinsurgency

Leaders must have a strong sense of the great responsibility of their office; the resources they expend on war are human lives.

Marine Corps Doctrinal Publication 1, 1997

There are leadership and ethical imperatives that are prominent and in some cases unique to counterinsurgency. The dynamic and ambiguous environment of modern counterinsurgency places a premium on leadership at every level, from sergeant to general. Combat in counterinsurgency is frequently a small unit leader's fight; however, the actions of commanders at the brigade and division levels are often more significant because the senior leader sets the conditions and the tone for all actions by subordinates. Today's Soldiers and Marines are required to be competent in a broad array of tasks. They must also rapidly adapt cognitively and emotionally to the perplexing challenges of counterinsurgency, and master new competencies as well as new contexts. Those in leadership positions must provide the moral compass for their subordinates as they navigate this complex environment. Underscoring these imperatives is the fact that exercising leadership in the midst of ambiguity requires intense, discriminating professional judgment.

LEADERSHIP IN COUNTERINSURGENCY

7-1. Military leaders, as members of American society, are expected to act ethically in accordance with shared national values and Constitutional principles, which are reflected in law and the military oaths of service. Military leaders have the unique professional responsibility of exercising military judgment on behalf of the American people they serve. They continually reconcile mission effectiveness, ethical standards, and thoughtful stewardship of the nation's precious resources—human and material—in the pursuit of national aims.

7-2. Military leaders work proactively to establish and maintain the proper ethical climate of their organizations. They serve as visible examples for every subordinate, demonstrating cherished values and military virtues in their decisions and actions. Army and Marine Corps leaders are responsible for ensuring the trying environment of counterinsurgency does not undermine the values of their Soldiers and Marines. Under all conditions, they must remain faithful to basic American, Marine Corps, and Army standards of proper behavior and respect for the sanctity of life.

7-3. Leaders educate and train their subordinates. They create standing operating procedures and other internal systems to prevent violations of legal and ethical rules. They check routinely on what Soldiers and Marines are doing. Effective leaders respond quickly and aggressively to indications of illegal or unethical behavior. The Nation's and profession's values are not negotiable. Violations of them are not just mistakes; they are failures in meeting the fundamental standards of the profession of arms.

LARGE AND SMALL UNIT LEADERSHIP TENETS

7-4. There are basic leadership tenets that are applicable for all levels of command and leadership in counterinsurgency (COIN), though their application and importance may vary.

7-5. Leaders are responsible for ensuring that Soldiers and Marines are properly trained and educated, including cultural preparation for the environment in which they will operate. In a COIN environment, it is often counterproductive to employ troops who are poorly trained or unfamiliar with operating in close proximity to the local populace. COIN forces aim to mobilize the good will of the people against the insurgents. Therefore, the populace must feel protected, not threatened, by COIN forces' actions and operations.

7-6. Proper training addresses many of the possible scenarios of the COIN environment. Education should prepare Soldiers and Marines to deal with the unexpected and unknown. Senior commanders should, at a minimum, ensure that their small unit leaders are inculcated with tactical cunning and mature judgment. Tactical cunning is the art of employing the fundamental skills of the profession in shrewd and crafty ways to out-think and out-adapt the adversaries. Developing mature judgment and cunning requires a rigorous regimen of preparation that begins before deployment and continues throughout. These skills are particularly important for junior leaders in COIN because of the decentralized nature of operations.

7-7. Senior leaders are responsible for determining the purpose of their operations. This entails, as discussed in chapter 4, a design process that focuses on learning about the nature of unfamiliar problems. Effective commanders know the people, topography, economy, history, and culture of their area of operations (AO). They know every village, road, field, population group, tribal leader, and ancient grievance within it. The COIN environment changes continually; good leaders appreciate that state of flux and constantly assess their situation.

7-8. Another part of analyzing a COIN mission involves assuming responsibility for everyone in the AO. This means that leaders feel the pulse of the local population, understand their motivations, and care about what the people want and need. Genuine compassion and empathy for the populace provides an effective weapon against insurgents.

7-9. Senior leaders exercise a leadership role for everyone within their AO. They exert a direct influence on those in the chain of command, but equally important is the indirect leadership provided to everyone within their AO. All elements engaged in COIN efforts look to the military for leadership. Therefore military actions and words must be beyond reproach. The greatest challenge for leaders may be in setting an example for the local population. Effective senior and junior leaders embrace this role and understand its significance. It involves more than just killing insurgents; it includes the responsibility to serve as a moral compass that extends beyond the COIN force and into the community. It is that moral compass that distinguishes Marines and Soldiers from the insurgents.

7-10. Senior commanders are responsible for maintaining the "moral high ground" in all deeds and words of their units. Information operations complement and reinforce actions, and actions reinforce the operational narrative. All COIN force activity is wrapped in a blanket of truth. Maintaining credibility requires immediately investigating all allegations of immoral or unethical behavior, and providing a prudent degree of transparency.

7-11. Military leaders emphasize to subordinates that, on the battlefield, the principles of honor and morality are inextricably linked. They do not allow subordinates to fall victim to the enormous pressures associated with the strain of prolonged combat against an elusive, unethical, and indiscriminate foe. While the environment that fosters insurgency is characterized by violence, immorality, distrust, and deceit, Army and Marine Corps leaders continue to demand and embrace honor, courage, and commitment to the highest standards. They know when to inspire and embolden their Soldiers and Marines, and when to enforce restraint and discipline. Effective leaders at all levels get out and around their units and among the people. Such leaders see what they are actually doing, engage in discourse, and most importantly, listen, in order to get a true sense of the complex situation in their AO.

7-12. Leaders at every level establish an ethical tone and climate that guards against the moral complacency and the frustrations that build up in protracted counterinsurgency. Leaders remain aware of the emotional toll that constant combat takes on their subordinates and the potential resulting psychoneurotic injury. Such injuries can result from cumulative stress over a prolonged period, witnessing the death of a comrade, or killing other human beings. Caring leaders recognize these pressures and provide emotional "shock absorbers" for their subordinates. It is critical that Soldiers and Marines have outlets to share their feelings and reach closure on traumatic experiences. These psychological burdens may be carried around

91 for a long time. Leaders watch for the signs of possible psychoneurotic injury within individuals and units.
92 These include—

93 ● Physical and mental fatigue.

94 ● Lack of respect for human life.

95 ● Loss of appetite; trouble with sleep; no interest in physical hygiene.

96 ● Lack of unit cohesion and discipline.

97 ● Depression and fatalism.

98 7-13. Combat requires commanders to be prepared to take some risk, especially at the tactical level.
99 Though this tenet is true for the entire spectrum of conflict, it is particularly important during COIN opera-
100 tions, where the insurgent seek to hide among the local populace.

Defusing a Confrontation

A small unit of American Soldiers was walking along a street in Najaf when hundreds of Iraqis poured out of the buildings on either side. Fists waving, voiced raised, they pressed in on the Americans, who glanced at one another in terror. The Iraqis were shrieking, frantic with rage. The source of this rage was unfathomable to the surprised soldiers. A journalist present with the soldiers expected that a shot would come from somewhere, the Americans would open fire, and the world would see the "My Lai massacre" of the Iraq war. At that moment, an American officer stepped through the crowd holding his rifle high over his head with the barrel pointed to the ground. Against the backdrop of the seething crowd, it was a striking gesture. "Take a knee," the officer commanded. The soldiers looked at him as if he were crazy. Then, one after another, swaying in their bulky body armor and gear, they knelt before the boiling crowd and pointed their guns at the ground. The Iraqis fell silent, and their anger visibly subsided. The officer then casually ordered his Soldiers to withdraw and continue on their patrol. Sometimes de-escalation, rather than generating overwhelming force, is the proper action to take.

---

118 7-14. Leaders prepare to indirectly inflict suffering on their Soldiers and Marines by sending them into
119 harm's way to accomplish the mission while attempting to avoid, at great length, injury and death to inno-
120 cents. This requirement gets to the very essence of what some describe as "the burden of command." The
121 steadfast ability to regularly lose Soldiers and Marines, yet continue day-in and day-out to close with the
122 enemy, requires a resoluteness and mental toughness in commanders and units that must be developed in
123 peacetime through study and hard training, and then maintained in combat.

124 7-15. Success in COIN operations requires small unit leaders agile enough to transition among many types
125 of missions and able to adapt to change. They must be able to shift through the spectrum of conflict from
126 unstable peace to combat and back again in the course of days or hours. Alert junior leaders recognize the
127 dynamic context of a tactical situation and can apply informed judgment to achieve the commander's intent
128 in a stressful and ambiguous environment. COIN operations are characterized by rapid changes in tactical
129 and operational environments and a high degree of ambiguity due to the presence of a civilian population
130 within which insurgents may disappear. Adaptable leaders scan the rapidly changing situation, identify its
131 key characteristics, ascertain what has to be done, and determine the most appropriate manner to accom-
132 plish the mission.

133 7-16. Cultural awareness has become an increasingly important competency for small unit leaders. Percep-
134 tive junior leaders learn how cultures affect military operations. They develop their knowledge of major
135 world cultures and learn the specifics of the new operational environment as a matter of priority when de-
136 ployed. Different solutions are required in different cultural contexts. Effective small unit leaders remain
137 aware of how their words and actions might be interpreted in different cultural settings and adapt to new
138 situations. Like all other competencies, cultural awareness requires self-awareness, self-directed learning,
139 and adaptability.

140
141
142
143
144
145
7-17. Self-aware leaders understand the need to continually assess their capabilities and limitations. They are humble, self-confident, and brave enough to admit their faults and shortcomings. More important, self-aware leaders work at improving and growing. After-action reviews, dialogues, and open discussion throughout a COIN force are essential to achieve understanding and improvement. Soldiers and Marines can become better, stronger leaders through a similar habit of self-examination, awareness, and focused corrective effort.

146
147
148
149
150
151
152
153
154
7-18. Commanders exercise initiative as leaders and fighters. Learning and adapting with appropriate deci-sion making authority are critical to gaining an advantage over insurgents. Effective senior leaders estab-lish a leadership climate that promotes decentralized modes of command and control, what the Army calls mission command. Under mission command, commanders create the conditions for subordinates' success by providing general guidance and the commander's intent, assigning authority commensurate with re-sponsibility to small unit leaders, and establishing control measures to monitor their actions and keep them within bounds without micromanaging everything they do. At the same time, it is important that the com-mander's presence be felt throughout the AO, especially at decisive points. The operation's purpose and commander's vision of resolution must be clearly understood throughout the force.

155
156
157
158
7-19. Sharing hardship and danger with subordinates builds confidence and esprit. Marines and Soldiers who know their leaders are committing them to schemes of maneuver based on firsthand knowledge of the point of contact are more confident in their chances of success. However, this fighter/leader concept does not absolve leaders from remembering their position and avoiding needless risk.

159
160
161
162
7-20. COIN operations require leaders to exhibit patience, persistence, and presence. Successful com-manders lead Soldiers and Marines, cooperate with the entire COIN force, leverage the capabilities of mul-tinational partners and nongovernmental organizations, and gain the confidence of the local population, while at the same time defeating and discrediting their adversary.

163
164

Patience, Presence, and Courage

165
166
167
168
169
The Marine platoon walked the streets in Iraq on foot, passing out candy, chocolates. and the occasional soccer ball to waving children. Their patrols weaved fearlessly around lines of cars and through packed markets. For the most part, their house calls began with knocks, not kicks. It was their strategy to win the respect of the city's Sunni population.

170
171
172
173
Until Monday. A night patrol had hit an improvised explosive device (IED) while pa-trolling in an armored Humvee. Five Marines were wounded initially, and two died shortly thereafter. A third Marine, a popular noncommissioned officer (NCO), subse-quently died of his wounds. The bomb had completely destroyed the vehicle.

174
175
176
177
178
The platoon was stunned. Some of the more veteran NCOs shrugged it off, but the younger Marines were keyed up, and wanted to make the elusive enemy pay a price. A squad leader, stood up in the squad bay and said, "When we go back out there tomorrow, I can guarantee you one thing, there will be a pile of dead Hajis out there. Take that to the bank."

179
180
181
182
183
184
185
186
187
188
189
Just then, the company commander walked in. He was widely respected within the command and generally short on words. He quickly sensed the unit's mood and rec-ognized the potential danger of their dark attitude. "You have to remember why we are here," he said, "a very small percentage of this country's people are out to create problems. They benefit from creating chaos, and they benefit from anything we do that detracts from our honor or our purpose here. They would love to get an overre-action to that IED last night. I know we lost some good people last night, and I know it can cause emotions to run high. Don't let that happen, don't get caught up in the an-ger of the moment and do something all of us will regret for a long time. When we go back out tomorrow, we all have to focus on what we're trying to accomplish here, and keep our mind on the mission. I know we've taken some hits, but escalating the vio-

190
191
192

> lence is not the way to win, it's falling for their strategy, instead of our game plan. You don't want to do something that you will regret forever or that will keep us here longer."

ETHICS

7-21. Article VI of the U.S. Constitution and the Army Values, Soldier's Creed, and Core Values of U.S. Marines all require obedience to the law of armed conflict. They hold Soldiers and Marines to the highest standards of moral and ethical conduct. Conflict brings to bear enormous moral challenges, as well as the burden of life and death decisions with profound ethical considerations. Combat, including COIN and other forms of irregular warfare, often obligates Soldiers and Marines to choose the riskier course of action to minimize harm to noncombatants. This risk-taking is an essential part of the Warrior Ethos. In conventional conflicts, balancing competing responsibilities of mission accomplishment at least friendly cost with protection of noncombatants is difficult enough. Complex COIN operations place even tougher ethical demands on Marines, Soldiers, and their leaders.

7-22. Even in conventional operations, Soldiers and Marines are not permitted to use force disproportionately or indiscriminately. Typically, more force reduces risk. But the American military values obligate Soldiers and Marines to accomplish their missions while taking measures to limit the destruction, particularly in terms of harm to noncombatants, caused by military operations. This restriction is based on the belief that it is wrong to harm innocents, regardless of their citizenship.

7-23. Limiting the misery caused by war requires combatants to consider certain rules, principles, and consequences that restrain the amount of force they may apply. At the same time, combatants are not required to take so much risk that they fail their mission or forfeit their lives. As long as their use of force is proportional to the gain to be achieved and discriminate in distinguishing between combatants and noncombatants, Soldiers and Marines may take actions where they knowingly risk, but do not intend, harm to noncombatants.

7-24. Ethically speaking, COIN environments can be much more complex than conventional ones. Insurgency is more than combat between armed groups; it is a political struggle with a high level of violence intended to destabilize and ultimately overthrow a government. COIN forces using excessive force to limit short-term risk alienate local residents. In doing this, they deprive themselves of the support or tolerance of the populace. This situation is what the insurgents want. It increases the threat they pose. Sometimes lethal responses are counterproductive. At other times, they are essential. The art of command includes knowing the difference and directing the appropriate action.

7-25. A key part of any insurgent's strategy is to attack their domestic and international opposition's political will. One of the insurgents' most effective means to undermine and erode political will is to portray their opposition as untrustworthy or illegitimate. These attacks are especially effective when insurgents can portray the opposition as unethical by their own standards. To combat these efforts, Soldiers and Marines treat noncombatants and detainees humanely and in accordance with America's values and internationally recognized human rights standards. In COIN, preserving noncombatant lives and dignity is central to mission accomplishment. This imperative creates a complex ethical environment that requires combatants to treat prohibitions against harming noncombatants as absolute. Further, it can sometimes require combatants to forego lethal solutions altogether. In practical terms, this consideration means that mission accomplishment sometimes obligates combatants to act more like police than warriors. That requirement imposes a very different calculus for the use of force.

WARFIGHTING VERSUS POLICING

7-26. In counterinsurgencies, warfighting and policing are dynamically linked. The moral purpose of combat operations is to secure peace. The moral purpose of policing is to maintain peace. In COIN operations, military forces defeat enemies to establish civil security; then, having done so, these same forces preserve it until indigenous police forces can assume responsibility for maintaining the civil order. When combatants conduct stability operations in a way that undermines civil security, they undermine the moral and practical

238
239
240
purposes they serve. There is a clear difference between warfighting and policing. COIN operations require that every unit be adept at both and capable of moving rapidly between one and the other, depending on circumstances.

241
242
243
244
245
246
247
7-27. In COIN operations, the environment frequently and rapidly shifts from warfighting to policing and back again. There are numerous examples from both Iraq and Afghanistan where U.S. forces drove insurgents out of urban areas but, because of the difficulty of maintaining civil security afterwards, the insurgents later returned and reestablished operations. The insurgents had to be dealt with as an organized combatant force all over again. To prevent this situation from occurring, forces that establish civil security need to be prepared to maintain it. And maintaining civil security entails very different ethical obligations than establishing it.

248
249
7-28. Civil security holds when institutions, civil law, courts, prisons, and effective police are in place and can protect the recognized rights of individuals. Typically this requires that—

250
251
- The enemy is defeated or transformed into a threat not capable of challenging a government's sovereignty.

252
253
- Institutions necessary for law enforcement, including police, courts, and prisons, are functioning.

254
- These institutions are credible, and people trust them to resolve disputes.

255
256
257
258
259
7-29. Where a functioning civil authority does not exist, COIN forces are obligated to work to establish it. Where U.S. forces are trying to build a host-nation (HN) government, the interim military government should transition to HN authority as soon as possible. Accomplishing this requires COIN forces to work within the framework of the institutions established to maintain order and security. In these conditions, COIN operations more closely resemble police work than combat operations.

260
261
262
263
264
265
7-30. The most salient difference between warfighting and policing is the moral permissibility of noncombatant and bystander casualties. In warfighting, noncombatant casualties are permitted as long as combatants observe the restrictions of proportionality and discrimination. In policing, bystanders may not be harmed intentionally under any circumstances. Failure to observe this rule undermines the peace—often tenuous in these circumstances—that military action has achieved. In maintaining the peace, police are permitted to use only the least force possible to achieve the immediate goal.

266
PROPORTIONALITY AND DISCRIMINATION

267
268
269
270
7-31. Because of the importance of gaining popular support and establishing legitimacy for the HN government, practicing proportionality and discrimination during COIN is an operational necessity as well as a legal and moral one. Proportionality and discrimination require combatants not only to minimize the harm to noncombatants, but also to make positive commitments to—

271
- Preserve noncombatant lives by limiting the damage they do.

272
- Assume additional risk to minimize potential harm.

273
274
275
7-32. Proportionality requires that the advantage gained by a military operation not be exceeded by the collateral harm. Combatants must take all feasible precautions in the choice of means and methods of attack to avoid and minimize loss of civilian life, injury to civilians, and damage to civilian objects.

276
277
278
279
280
281
282
283
284
7-33. In conventional operations, proportionality is usually calculated in simple utilitarian terms: civilian lives and property lost versus enemy destroyed and military advantage gained. But in COIN operations, advantage is best calculated not in terms of how many insurgents are killed or detained, but rather which enemies are killed or detained. If certain key insurgent leaders are essential to the insurgents' ability to conduct operations, then military leaders need to consider their relative importance when determining how best to pursue them. In COIN environments, the number of civilian lives lost and property destroyed needs to be measured against how much harm the targeted insurgent could do if allowed to escape. If the target in question is relatively inconsequential, then proportionality requires combatants to forego severe action, or seek noncombative means of engagement.

285
286
7-34. Further, when conditions of civil security exist, COIN forces are prohibited from taking any actions in which noncombatants will knowingly be harmed. However, this does not mean COIN forces are prohib-

287 ited from taking risks that might place civilian lives in danger. But those risks are subject to the same con-
288 siderations of proportionality. The benefit anticipated must outweigh the risk taken.

289 7-35. Discrimination requires combatants to differentiate between enemy combatants who represent a
290 threat and noncombatants who do not. In conventional operations, this restriction means that combatants
291 cannot intend to harm noncombatants, though proportionality permits them to act knowing some noncom-
292 batants may be harmed. In policing situations, combatants cannot act in any way in which they know by-
293 standers may be harmed.

294 7-36. In insurgencies, it is not only difficult to distinguish combatant from noncombatant; it is also diffi-
295 cult to determine whether the situation permits harm to noncombatants. Two levels of discrimination are
296 necessary:

297 ● Deciding between targets.

298 ● Determining an acceptable risk to noncombatants and bystanders.

299 7-37. Discrimination also applies to the means by which combatants engage the enemy. The law and mo-
300 rality of war prohibit the use of certain weapons, such as chemical munitions, soft lead bullets, and others
301 against combatants. (See FM 27-10.) In COIN environments, combatants must not only discriminate
302 among the kinds of weapons used but also whether lethal means are permitted in the first place. Where
303 civil security exists, even tenuously, COIN forces are obligated to pursue noncombat and nonlethal means
304 first, using lethal force only when necessary.

DETENTION AND INTERROGATION

306 7-38. Detentions and interrogations are critical components to any military operation. But the nature of
307 COIN operations produces the potential for complex detainee situations due to insurgents' deliberate min-
308 gling with the civilian population and lack of distinctive uniforms. Interrogators are often under extreme
309 pressure to get information that can save the lives of noncombatants, Soldiers, and Marines. While enemy
310 prisoners in conventional war are considered moral and legal equals, the moral and legal status of insur-
311 gents is ambiguous and often contested. What is not ambiguous is the legal obligation of Marines and Sol-
312 diers to treat all prisoners and detainees according to the law. (Appendix D provides more guidance on the
313 legal issues concerning detention and interrogation.)

LIMITS ON DETENTION

315 7-39. Mistreatment of noncombatants, including prisoners and detainees is illegal and immoral. It will not
316 be condoned. The Detainee Treatment Act of 2005 makes the standard clear:

317 *No person in the custody or under the effective control of the Department of Defense or*
318 *under detention in a Department of Defense facility shall be subject to any treatment or*
319 *technique of interrogation not authorized by and listed in the United States Army Field*
320 *Manual on Intelligence Interrogation.*

321 *No individual in the custody or under the physical control of the United States Govern-*
322 *ment, regardless of nationality or physical location, shall be subject to cruel, inhuman,*
323 *or degrading treatment or punishment.*

324 §§ 1002(a); 1003 (a), Pub. L. No. 109-148, enacted at Title X of the Defense Appropriation
325 Act, H.R. 2863, 109th Cong., 1st Sess. (Dec. 30, 2005).

326 7-40. In COIN environments, distinguishing an insurgent from a civilian is difficult and often impossible.
327 Treating the second like the first, however, is a sure recipe for failure. Individuals suspected of insurgent or
328 terrorist activity may be detained for two reasons:

329 ● To prevent them from conducting further attacks.

330 ● To gather information in order to prevent other insurgents and terrorists from conducting at-
331 tacks.

332 These reasons allow for two classes of persons to be detained and interrogated:

333 ● Persons who have engaged in, or assisted those who engage in, terrorist or insurgent activities.

334
335
• Persons who have incidentally attained knowledge regarding insurgent and terrorist activity, but who are not guilty of associating with such groups.

336
337
338
339
340
341
By engaging in such activities, persons in the first category may be detained as criminals or enemies, depending on the context. Persons in the second category may be detained and questioned for specific information. However, since they have not, by virtue of their activities, represented a threat, they may be detained only long enough to obtain the relevant information. Since persons in the second category have not engaged in criminal or insurgent activities, they must be released even if they refuse to provide information.

342
343
7-41. At no time is it permissible to detain family members or close associates in order to compel suspected insurgents to surrender. This kind of hostage taking is both unethical and illegal.

344
LIMITS ON INTERROGATION

345
346
347
348
349
350
351
7-42. Abuse of detained persons is immoral, illegal, and unprofessional. Those who engage in cruel or inhuman treatment of prisoners betray the standards of the profession of arms and the laws of the United States. They are subject to punishment under the Uniform Code of Military Justice. The Geneva Conventions as well as the Convention against Torture and Other Cruel, Inhuman or Degrading Treatment or Punishment agree on what is unacceptable for interrogation. Torture and cruel, inhumane, and degrading treatment is never a morally permissible option, even in situations where lives depend on gaining information. No exceptional circumstances permit the use of torture and other cruel, inhuman or degrading treatment.

352
353

Lose Moral Legitimacy, Lose the War

354
355
356
357
358
359
During Algerian war of independence between 1954 and 1962, French leaders decided to permit torture against suspected insurgents. Though they were aware that it was against the law and morality of war, they argued that (1) this was a new form of war and these rules did not apply; (2) the threat the enemy represented, communism, was a great evil that justified extraordinary means; and (3) the application of torture against insurgents was measured and nongratuitous.

360
361
362
363
364
365
366
367
368
369
This official condoning of torture on the part of French Army leadership empowered the moral legitimacy of the opposition, undermined the French moral legitimacy, and caused internal fragmentation among serving officers that led to an unsuccessful coup attempt in 1962. In the end, failure to comply with moral and legal restrictions against torture severely undermined French efforts and contributed to their loss despite a number of significant military victories. Illegal and immoral activities made the counterinsurgents extremely vulnerable to enemy propaganda inside Algeria among the Muslim population, as well as in the United Nations and the French media. These actions also degraded the ethical climate throughout the French Army. France eventually recognized Algerian independence in July, 1963.

370
371
372
373
374
375
376
377
378
7-43. The ethical challenges posed in COIN require commanders' attention and action. Proactive commanders establish procedures and checks to ensure subordinate leaders do not allow apparent requirements of the moment to result in violations of proper detainee-handling procedures. Prohibitions against mistreatment may sometimes clash with leaders' moral imperative to accomplish their mission with minimum losses. Such situations place leaders in the difficult situation where they must choose between obedience to the law and the lives of their troops. However, American law and professional values compel commanders to forbid mistreatment of noncombatants in their hands, including captured enemies. It is essential that senior commanders make the limits of acceptable behavior clear to their subordinates and take positive measures to ensure their standards are met.

379
380
381
7-44. To the extent that the work of interrogators is indispensable to fulfilling the state's obligation to secure its citizens' the lives and liberties, conducting interrogations is a moral obligation. The methods used, however, must reflect the Nation's commitment to human dignity and international humanitarian law. A

commander's need for information remains valid and can be met while observing relevant regulations and ethical standards. Acting morally does not necessarily mean that leaders give up obtaining critical information. Acting morally does require that leaders relinquish certain ways of obtaining information, even if that means that members of the military and intelligence professions must take greater risk.

THE LEARNING IMPERATIVE

7-45. Today's operational environment requires military organizations at all echelons to prepare for a broader range of missions than ever before. The Services are preparing for preventative stability operations or postconflict reconstruction tasks with the same degree of professionalism and study given to the conduct of combat operations. Similarly, COIN operations are receiving the attention and study merited by their frequency and potential impact. This broader mission set implies significant leadership development, education, and training implications, especially for land forces. The next chapter highlights how COIN requirements affect logistic operations.

7-46. Army and Marine leaders need to visualize the operational impact of many tactical actions and relate their operations to larger strategic purposes. Effectively blending traditional military operations with other forms of influence is necessary. Effective leaders are placing a stronger emphasis on organizational change, developing subordinates, and empowering them to execute critical tasks in consonance with broad guidance. Leaders are required to directly and indirectly influence the behavior of others outside their chain of command. Finally, leaders are increasingly responsible for creating environments in which individuals and organizations learn from their experiences and for establishing climates that tap the full ingenuity of subordinates. Open channels of discussion and debate are needed to encourage growth of a learning environment in which experience is rapidly shared and lessons adapted for new challenges. The velocity of organizational adaptation must outpace the enemy's efforts to identify and exploit weaknesses or develop countermeasures. (For more on the importance and application of learning in counterinsurgency, see Appendix G.)

7-47. Effective individual professional development programs develop and reward initiative and adaptability in junior leaders. Self-development, life-long learning, and reflection on experience should be encouraged and rewarded. Cultural sensitivity, development of nonauthoritarian interpersonal skills, and foreign language ability must be mastered. Institutional professional development programs must develop leaders' judgment to help them recognize when situations change from combat to policing. Leaders must be as skilled at limiting lethal force as they are in concentrating it. Indeed, they must learn that nonlethal solutions may often be preferable to "breaking things and killing people."

SUMMARY

7-48. Senior leaders must model and transmit to their subordinates appropriate respect for professional standards of self-discipline and adherence to ethical values. Effective leaders create command climates that reward professional conduct and penalize unethical behavior. They also are comfortable with delegating authority. However, as always, accountability for the overall behavior and performance of a command cannot be delegated. Commanders remain accountable for the attainment of objectives and the manner in which they are attained.

Chapter 8

Building and Sustaining Capability and Capacity

...the object [is to]...restore government authority in an area and establish a firm security framework.... "Winning" the population can tritely be summed up as good government in all its aspects. ...More desirable than outright gifts are schemes which are self-perpetuating or encourage a chain reaction. For example, building plans should stimulate the production of local building material. ...All this helps to give the impression not only that the government is operating for the benefit of the people, but that it is carrying out programs of a permanent nature...It gives people a stake in stability and hope for the future....

Sir Robert Thompson
Defeating Communist Insurgency: The Lessons of Malaya and Vietnam, 1966

This chapter begins with a general discussion and analysis of how logistics in counterinsurgency (COIN) operations are different from logistics in conventional operations. This is followed by a survey of COIN-specific factors that affect how commanders can leverage available logistic assets and assign logisticians to meet special requirements needed to support different COIN logical lines of operations. The discussions that follow acknowledge that COIN operations may be entered into from a variety of military conditions ranging from unstable peace to general war. Furthermore, due to the long-term commitment usually required for successful COIN, logistic units need to become more adept at relief-in-place operations that maintain continuity of progress between rotations.

LOGISTIC CONSIDERATIONS IN Counterinsurgency OPERATIONS

8-1. In counterinsurgency (COIN) operations, logistic units are asked to not only support combat efforts as they always have, but also to provide decisive or shaping support as well. Logistic providers are now no longer the tail but the nose of a counterinsurgent force. Among the most valuable services that military logisticians can provide to COIN operations are the means and knowledge required for setting up or restarting self-perpetuating sustainment schemes that give the people a stake in stability and hope for the future. One of the paradoxes of counterinsurgency is that many of the logisticians' best weapons for countering an insurgency do not shoot. In this respect, logistic units can provide some of the most versatile and effective nonlethal resources available to U.S. military forces. Logisticians supporting COIN operations prepare to provide support across all logical lines of operations (LLOs) visualized and articulated by the commander. In many cases, logisticians already supporting COIN combat operations may be the only available source able to provide essential knowledge, capabilities, and materials in the most timely manner. While this chapter focuses heavily on the capabilities and responsibilities of logistic units and logisticians, commanders of all types of units at all levels should also be aware of the distinct characteristics of COIN support.

WHAT IS DIFFERENT ABOUT LOGISTICS IN COIN OPERATIONS

8-2. In COIN operations, logistic units and other logistic providers find themselves performing all the functions they normally do in conventional operations, as well as some new ones. Conventional operations usually involve two recognizable military-like organizations engaged in force-on-force operations in contiguous areas of operations. What is different in COIN operations is that all the usual logistic functions, as well as the COIN-specific new activities, must now all be performed in a frequently disorienting environ-

43
44
45

ment in which security conditions can rapidly change from moment to moment and every few hundred yards over a wide variety of terrain conditions.

	Conventional Operations	CounterInsurgency Operations
Mission	Support combat unit missions Sustain and build combat power	Same as conventional operations plus support COIN-specific LLOs
	Supporting a mobile force with a clear organization and structure Typically in direct support	Supporting both a static force and mobile force Increased requirements of area support operations
	Logistic units and assets conduct only sustaining operations (focused on the force)	Logistic units and assets can be assigned as decisive and shaping operations (focused on the environment)
Enemy	Enemy forces have supply trains and support echelons	Insurgents use nonstandard, covert supply methods that are difficult to template
	Friendly operational surprise (masking possible) Difficult for enemy to conduct pattern analysis	Limited operational surprise Easy for enemy to observe patterns in friendly logistic operations
	Targeting logistic units is the enemy's shaping effort and considered a second front	Insurgents place a high value on attacking logistic units and other less formidable "softer/high-payoff targets"
Terrain	Fought in a definable AO Focus destruction of enemy combat forces Few constraints	Operational environment poorly defined with multiple dimensions HN population is center of gravity Constrained time to achieve results, yet many COIN tasks are inherently time consuming
	Contiguous areas of operations, echeloned formations, and discernable hierarchical logistic organizations supporting well defined deep, close, rear areas	Noncontiguous and wide dispersion of units No front; everything is potentially close, yet far
	Relatively secure LOCs facilitate distribution operations from theater to corps to division to brigade	Need to maximize multiple LOC capacity/greater complexity Potentially decreased throughput capabilities Increased area support requirements LOCs very vulnerable
Troops and Support Available	Uniformed personnel: always suitable Contractor personnel: Suitable for secure areas only	Uniformed personnel: Usually suitable Contractor personnel: Case by case; Task and location dependent; must be part of economic pluralism promotion plan
Time Available	Tempo quicker Geared toward decisive major combat	Long duration operations Continuity/logistics hand-off planning often required
Civil Considerations	Secondary to considerations of how to defeat the enemy	Primarily considered as the center of gravity May figure prominently in logistic planning
AO area of operations LOC line of communications		LLO logical line of operations

46

Figure 8-1. Conventional and counterinsurgency operations contrasted

8-3. These differences between the characteristics of logistic support to conventional operations and COIN operations are listed in figure 8-1 (above).

8-4. Differences in COIN logistics are summarized into the following major considerations:

- Logistic units are an essential part of the COIN fight.
- Logistic units are perceived by insurgents as high-payoff targets and potential sources of supplies; COIN lines of communications (LOCs) are a main battle area for insurgents. (See the following case study.)

What is Different: Insurgent Perceptions of Military Logistics

Insurgents have a long history of exploiting their enemies' LOCS as a source of supply. During the Revolutionary War, American Soldiers significantly provisioned themselves from the British Army's overindulgent and carelessly defended logistic tail. In the 1930s, Mao Zedong codified a doctrine for insurgency logistics during the fight against the Japanese occupation of China. Without exaggerating, Mao stated, "We have a claim on the output of the arsenals of [our enemies]...and, what is more, it is delivered to us by the enemy's transport corps. This is the sober truth, not a jest." For Mao's forces, his enemy's supply trains provided a valuable source of supply. Mao believed the enemy's rear was the guerrillas' front, and the guerrillas' advantage was that they had no discernable logistic rear.

This relative lack of logistic capacity was not an insurmountable problem for Mao or one of his logistic theorists, Ming Fan. Ming argued that, "ammunition can be obtained in the following ways:...given by friendly troops; purchased or appropriated from the people; captured by ambushing enemy supply columns; purchased under cover from the enemy army; from salvage in combat areas; from the field of battle; and self made [or adapted] by guerrilla organizations, especially items such as grenades..." Beyond these specifics, this doctrine prescribes a mindset of actively seeking parasitic logistic relationships with not only the conventional enemy forces that the insurgents seek to co-opt and defeat, but also active linkages to local black market activities and the cultivation of host-nation sympathizers.

For these reasons forces conducting COIN operations must vigorously protect their LOCs, scrupulously collect and positively control dud munitions and access to other convertible materiel, and actively seek ways to separate the insurgent from black market activities, all of which are potentially significant sources of insurgent supplies.

In one moment in time, our Service members will be feeding and clothing displaced refugees—providing humanitarian assistance. In the next moment, they will be holding two warring tribes apart—conducting peacekeeping operations. Finally, they will be fighting a highly lethal mid-intensity battle. All in the same day, all within three city blocks. It will be what we call the three block war.

General Charles C. Krulak
"The Three Block War: Fighting in Urban Areas"

8-5. In a COIN environment, units tasked with providing logistic support can potentially be involved in more complicated tasks than even those of the three block war metaphor described in the quote above. The significant difference in COIN operations is that logistic units must be prepared to provide conventional logistic support to highly lethal, mid-intensity combat operations while simultaneously supporting humanitarian operations (until conditions stabilize locally and civilian organizations can assume those duties).

8-6. This complex operational environment, with numerous competing demands, means that logisticians need to creatively seek distribution efficiencies wherever possible. In the

94
95
96
97
98
99
8-7. COIN environment, it is highly desirable to eliminate backtracking and unnecessary distribution traffic. Methods of logistic throughput that bypass, either on the ground or by air, population centers and heavily used civilian transportation nets are desirable. The benefits derived from such practices are especially valuable in COIN operations because they improve logistic security, speed delivery of required support, and minimize adverse effects and stress on the civilian population that is supposed to benefit from the COIN operation.

100
101
102
103
104
8-8. Because of the diversity of requirements for COIN operations, logisticians must be involved from the very beginning of the planning process, and they must begin planning in detail as early as possible. Because of the complexity and diversity of logistic requirements as well as the variety of conditions under which COIN operations are pursued, a careful logistic preparation of the area of operations (AO) is required.

105
LOGISTIC PREPARATION OF THE COUNTERINSURGENCY AREA OF OPERATIONS

106
107
108
109
110
111
8-9. Logistic preparation of the AO is related to and can be treated as a COIN-specific logistic preparation of the theater. (See FM 4-0, paragraphs 5-34 through 5-55.) In COIN operations, detailed analysis of civil logistic and economic assets takes on an even greater importance because of their potential significance for supporting the insurgency as well as their potential value to the development and sustainment of host-nation (HN) security forces and the restoration of other essential services. Some examples of essential information for COIN logistic planning include the following:

112
113
- Analysis of the HN conventional force's existing logistics as source of supply for the developing HN security forces and the potential insurgents/black market.

114
- Effects of requirements generated by combat operations/collateral damage.

115
- Effects of multinational distribution requirements on HN lines of commerce.

116
- HN economic base (industry/manufacturing/agricultural).

117
118
- HN lines of commerce (such as main supply routes, industrial cities, technical cities, pipelines, rail, and air and maritime ports).

119
- HN public works/utilities/health/transportation/legal/justice systems.

120
- Potential or existing displaced person populations/refugee requirements.

121
ANALYSIS OF INSURGENT LOGISTIC CAPABILITIES

122
123
124
125
126
127
128
8-10. In COIN operations, analysis of the logistic capabilities/shortfalls of the insurgent forces is especially significant. Logisticians, together with intelligence personnel perform what was formerly known as reverse-BOS (battlefield operating systems) analysis. The purpose of this analysis is not just the targeting of enemy logistic capabilities and LOCs; it also includes an assessment of the suitability of various sources of supply for the development and sustainment of insurgent forces. This analysis includes an assessment of black market material, including salvage goods that might still be useful to the insurgents in an improvised manner.

129
LOGISTIC SUPPORT TO LOGICAL LINES OF OPERATIONS

130
8-11. Although logisticians support all LLOs, logistic support during COIN focuses on the following:
131
- Combat operations.
132
- Training and employment of HN security forces.
133
- Essential services.
134
- Governance.
135
- Economic development.

136
SUPPORTING COMBAT OPERATIONS

137
138
8-12. The combat operations LLO is the line of operations most familiar to logisticians and nonlogisticians alike. The paramount role of logistic units will always remain to support Soldiers and Marines in

139 accomplishing the commander's intent and other Title 10 responsibilities. The use of logistic units to aug-
140 ment civil programs supporting other lines of operations must not take away from the logistic system's
141 capability to support maneuver forces engaged in combat operations.

Support to and from Operating Bases

143 8-13. Logistic support to COIN combat operations is often accomplished from bases (see FM 3-90, ap-
144 pendix E-1) or forward operating bases (see FM 100-25). Operating bases provide combat units with rela-
145 tively secure locations from which to plan and prepare for operations. As a result, these bases take on new
146 significance in operations executed in a noncontiguous COIN environment.

Considerations for Situating Operating Bases

148 8-14. In COIN, base site selection becomes extremely important for more reasons than providing opti-
149 mal support to combat operations. Under certain geographic conditions, such as in rugged mountains with
150 few passes or desolate desert terrain, placement of secure operating bases astride or near the insurgents'
151 LOCs can improve the COIN force's interdiction and disruption capabilities. In other environments, such
152 as urban areas and jungles, the advantages of such an operating base position may be negated by the in-
153 surgents rerouting their LOCs around the base, as was often the case with U.S. efforts to interdict insur-
154 gent supplies on the Ho Chi Minh Trail in Vietnam.

155 8-15. Other reasons for carefully considering base placements in COIN operations involve the sensitivi-
156 ties and concerns of the local population. The potential for ill-considered bases to be substantially disrup-
157 tive and to produce unintentionally negative effects on the daily lives and operations of the population is
158 significant, even if the COIN force arrives with the most positive of intentions. Other related concerns in-
159 clude taking care to set up bases so that they do not project an image of undue permanency or a posture
160 suggestive of a long-term foreign occupation. Similarly, logistic postures that project an image of unduly
161 luxurious living by foreign forces while the HN population suffers in poverty undermines the COIN mes-
162 sage and mission. Such perceptions derived from logistic practices can be construed and twisted by insur-
163 gent propaganda into evidence of malign intentions by COIN forces. While all these considerations take
164 on special significance in COIN, none of them override the primary concern that operating bases be se-
165 curable and defendable. (See FM 3-90, paragraph 9-44.)

166 8-16. Operating base site selection and development in support of COIN operations also requires the
167 careful consideration and balancing of several other factors. In COIN operations, it is desirable to provide
168 support through a carefully optimized mix of supply-based or supply-point ("just-in-case") practices with
169 distribution-based or unit distribution ("just-in-time") logistic methods. (See FM 4-0, paragraph 1-39;
170 MCDP 4, page 67) In COIN operations, situations can swiftly develop that require equally rapid logistic
171 responses to prevent further deterioration of security conditions. Under these COIN-specific circum-
172 stances, "just-in-time" practices may still not be quick enough and use of "just-in-case" capabilities may
173 be much more appropriate, effective, and timely, while economizing resources. This dilemma of COIN
174 logistics is best illustrated by a fire-fighting analogy. When a fire is small and confined to a pan on the
175 stove, the five-pound extinguisher hanging on the wall in the kitchen can easily put it out. But when this
176 small but important resource is not immediately available to put out the fire, half of the house may be
177 consumed by the time the fire department arrives. The problem of the fire now requires thousands of gal-
178 lons of water, trucks, and hoses, not to mention construction materials to restore the house to its former
179 state. The challenge for logisticians supporting COIN operations is to correctly identify which materials
180 are the COIN forces' "five-pound extinguishers" and to make sure these items are supply-based at the
181 most appropriate operating bases.

182 8-17. This carefully considered balancing between distribution- and supply-based methods supports the
183 goal of minimizing operating base footprints. Not only are minimal footprint bases easier to relocate as
184 required; they are also less intrusive and antagonizing to the local population, all characteristics which
185 significantly assist COIN operations.

186 8-18. Another consideration in operating base selection is the base's purpose. If, for instance, extensive
187 logistic throughput is anticipated, planners pay close attention to entry and exit points. Ideally, there
188 should be more than one entry/exit point. Where possible, at least one of these points should not require

189 convoys to travel though a heavily populated area. Space should be allotted near an entry point for stag-
190 ing areas for logistic convoys so that they do not have to transit the base to form up.

191 8-19. Due to the noncontiguous nature of COIN operations, logisticians give special care to the devel-
192 opment of web-like LOCs and main supply routes between operating bases and logistic bases. The advan-
193 tage of web-like linkages between bases is twofold. It minimizes by means of dispersion the intrusive ef-
194 fects of COIN logistic operations on the population. It also provides redundancy in distribution
195 capabilities, making the system more robust and impervious to the effects of insurgent interdiction of any
196 one LOC. Additionally, more ground LOC routings provide more opportunities to observe the population
197 and gather information from them. For these reasons, wherever possible in COIN operations, multiple
198 LOCs between bases should be identified.

199 **Force Protection**

200 8-20. Force protection of all logistic activities takes on greater significance during COIN operations.
201 Historically, insurgents deliberately seek out and engage logistic units, particularly those that they assess
202 as poorly defended and easy targets. Because of the intensive manpower requirements and dispersed na-
203 ture of COIN operations, logistic units cannot assume that combat arms units will be available to assist
204 them with force protection. For this reason, logistic units play a much greater role in base and LOC de-
205 fenses and also must assume responsibility for protecting civilian logistic augmentees, whether Depart-
206 ment of Defense (DOD) civilians or contractors, working in their AOs.

---

207 ### Vietnam: Meeting the Enemy and Convoy Security
208

209 The year 1968 proved a turning-point for units of the 48th Transportation Group as-
210 signed the task of transporting supplies to the 25th Infantry Division operating in the
211 Cu Chi province of Vietnam. In August, a supply convoy from the group was caught
212 in a North Vietnamese Army and Vietcong (NVA/VC) ambush. The combat brigade
213 responsible for clearing that section of main supply route had recently been assigned
214 other missions, and its resources were spread thin. The ambushers chose their mo-
215 ment to attack well: monsoon conditions prevailed, the site was outside the range of
216 supporting indirect fire, and the supplies were destined for the unit tasked with re-
217 sponding to such attacks. Under dangerous weather conditions, two UH-1C "Huey"
218 gun ships were the first assets to arrive to assist the beleaguered convoy. From the
219 air, the aviators witnessed NVA soldiers unloading supplies from the U.S. vehicles
220 and onto NVA trucks hidden in the tree line off the road. Almost three hours later, the
221 first relief force arrived on the ground with barely enough capability to continue a
222 minimal defense of the remaining convoy assets and surviving personnel. Finally,
223 seven hours later, a U.S. armored cavalry force arrived and forced the NVA/VC at-
224 tackers to withdraw.

225 Thirty U.S. Soldiers were killed, 45 were wounded, and 2 were taken prisoner in this
226 devastating ambush. This event forced the 48th Group and the 25th Infantry Division
227 to rethink their convoy operations. As a result, the two units started to conduct de-
228 tailed convoy planning meetings, renewed their enforcement of Soldier discipline,
229 placed security Soldiers on every vehicle, hardened the cabs of supply trucks with
230 steel plates, and mounted M60 machine guns on every vehicle possible. Not the
231 least of the improvements was the clarification of the command and support relation-
232 ships and responsibilities between the 48th Group and the 25th Infantry Division, in-
233 cluding publication of joint convoy standing operating procedures. With these new
234 practices in place, NVA/VC convoy ambushes soon had very different endings. A
235 change in thinking about a logistic problem converted the perception of convoy op-
236 erations from an unglamorous defensive activity into a valuable opportunity to offen-
237 sively engage elusive insurgents.

---

238 **Combat Logistic Patrolling**

239 8-21. During COIN operations, every logistic package (LOGPAC) or resupply operation becomes a
240 mounted combat operation, or combat logistic patrol. From the insurgents' perspective, attacks on resup-
241 ply operations are not only a potential source of dramatic propaganda but can also be a source of supplies
242 and materiel. For this reason LOGPAC convoys should project a resolute ("hard and prickly") image that
243 suggests that they will not be an easy ("soft and chewy") target. Logistic convoys should project their
244 available combat power to the maximum extent possible as would any other combat patrol. Under these
245 conditions, logistic units, or anyone else involved in resupply operations, conducts a detailed intelligence
246 preparation of the battlefield and prepares a fire support plan as well as identifying usable intelligence,
247 surveillance, and reconnaissance assets. Additionally, combat logistic patrols should gather information,
248 report on road statuses, and be a valuable source of contributions to intelligence collection plans. Logisti-
249 cians should be mindful that, in the COIN environment, distribution-based practices may actually provide
250 insurgents with more opportunities to target resupply activities, due to the large blocks of time materiel is
251 in motion towards the customer. Correspondingly, logisticians remain aware that insurgents constantly
252 and deliberately seek out adaptive countermeasures to logistic activities, such as development and prolif-
253 eration of improvised explosive devices (IEDs)—a natural counter to U.S. distribution-based doctrine.
254 For these reasons, it is essential that logisticians conduct careful analysis of conditions and thorough
255 combat preparations before launching combat logistic patrols.

256 **Unit Equipment for COIN**

257 8-22. Due to the rapidly shifting nature of COIN operations, logistic units and the combat units they
258 support usually deploy with at least some of their equipment unsuitable for the prevailing operational and
259 tactical trends when they arrive. This dynamic of COIN operations causes logisticians to actively seek
260 equipment modifications and new items on behalf of customer and logistic units. For this reason, COIN
261 operations may be particularly well suited to the adoption of improved and new procurement programs,
262 such as rapid fielding initiatives (RFIs) and the purchase of commercial off-the-shelf (COTS) items.
263 These approaches make sense when one realizes that, in many cases, the source of the insurgents' surpris-
264 ing capabilities also comes from their creative exploitation of commercially available technologies and
265 materials as well as their lack of bureaucratic encumbrance.

266 8-23. By utilizing more streamlined materiel procurement procedures, COIN forces can benefit from
267 closer to real-time satisfaction of previously unforeseen needs generated by specific and localized envi-
268 ronmental and cultural conditions. Examples of COIN requirements that fit into these categories for
269 COTS are—

270 ● Public address systems.
271 ● Language translation devices.
272 ● Nonlethal weapons.
273 ● Back-pack drinking water systems.
274 ● "Gator" mobility systems.

275 8-24. Examples of RFIs are—
276 ● Up-armor kits for light wheeled vehicles.
277 ● Body armor improvements.
278 ● Improved explosive detectors.
279 ● IED signal jammers.

280 8-25. A potential drawback when adopting COTS equipment is that maintenance support packages and
281 repair parts may be inadequate or difficult to obtain in theater. Many civilian commercial product manu-
282 facturers have no experience or infrastructure in place to support their equipment under military condi-
283 tions and in quantity, operating in hostile austere theaters, far from their regular markets and customer
284 base. It may take time to get needed parts into normal supply channels and trained personnel into theater
285 to fix such novel equipment. Actions to be considered to permit more continuous operation of newly vital
286 equipment are the establishment of pools of COTS "floats" and allocating time for floated equipment to
287 be evacuated to locations where maintenance and repairs can be performed.

288   8-26. Units conducting COIN operations may be required to temporarily draw additional or specialized
289   equipment in theater. For long-term COIN operations, theater property books may be established for the
290   maintenance and accountability of rotationally issued additions to standard equipment, as well as special-
291   ized or specially modified equipment. In-theater special issues and fieldings may include materiel and
292   equipment procured through military channels, RFIs, or COTS sources. Depending on the unit's non-
293   COIN primary function, supplementary or modified equipment might be drawn to a greater or lesser de-
294   gree. For example, an artillery unit normally equipped with self-propelled howitzers would have to draw
295   a substantial number of hardened HMMWVs to conduct security missions and would probably leave
296   many of their howitzers at home station. Conversely a military police unit might already be well equipped
297   to conduct security activities and might only have to draw a few pieces of the latest specialized equip-
298   ment. Other examples of items that might be provided as required to COIN forces by this method in-
299   clude—

300   • Up-armored vehicles.
301   • Cargo trucks.
302   • "Gators."
303   • IED jammers.
304   • Body armor.

305   8-27. Units in COIN operations can expect somewhat different maintenance requirements than in con-
306   ventional operations. For instance, units may put very high mileage on their wheeled vehicles and there-
307   fore need more frequent servicing. Armor packages may wear out shock absorbers and springs much
308   faster; these would require replacement sooner than normal (compared to conventional operations). Due
309   to the mission and the remoteness of many operating bases, unit leaders should expect their maintenance
310   sections to have to perform higher echelons of maintenance than normal and therefore to need greater or-
311   ganic capability.

312   **Unit Basic Loads and Operational Reach**

313   8-28. As seen during operations in Somalia, Sadr City during Operation Iraqi Freedom, and Operation
314   Anaconda during Operation Enduring Freedom, Soldiers and Marines conducting COIN operations are at
315   risk of being cut off from bases and forward operating bases due to weather changes, enemy action, or
316   civil protests. Because these types of rapidly developing situations can keep units without access to re-
317   supply for extended periods, units conducting activities away from supporting bases maximize the
318   amount of basic supplies (for example, water, food, ammunition, first aid, and equipment batteries) that
319   they carry with them on their vehicles. Additionally, some COIN operations can consume surprisingly
320   high quantities of ammunition (specifically small arms) because of combined defensive and offensive ac-
321   tions. Logisticians supporting these types of operations adjust stockage levels for unit basic loads and
322   other sustainment commodities. In turn, logisticians and their supported units should rethink how their
323   supporting vehicles can best be configured as supply platforms that meet these COIN-specific needs.
324   Successful solutions should be validated by competent authorities and standardized across formations to
325   ensure safety and to support planning for effective employment.

326   8-29. Unit leaders, when developing their operating base requirements, should plan for ammunition/ ex-
327   plosive storage areas. Ideally, units are issued ammunition and explosives, anticipating that it may be
328   some time before resupply is affected. Units normally carry only their basic load. The rest should be
329   staged appropriately.

330   **Aerial Distribution**

331   8-30. During COIN operations, intratheater aerial resupply should be utilized to the maximum extent
332   possible. This practice not only reduces the vulnerability of resupply activities to ground-based attacks by
333   insurgents, but also has the added benefit of minimizing the negative effects of COIN logistic activities
334   on public roadways and reduces the potential for alienating the population. Site selection for bases and
335   forward operating bases includes assessment of aircraft support capabilities and seeks to maximize the
336   possible options for aerial delivery (that is, by rotary- and fixed-wing aircraft, air drops, and landings).

337
338

Additionally, most bases have some medical capability (level II or higher) and require a helicopter pad near the medical area.

339
340

Air Delivery in Iraq: Maximizing Counterinsurgency Potential

341
342
343
344
345
346
347

For almost five months in 2004, two Marine battalions with attached units operating in remote areas of Iraq were resupplied by air drops from rotary- and fixed-wing aircraft. Until then, cargo air drops from helicopters had been suspended since the mid-1990s due to technical difficulties that posed unacceptable peacetime risks. The high tempo of operations in Iraq, coupled with the severe challenges of using ground supply methods to reach remote locations, made circumstances opportune to reexamine helicopter airdrop possibilities.

348
349
350
351
352
353
354
355
356
357

A careful analysis of the earlier challenges resulted in a clarification of flying procedures during drops. Furthermore, in the intervening decade, technological advances made it possible to drop bundles by parachute with some directional adjustment after release. This new ability greatly increased the accuracy and utility of this procedure. Through this combination of improvements, dangerous nine-hour convoys were replaced by quicker and more secure flights covering three-quarters the ground distance. Overall, during this period, more than 103 tons of supplies were delivered this way to the Marines. A logistic method that had been set aside as unworkable found new utility when operational conditions changed and technological shortcomings were addressed.

358
359
360
361

With the need to quickly deliver a wide variety of supplies to diverse locations, COIN forces discovered that adopting a variety of air delivery procedures significantly improved their logistic posture under challenging conditions, while minimizing the risk of negative encounters with the HN population or the insurgents.

362

SUPPORTING THE TRAINING AND EMPLOYMENT OF HOST-NATION SECURITY FORCES

363
364
365
366
367
368
369
370

8-31.   One of the most important missions and LLOs for U.S. forces engaged in COIN operations is the establishment and employment of HN security forces. Security sector reform is done through various support and training activities, all of which may substantially involve military logisticians. The development and support of police forces and their training normally falls under the auspices of non-DOD agencies, such as the Department of State and the Department of Justice, or UN mandated missions. The development and support of HN military forces is a COIN mission that military logisticians prepare to undertake, from planning at the strategic level to practical implementation on the ground. (Chapter 6 covers the support of HN security forces in more detail.)

371
372
373
374
375
376
377
378

8-32.   Some tasks required to establish HN security forces may initially fall to military logistics units until other government agencies' programs start, other logistic support can be contracted, or HN logistic organizations are in operation. They are—
- Providing operating base space or establishing another supportable secure location for the recruiting, reception, and training of HN security forces.
- Providing initial logistic support to forming HN security forces, to include possibly equipping, arming, feeding, billeting, fueling, and medical support.
- Providing logistic training to newly formed HN security force logistic organizations.

379

Equipping and Sustaining Host Nation Security Forces

380
381
382
383

8-33.   Logisticians involved in the development of plans and programs for the sustainment of HN security forces should take care to ensure that equipment selected is suitable to and sustainable through the host nation's capabilities. Logisticians remain mindful that equipment and support programs must be within the host nation's resources, including budget and technological capabilities. In many cases "good

384   enough to meet standards" equipment that is indigenously sustainable is preferable to "high-tech, best
385   available" that requires substantial foreign assistance for maintenance support for years to come, all the
386   while providing the insurgent movement with a valuable propaganda point that could negate any potential
387   technological advantages.

388   8-34.   One of the acknowledged difficulties in establishing HN security forces is identifying where suit-
389   able materiel and equipment is to come from. Often, plans developed in a joint or multinational environ-
390   ment include provisions for such forces to be initially equipped from multiple donor nations and agency
391   sources. As a result, logisticians may be compelled to familiarize themselves with the capabilities of these
392   agencies and nations' supply and maintenance systems, though support packages may not be included
393   with the donation.

394
395

396
397
398
399
400
401
402
403
404
405
406
407
408
409
410
411
412
413
414

Building a Military: Sustainment Failure

By 1969, pressure was on for U.S. forces in Vietnam to turn the war over to the host nation in a process that became known as Vietnamization. In the process of growing the South Vietnamese military forces, the United States armed and equipped them with modern small arms, communications, and transportation equipment, all items produced by and sustained from the American industrial base. This modern equipment required an equally sophisticated maintenance and supply system to sustain it, something the mostly agrarian South Vietnamese were hard-pressed to do both economically and culturally, despite the training of several thousand South Vietnamese in American supply and maintenance practices. In short, the American way of war was not indigenously sustainable and was incompatible with their material culture and economic capabilities. South Vietnam's predominately agrarian-based economy was unable to sustain the high-technology equipment and computer-based systems established by American forces and contractors. Consequently, the South Vietnamese military transformation was artificial and superficial. Many of the South Vietnamese involved in running the sustainment systems had little faith in them. Such attitudes encouraged poor administration and rampant corruption. After U.S. forces left and most American support was cut off, the logistic shortcomings of the supposedly "modern" South Vietnamese military contributed to its rapid disintegration in the face of the North Vietnamese advance in 1975.

Host-Nation Security Forces Logistics

416   8-35.   Logisticians involved in the training of HN security force logistic personnel need to be aware that
417   a large part of the problem with previously dysfunctional military cultures in many developing countries
418   was that pervasive climates of corruption and graft crippled their attempts to develop effective support
419   services. Logisticians conducting such training should expect to find themselves repeatedly emphasizing
420   the long term benefits of supply discipline and materiel accountability and the importance of those prac-
421   tices to the security and development of the host nation. For this reason, emphasis should be placed on
422   inventory procedures. Simultaneously the black market should be monitored for the presence of pilfered
423   military equipment and as a means of determining the effectiveness of logistic procedures and account-
424   ability training. Of all the capabilities being developed for HN security forces, logistic functions may take
425   the longest to impart due to their inherent complexity and potential cultural challenges. For these reasons,
426   it may be a long time before HN security forces are able to operate independently of U.S. or multinational
427   logistic support.

428   8-36.   HN produced materiel should be procured and used to support HN security forces if it is reasona-
429   bly and reliably available and deemed adequate to meet requirements. Not only does this help stimulate
430   the HN economic base; it also promotes an attitude of self-sufficiency on the part of HN forces and rein-
431   forces the important political message that the HN security forces are of the population and not the agents
432   of foreign powers. When promoting these practices, logisticians may find themselves stepping into roles

433 somewhat beyond the normal scope of their duties when they assess the suitability of locally available
434 materials and provide advice on how such materials might be made suitable for self-sustainment. In this
435 case, the most valuable lesson logisticians may be communicating to HN security forces and those tasked
436 with supporting them is not "what to do" but "how to think about the problem of sustainment" and its
437 linkage to security effectiveness.

438 ## ESSENTIAL SERVICES

439 8-37. In general, according to existing U.S. military logistic doctrine, there is no provision for U.S.
440 forces to become decisively or exclusively engaged in providing essential services to the HN population
441 during COIN operations. However, this doctrinal position does not prohibit units from using applicable
442 skills and expertise resident in their military organizations to help assess essential HN service needs. In
443 conjunction with these assessments, logistic and other units can also be used to meet immediate needs
444 where possible and in the commander's interest, and to assist in the handoff of essential service functions
445 to appropriate U.S. government agencies, HN agencies, and other civil support organizations.

446 ### Assessing Essential Services Requirements

447 8-38. Military logisticians should already have good insights into capabilities, requirements, and short-
448 falls from their logistic preparation of the theater and more detailed assessment of COIN-specific issues.
449 (See paragraph 8-9.) As a work in continuous progress, this assessment should be merged later with in-
450 formation from civil affairs area assessments. Areas of civil affairs concern where logisticians and logis-
451 tic branches can contribute to the area assessment include, for example—

452 - **Sanitation.**
453   - *Quartermaster.* Water specialists.
454   - *Engineer.* Earth moving specialists, plumbing construction, soil analysis, concrete casting,
455 and sanitary landfill management.
456   - *Ordnance.* Heavy wheeled vehicle, pump, and mechanical repair
457 - **Water.**
458   - *Quartermaster.* Water purification.
459   - *Medical service.* Preventive medicine and sanitation specialists.
460   - *Ordnance.* Pumps and mechanicals repair.
461 - **Electricity.** *Engineer.* Power generation specialists.
462 - **Academic.**
463   - *Engineer.* Vertical construction specialists.
464   - *All.* Training and education.
465 - **Transportation.**
466   - *Engineer.* Assessment of rail/bus/port facilities, roadway/bridge assessments, and airfield
467 capabilities/upgrades.
468   - *Transportation.* Assessment of rail/bus/ferry capacities.
469   - *Ordnance.* Assessment of mechanical maintenance of rail/bus/truck/ferry operating equip-
470 ment.
471   - *Military police.* Roadway and traffic flow assessments.
472 - **Medical.** *Medical service/medical corps.* Assess local health care capabilities/needs
473 - **Security.** *Military police.* Security survey, crowd control
474 - **Food Supply.**
475   - *Veterinary corps.* Food source/quality, vector control.
476   - *Quartermaster.* Food packaging and distribution
477 - **Fuel.** *Quartermaster.* Fuel specialists, testing of locally procured fuel supplies
478 - **Financial.** *Finance.* Requirements to reestablish accountability/security of HN funds/captured
479 funds, assess financial support requirements banking/currency access, and so forth.

Time as a Logistic Commodity

8-39. Success in COIN operations and buy-in of the population into COIN efforts, may hinge on there being as small a gap as possible between the time assessments of essential services needs are determined and the time that initial remediation efforts are begun. In order to keep this time gap as small as possible and manage the development of popular expectations, COIN logistic units may be compelled to begin to remediate essential services until civilian authorities and agencies can assume these functions. The rationale for logistic units taking up this mission is similar to that applied to a metaphor of lifesaving trauma care. If one thinks of the populace as a patient destabilized by the trauma of insurgency, in COIN operations, logistic and other units may need to function much like the first-responder medic on the scene, conducting initial assessments of patient's needs, providing lifesaving first aid, and letting the hospital know what more specific higher-level care is required. Both medical and COIN first responders are most effective when they can assess and initiate life support treatment immediately. In COIN operations though, "immediately" may last weeks or months and is harder to determine without obvious calibrated vital signs, such as blood pressure, pulse, and temperature. For this reason, COIN logistic units may be required to take whatever measures they are able to immediately provide for stabilizing essential services and preventing deteriorating conditions.

8-40. The ways in which logistic and other units can be used to bridge the essential services gap is an extension of the assessment capabilities of these units. Some examples of how military logistic assets and capabilities can be used to meet immediate and essential service needs are—

- **Contingency contracting officer** procures commercial public utility equipment with Title 22 funds, and then employs theater contractors and external theater subject matter experts/trainers to maintain the asset.
- **Reverse osmosis water purification unit** provides immediate source of potable water until the water pumps at a purification plant can be restored.
- **Distribution companies and supply support activities** provide temporary storage and distribution of foreign humanitarian assistance.
- **Explosive ordnance disposal**, supported by logistic transportation, disposes of munitions in populated areas.
- **Combat logistic patrols** provide security for nongovernmental organization (NGO) transportation of critical humanitarian assistance. (Not all NGOs agree to this.)
- **Medical units** provide a medical civic action program (MEDCAP) team to conduct a visiting clinic at a small or remote village; may augment an NGO (for example, Doctors without Borders).
- **Brigade surgeon/brigade engineer** work with the contracting officer representative to restore a clinic/hospital service.
- **Medical personnel/units** assist with upgrading/restoring HN medical training programs to meet civil healthcare provider critical shortfalls.
- **Senior power generation technicians** provide advice and troubleshoot for a municipal power source.
- **Class I rations section** accounts for, preserves, and distributes humanitarian daily rations (a State Department–controlled item).
- **Medium truck companies** move internally displaced persons (IDPs).
- **Logistic units** tasked with providing life support to IDP (or refugee) camps, that is, billeting, food service, personnel (biometric) accountability, and work placement).
- **Preventive medicine team,** in conjunction with veterinary support, conducts vector/parasite analysis on farm livestock (HN food source).
- **Engineers** repair a critical highway, renovate a bridge, or build a building (such as a clinic or school).

Handoff of Essential Services

8-41. Frequently logisticians who have provided stopgap essential services may be the only personnel with accurate knowledge of essential services needs and priorities. For this reason, logisticians providing these services should expect to be actively involved in the handoff process to other government agencies and designated civil organizations until those agencies and activities are reasonably functional and adequately able to meet essential services needs. A poor handoff can provide the insurgents with propaganda opportunities and evidence of the "insincerity" of COIN efforts.

8-42. Through the restoration and transition of essential services to the HN government, one of the principal causes exploited by the insurgents is removed. This action greatly assists the HN government in its struggle for legitimacy. For this reason, the insurgents can be expected to conduct attacks against restored services. During this handoff period, multinational logistic assets may need to maintain a logistic quick reaction force to ensure the continuity of services and marginalization of the insurgents' counteractions and messages.

Public Transportation, Population Movement, and Life Support to Internally Displaced Persons and Refugees

8-43. Under conditions of national crisis or insurgency, public transportation systems often fall into disarray. COIN forces may be faced with the task of recovering stolen or misappropriated buses, trucks, cars, and other government vehicles (including former military equipment), and restoring them to public service. This action not only helps alleviate urgent requirements for public transportation; it sends an unmistakable message of resumption of governmental authority, and it can substantially reduce the amount of replacement equipment that must be procured from other sources. Logistic units and personnel can expect to be significantly involved in this process, from reestablishing accountability procedures to assessing repair and maintenance needs, until competent public or government authorities can resume these duties.

8-44. While not a specified essential services task or an LLO, one of the most common problems arising on short notice during an insurgency is the creation of substantial groups of IDPs and refugees. Attending to IDP and refugee needs can quickly become an urgent logistic requirement, drawing on all essential services in the process of providing secure emergency shelter (IDP/refugee camps) and life support (food, water, medical care). While it is normally the mission of NGOs and other civilian agencies to furnish this type of support to IDPs and refugees, conditions may prevent them from providing these services in a timely manner. Furthermore, in COIN operations, IDP and refugee security may take on heightened military importance because traumatized and dislocated persons may become vulnerable to insurgent threats and recruitment. Restoration and maintenance of public transportation services can also be helpful for IDP and refugee support. Figure 8-2 (below) shows that, as essential services projects take root and start to provide tangible benefits for the HN population, management of these activities can be handed-off from military forces to civilian aid agencies and ultimately to HN authorities. Progress in these individual endeavors may experience individual setbacks as programs and projects are calibrated to specific localized needs, but overall progress should be measurable before agency transfers are implemented. As essential services become more effective, insurgent activities are marginalized and generate less popular support.

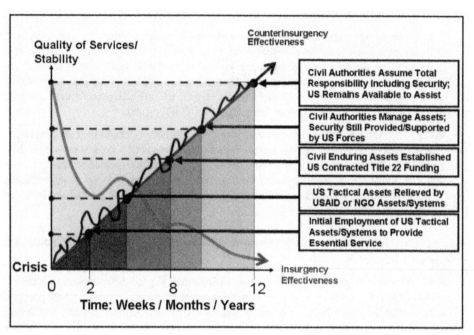

568   **Figure 8-2. Comparison of essential services availability to insurgency effectiveness**

569   ## GOVERNANCE SUPPORT

570   8-45. One of the main objectives of COIN operations is the restoration of the rule of law, order, and
571   civil procedures to the authority of the HN government. All actions of Soldiers and Marines conducting
572   COIN operations must be congruent with those of agents of a legitimate and law-abiding HN govern-
573   ment. Multinational and U.S. military units brought in to support this objective remain mindful that all of
574   their actions will be scrutinized by the population to see if they are indeed consistent with this avowed
575   purpose. For this reason, legal support to COIN operations is of particular significance and sensitivity. In-
576   consistencies in this area can furnish insurgent forces with valuable issues for manipulation in propa-
577   ganda.

578   ### Legal Support to Operations

579   8-46. Legal support to COIN operations can cover many areas. (See FM 4-0, chapter 12; MCWP 4-12,
580   page1-7, and appendix D of this manual.) One of the most significant areas of legal support in COIN op-
581   erations is the continuous monitoring and evaluation of rules of engagement as they are applied in a con-
582   stantly changing environment. This legal status may also affect the conduct of contractors and their re-
583   quirements for protection. Another area of legal support is status of forces agreements, which will need to
584   be negotiated and revised as the HN government is able to responsibly assume and exercise sovereignty.
585   Status of forces agreements affect how legal disputes between forces and local nationals are handled, in-
586   cluding those disputes emerging from contracting and other commercial activities. Contracts and claims re-
587   quire sensitive and fair construction and execution, so perceptions of exploitation and favoritism do not
588   undermine overall COIN initiatives. Since COIN operations often depend upon a variety of complex fund-
589   ing sources, judge advocate legal advice on fiscal law is particularly important to ensure compliance with
590   domestic statutes governing the funding of military and nonmilitary activities. In these contexts, judge ad-
591   vocates may also be asked to advise the HN government at all political levels, about how to establish and
592   administer appropriate legal safeguards.

Legal Aspects of Contracting and Claims

8-47. In COIN operations, two circumstances may require extensive civil law support. The first situation is when COIN forces engage in commercial contracts with host nationals for provision of goods and or services. The second occasion is when host nationals seek compensation for damages, injuries, or deaths that individuals or their relatives claim to have suffered due to the actions of COIN forces.

8-48. Legally reviewing COIN contracts negotiated with host nationals establishes several important conditions. First the process makes clear to the host nationals that there are indeed established procedures rooted in law that govern such transactions, and it sends the message that favoritism and partisanship are not part of the process in a legitimate government. Second, this review potentially forestalls contracts going to individuals who may be part of the insurgency and may already be named or identified as subjects of other ongoing investigations or legal actions.

8-49. In the case of claims for damages allegedly caused by COIN forces, legal reviews show genuinely wronged host nationals that their grievances are indeed taken seriously. In the case of insurgents or opportunists misrepresenting the terms or conditions under which "damages" occurred, legal reviews provide an effective method of assessing the validity or falsehood of such claims and thereby prevent COIN forces from squandering resources or at worst inadvertently supporting the insurgents.

Restoration of Civil Judicial Functions

8-50. In periods of extreme unrest and insurgency, HN legal structures (such as courts, prosecutors, defense assistance, prisons) may fail to exist or function at any level. Under these conditions, in order to establish legal procedures and precedents for dealing with captured insurgents and common criminals, provisions may be made for the establishment of special tribunals under the auspices of either a provisional authority or a United Nations mandate. While legal actions are being handled under these provisions, COIN forces can expect to be substantially involved in providing sustainment and security support as well as legal support and advice on these functions.

8-51. Even when judicial functions are restored to HN authorities, COIN forces may have to provide logistic and security support to judicial activities for a prolonged period if insurgents continue to demonstrate interest in disrupting all activities supporting the legitimate rule of law. With restoration of judicial functions to the HN government, COIN forces must recognize and acknowledge that not all laws passed by the popular legislative or parliamentary branches of the HN government will be consistent with those experienced by multinational forces in their home countries. Under such conditions, COIN forces need to look to their legal advisors, commanders, and diplomatic representatives for appropriate guidance on dealing with these sensitive matters.

ECONOMIC DEVELOPMENT

8-52. Many commanders are unfamiliar with the tools and resources required for promoting economic pluralism. In COIN operations, economic development is probably the LLO with the greatest logistic significance. This LLO is usually exercised by military commanders using resource managers (comptrollers) and contingency contracting officers. The challenge for these staff officers is to deliver financial resources such that they—
- Maximally benefit the HN population.
- Support the COIN force's other LLO objectives.
- Ensure the funds are not diverted into insurgent hands.

Achievement of these objectives depends upon the efforts of logisticians maintaining a thorough and accurate logistic preparation of the theater and commanders and contracting officers obtaining goods and services consistent with its assessments. Such purchasing must also promote vendors and businesses whose practices are supportive of widespread job stimulation and local investment. In addition to the logistic preparation of the theater issues discussed in paragraph 8-9, some other areas for assessment and analysis are—
- HN economic capabilities and shortfalls suitable for filling by external means.

641
642
- Methods of determining land and other real property ownership, means of transfers, and dispute resolution.

643
644
- Methods for promoting and protecting property and asset rights, and open access to trade goods and services.

645
646
- Prevailing wage rate standards and correlation to occupational category (unskilled/skilled/ professional labor).

647
- Historic market demographics.

648
- Identification of potential vendors with local sources of supply in the AO.

649
Sources of Funding

650
651
8-53. COIN operations are usually supported by a variety of funding sources. The two types of funds that U.S. forces most commonly operate under are—

652
653
- **Title 10** funds, which are strictly for the supply, support, and sustainment of DOD service members and employees.

654
655
- **Title 22** funds, which are appropriated for foreign relations purposes and used solely for the benefit and support of the HN government and population.

656
657
658
659
660
661
662
8-54. Other sources of funding that may be encountered in COIN operations are those provided by other government agencies (such as the U.S. Agency for International Development and State Department), other donor nations and agencies, the United Nations or even the host nation itself. In some cases misappropriated or illicit funds may be seized or captured by HN government or COIN forces and redistributed to fund COIN activities. Under these complex fiscal circumstances, resource managers and staff judge advocates are the best sources of guidance on the legal use of different types of funds. (This subject is covered in more detail in appendix D.)

663
Contracted Logistic Support

664
665
8-55. In COIN operations, U.S. forces can expect to be supported by contracted logistic support. Contractor activities fall into three different categories:

666
- Theater support contractors.

667
- External support contractors.

668
- System contractors.

669
(See FM 4-0, paragraphs 5-92 through 5-95; MCWP 4-12, page 4-8.)

670
671
672
673
674
675
676
8-56. For the purposes of promoting economic pluralism in COIN operations, theater support contractors are the most significant because this type of contractor is the most reliant on HN employees and vendors. External support contractors and the Logistic Civilian Augmentation Program (LOGCAP) are designed to provide logistic services, usually through large-scale prearranged contracts with major contractors, who may in turn subcontract various components of their large contracts to smaller theater-based providers. Systems contracts are designed by systems program managers to support special or complex equipment and generally have little influence on promoting economic pluralism.

677
THEATER SUPPORT CONTRACTORS

678
679
680
681
682
8-57. Theater support contractors can be obtained either under prearranged contracts or by contracting officers serving under the direct authority of the theater principle assistant responsible for contracting. Theater support contractors usually obtain most of their materials, goods, and labor from the local manufacturing and vendor base. Some examples of goods and services that can often be obtained from theater support contractors are—

683
684
- Construction, delivery, and installation of concrete security barriers for the defense of COIN force bases and HN public buildings.

685
- Construction of security fencing.

686
687
- Public building construction and renovations (for example, site preparation, structure construction, electrical and plumbing installation, and roofing).
688
- Sanitation services.
689
- Maintenance augmentation in motor pools.
690
- Road construction and repair.
691
- Trucking and cartage.
692
- Manual labor details (for example, grounds maintenance, and sandbag filling).
693
- Housekeeping (such as warehouses).

694 ## COIN CONTRACTOR CONSIDERATIONS

695
696
697
698
699
700
701
702
703
704
705
706
707
708
709
8-58. In a COIN environment, the employment of theater support contractors and host nationals must be carefully considered and supervised so as not to undermine larger COIN objectives. Due to the subversive nature of many insurgent activities, all contractors and their employees require vetting through the G-2/S-2 as well as tamperproof photograph biometric-tagged identification, coded to indicate access areas and level of security and supervision required. Particularly in the case of HN employees, "badging" can also function as a valuable accountability tool if the badges are issued and returned at entry control points on a daily basis. While contractor security breaches are one concern, so is the security and safety of the contractor's employees. Though they may be targeted by insurgents, logistic contractors and their employees are not combatants. They are classified as "civilians accompanying the force." This status must not be jeopardized and the military units with which they work are responsible for their security in the workplace. Units employing HN contractors and employees need to be on the lookout for signs of exploitive or corrupt business practices that may alienate segments of the local population and inadvertently undermine COIN objectives. Treated fairly and respectfully, HN employees can become good sources of insights into the local language, culture, and perceptions of COIN activities as well as other issues affecting communities in the AO.

710
711
COIN Host-nation Contracting: A Potential Double-edged Sword

712
713
714
715
716
717
718
Early in Operation Iraqi Freedom, a brigade from the 101st Airborne Division was assigned a large AO near Tal Afar, in northern Iraq. The terrain the unit was required to cover and support exceeded the distribution capabilities of its ground transportation assets. Logistic officers supporting the brigade sought out and found a local business leader with a family-owned transportation company. He was positive towards U.S. aims for improving Iraq and willing to work with U.S. forces by providing a wide variety of truck and bus services.

719
720
721
722
723
724
725
726
727
After two months of ad hoc daily arrangements for services at the U.S. forces' compound entry point, the unit established a six-month contract to make this transportation support more regular. As the working relationship became more solid, the contractor and his employees also furnished insights into the effectiveness of U.S. information operations as well as information on the presence and activities of suspicious persons possibly affiliated with the insurgency. The arrangement worked exceptionally well, effectively supported COIN activities, and maintained peace and security—as long as the original unit that established the services was stationed in the area.

728
729
730
731
732
733
In time though, the unit that made the arrangement with the transportation contractor was replaced by a smaller task force, and the security situation in the area began to deteriorate. The insurgents, upon detecting this change in security posture, were quick to find the transportation contractor and kill him. No doubt their intent was not only to degrade the U.S. forces' logistic posture, but also to use terror to send the message to other local vendors that doing business with the Americans was costly.

734
735
| Eventually the contactor's brother took over operations, but understandably support deteriorated.

736
737
738
739
740
741
742
743
| When setting up logistic contracting arrangements with host nationals in a COIN environment, U.S. logisticians and contracting officers must be mindful of the grave risks host nationals may be taking in accepting these responsibilities. Insurgents are exceptionally attuned to finding ways to attack COIN logistics; there is an added terror or political message benefit for the insurgents if this can be done against host nationals branded as traitors. Inadequate or shifting U.S. security arrangements can provide openings for insurgents to more easily attack HN contractors and logistic providers.

744 ## EXTERNAL SUPPORT CONTRACTORS

745
746
747
8-59. Many of the same considerations that apply to theater contractors apply to external support contractors and their subcontracted employees, particularly if they are local or in-theater hires who are not U.S. or multinational partner citizens.

748 ## SYSTEM CONTRACTORS

749
750
751
752
753
8-60. System contractors generally work on technologically complicated military systems, such as vehicles, weapons systems, aircraft, and information systems. They are provided under prearranged contracts negotiated by program executive officers and program managers. These contractors provide systems expertise. Most are U.S. citizens and many of them are former U.S. military members. These contractors generally have to meet similar deployment and security requirements as U.S. government employees.

754 ## CONTINGENCY CONTRACTING OFFICERS AND OTHER AGENTS

755
756
757
8-61. In COIN operations the timely and well-placed distribution of funds at the local level can serve as an invaluable force multiplier. Challenges to accomplishing payments and purchases in the COIN environment are many, including the following—

758
- Problems with the security of financial institutions, agents, and instruments.

759
- Potential for sudden volatility in the HN economy.

760
- Reliability issues with local supplier and vendors.

761
- Peculiarities of local business cultures.

762
763
764
8-62. The challenge for contracting officers and other agents authorized to make payments to support COIN activities is that it is often difficult to obtain reliable information upon which to make decisions and conduct negotiations. Military means for accomplishing this type of purchasing are found at two levels:

765
766
- The contingency contracting officer, who acts upon unit-generated purchasing request and committals.

767
- The ordering officer for smaller purchases.

768
769
8-63. Because of the legal requirement to keep U.S. funds for different purposes separate and distinct, two types of purchasing officer teams need to be maintained by COIN units:

770
- Field ordering officer teams for Title 10 funds.

771
- Project ordering officer teams for Title 22 funds.

772
773
(Figure 8-3 (below) outlines the relationships and roles among different contracting and ordering officers and the types of funds that they manage.)

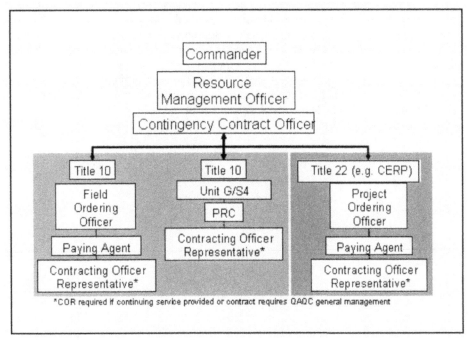

774   **Figure 8-3. Tactical financial management organizations**

775 **CONTINGENCY CONTRACTING**

776   8-64. Because contingency contracting officers are able to set contracts for larger amounts than ordering
777   officers, they normally place purchasing request and committal requirements out for bid to local vendors.
778   During COIN operations, it is especially important to spread contracts across different vendors to forestall
779   any appearance of partiality.

780 **FIELD ORDERING AND PROJECT ORDERING OFFICER TEAMS**

781   8-65. Both field ordering and project ordering officer teams consist of the respective contracting officer
782   agent, a paying agent, and a security detail, and they operate under similar regulatory constraints. These
783   officers' duties differ with respect not only to the type of funds they disburse, but also in the increment
784   caps applied. Field ordering officers with Title 10 Funds are limited to individual payments not to exceed
785    $2,500. Project ordering officers can make individual contract payments of up to $20,000 due to the
786   higher costs and scale associated with projects. In both cases, these teams provide an invaluable asset for
787   reaching into the HN communities and promoting economic pluralism while assessing the economic ef-
788   fects of purchasing activities and economic stability initiatives.

789   8-66. Due to their activities throughout HN communities and their cultivation of local business connec-
790   tions, field ordering and project ordering officer teams should not be overlooked as a potential informa-
791   tion gathering and distribution sources. Vendor observations and actions may also reveal much about the
792   real status of the COIN effort. These indicators may be as simple as vendor comments about outsiders
793   moving into the area, contractors failing to show up for work and deliver goods, or HN employees asking
794   to leave early because they have been tipped off about an impending attack. By doing business with
795   COIN forces, local contractors and vendors may be putting themselves at great risk. Protection of their
796   activities can pose a great challenge and must be seriously considered when doing business with them.

797 # SUMMARY

798   8-67. Logistic activities are an integral part of counterinsurgency operations. These activities take on both
799   the traditional form of support to combat and security forces as well as the unconventional form of provid-

800   ing mixes of essential and timely support to a wide variety of HN security and stability-enhancing activities
801   that may seem to be purely civil in character. Initially this support may have to be conducted by uniformed
802   military logistic providers, but one of the logistic objectives of COIN operations should be to encourage
803   and promote HN providers as soon as security conditions make this reasonably feasible. This transition is a
804   delicate one. Logistic providers must be continuously aware of how their practices are or are not contribut-
805   ing to their long-term objectives and adjust their methods accordingly. If there is a final paradox in coun-
806   terinsurgency, it is that logistic postures and practices are a major part of the fight and may well determine
807   if COIN forces are able to deliver the desired political end state.
808
809

1 **Appendix A**

2 # A Guide for Action: Plan, Prepare, Execute, and Assess

3 Translating the lessons of this manual into practice begins with planning and prepara-
4 tion for deployment. Successful counterinsurgents execute wisely and continually as-
5 sess their area of operations, the impact of their own operations, and their enemy's
6 strategy and tactics to adapt and win. This appendix discusses several techniques that
7 have proven successful during counterinsurgency operations. They are discussed
8 within the framework of the operations process. However, this does not limit their use
9 to any operations process activity. Successful counterinsurgents assess the operational
10 environment continuously and apply the appropriate techniques when they are
11 needed.

12 # PLAN

13 A-1. *Planning* is the process by which commanders (and staff if available) translate the commander's
14 visualization into a specific course of action for preparation and execution, focusing on the expected results
15 (FMI 5-0.1). Planning for counterinsurgency (COIN) operations is no different than for conventional op-
16 erations. However, effective counterinsurgency planning requires paying at least as much attention to as-
17 pects of the environment as to the enemy force.

18 ## ASSESS DURING PLANNING: PERFORM MISSION ANALYSIS

19 A-2. Learn about the people, topography, economy, history, religion, and culture of the Area of Opera-
20 tions. Know every village, road, field, population group, tribal leader, and ancient grievance. Become the
21 world expert on these topics. If the precise destination is unknown, study the general area. Focus on the
22 precise destination when it is determined. Understand factors in adjacent AOs and the information envi-
23 ronment that can influence AO. These can be many, particularly when insurgents draw on global griev-
24 ances.

25 A-3. Read the map like a book. Study it every night before sleep, and redraw it from memory every morn-
26 ing. Do this until its patterns become second nature. Develop a mental model of the AO. Use it as a frame-
27 work into which to fit every new piece of knowledge.

28 A-4. Study handover notes from predecessors. Better still, get in touch with the unit in theater and pick
29 their brains. In an ideal world, intelligence officers and area experts provide briefings. This may not occur.
30 Even if it does, there is no substitute for personal mastery.

31 A-5. Require each subordinate leader, including noncommissioned officers, to specialize on some aspect
32 of the AO and brief the others.

33 ## ANALYZE THE PROBLEM

34 A-6. Mastery of the AO provides a foundation for analyzing the problem. Who are the insurgents? What
35 drives them? What makes local leaders tick? An insurgency is fundamentally a competition among many
36 groups, each seeking to mobilize the population in support of its agenda. Thus, COIN is always more than
37 two-sided.

38 A-7. Understand what motivates the people and how to mobilize them. Knowing why and how the insur-
39 gents are getting followers is essential. This requires knowing the real enemy, not a cardboard cut-out. In-
40 surgents are adaptive, resourceful and probably grew up in the AO. The locals have known them since they

41 were young. U.S. troops are the outsiders. The worst opponents are not the psychopathic terrorists of the
42 movies; rather, they are charismatic warriors who would do well in any armed force. Insurgents are not
43 necessarily misled or naive. Much of their success may be due to bad government policies or security
44 forces that alienate the population.

45 A-8. Work the problem collectively with subordinate leaders. Discuss ideas, and explore possible solu-
46 tions. Once all understand the situation, seek a consensus on how to address it. If this sounds un-military,
47 get over it. Such discussions help subordinates understand the commander's intent. Once in theater, situa-
48 tions requiring immediate action will arise too quickly for orders. Subordinates will need to exercise sub-
49 ordinates initiative and act based on the commander's intent informed by whatever knowledge they have
50 developed. Corporals and privates will have to make quick decisions that may result in actions with strate-
51 gic implications. Such circumstances require a shared situational understanding and a leadership climate
52 that encourages subordinates to assess the situation, act on it, and accept responsibility for their actions.
53 Employing mission command is essential in this environment. (Mission command, commander's intent,
54 and subordinates' initiative are defined in the glossary. See FM 6-0 for discussions of the principles in-
55 volved.)

PREPARE

57 A-9. *Preparation* consists of activities by the unit before execution to improve its ability to conduct the
58 operation including, but not limited to, the following: plan refinement, rehearsals, reconnaissance, coordi-
59 nation, inspections, and movement (FM 3-0). Compared with conventional operations, preparing for COIN
60 operations requires greater emphasis on organizing for intelligence and for working with nonmilitary or-
61 ganizations, preparing small unit leaders for increased responsibility, and maintaining flexibility.

ORGANIZE FOR INTELLIGENCE

63 A-10. Intelligence and operations are always complementary, especially in COIN operations. COIN opera-
64 tions are intelligence-driven, and units often develop much of their own intelligence. Commanders must
65 organize their assets to do that.

66 A-11. Each company may require an intelligence section, including analysts and an individual designated
67 as the "S-2." Platoon leaders may also have to designate individuals to perform intelligence and operations
68 functions. A reconnaissance and surveillance element is also essential. Augmentation for these positions is
69 normally not available, but the tasks must be performed. Put the smartest troops in the intelligence section
70 and the reconnaissance and surveillance element. Doing these things results in one less rifle squad, but in-
71 telligence section pays for itself in lives and effort saved.

72 A-12. There are never enough linguists available. Commanders consider carefully where best to employ
73 them. Linguists are a battle-winning asset, but like any other scarce resource must be allocated carefully.
74 During predeployment, the best use of linguists may be to train troops in basic language skills.

ORGANIZE FOR INTERAGENCY OPERATIONS

76 A-13. Almost everything in COIN is interagency. Everything from policing to intelligence to civil-military
77 operations to trash collection involves working with interagency and indigenous partners. These agencies
78 are not under military control, but their success is essential to accomplishing the mission. Train troops in
79 conducting interagency operations. Get a briefing from the State Department, aid agencies and the local po-
80 lice or fire department. Designate interagency subject matter experts in each subordinate element and train
81 them. Realize that many civilians find rifles, helmets, and body armor intimidating. Learn how not to scare
82 them. Seek advice from those who come from that country or culture. Look at the situation through the
83 eyes of a civilian who knows nothing about the military. Most importantly, know that military operations
84 create temporary breathing space, but that long-term development and stabilization by civilian agencies are
85 required to prevail.

86 ## TRAVEL LIGHT AND HARDEN YOUR SUSTAINMENT ASSETS

87 A-14. A normal combat load includes body armor, rations, extra ammunition, communications gear, and
88 many other things—all of which are heavy. Insurgents may carry a rifle or rocket-propelled grenade, a
89 head scarf, and a water bottle. This situation requires ruthlessly lightening troops' combat load and enforc-
90 ing a habit of speed and mobility. Otherwise, the insurgents consistently outrun and outmaneuver them.
91 However, make sure troops can always reach back for fires or other support.

92 A-15. Also, remember to harden sustainment bases. Insurgents often consider them weak points and attack
93 there. Most attacks on coalition forces in Iraq in 2004 and 2005, other than combat actions, were against
94 sustainment installations and convoys. Ensure sustainment assets are hardened and have communications.
95 Make sure the troops there are trained in combat operations. They may do more fighting than some rifle
96 squads.

97 ## FIND A POLITICAL/CULTURAL ADVISOR

98 A-16. A force optimized for COIN would have political/cultural advisors at company level. Current force
99 structure gives corps and division commanders a political advisor. This situation requires lower-echelon
100 commanders to improvise. They must select a political/cultural advisor from among their troops. This per-
101 son may be a commissioned officer, but may not. The position requires someone with "people skills" and a
102 feel for the environment. Commanders should not try to be their own cultural advisor. They must be fully
103 aware of the political and cultural dimension, but this is a different role. Also, this position is not suitable
104 for intelligence professionals. They can help, but their task is to understand the environment. The political
105 advisor's job is to help shape it.

106 ## TRAIN THE SQUAD LEADERS—THEN TRUST THEM

107 A-17. COIN is largely executed by squads and platoons. Small-unit actions in a COIN environment often
108 have greater impacts than similar actions during major combat operations. Engagements are often won or
109 lost in moments; whoever can bring combat power to bear in seconds wins. The on-scene leader controls
110 the fight. This situation requires mission command and subordinates' initiative. (See FM 6-0, paragraphs 1-
111 67–1-80 and 2-83–2-92.) Leaders at the lowest echelons must be trained to act intelligently and independ-
112 ently without orders.

113 A-18. Training should focus on basic skills: marksmanship, patrolling, security on the move and at the halt,
114 and basic drills. When in doubt, spend less time on company and platoon training, and more time on
115 squads. Ruthlessly replace ineffective leaders. Once troops are trained, give them a clear commander's in-
116 tent and trust them to exercise subordinates' initiative within it. This allows subordinates to execute COIN
117 operations at the level at which they are won.

118 ## IDENTIFY AND USE TALENT

119 A-19. Not everyone is good at COIN. Many leaders don't understand it, and some who do can't execute it.
120 COIN is difficult and anyone can learn the basics. However, people able to intuitively grasp, master, and
121 execute COIN techniques are rare. Learn how to spot these people and put them into positions where they
122 can make a difference. Rank may not indicate the required talent. In COIN, a few good troops under a
123 smart junior noncommissioned officer doing the right things can succeed, while a large force doing the
124 wrong things will fail.

125 ## CONTINUE TO ASSESS AND PLAN DURING PREPARATION: BE FLEXIBLE

126 A-20. *Commander's visualization* is the mental process of developing situational understanding, determin-
127 ing a desired end state, and envisioning how to move the force from its current state to that end state (FMI
128 5-0.1). It begins with mission receipt and continues throughout any operation. The commander's visualiza-
129 tion forms the basis for conducting (planning, preparing for, executing and assessing) an operation.

130 A-21. Commanders continually refine their visualization based on their assessment of the operational envi-
131 ronment. They describe and direct any changes they want made as the changes are needed. The do not wait

132 for a set point in any process. This flexibility is essential during preparation for COIN operations. Some are
133 tempted to try and finalize a plan too early. They then prepare to execute the plan rather than what changes
134 in the operational environment require. However, as commanders gain knowledge, their situational under-
135 standing improves. They get a better idea of what needs to be done and of their own limitations. This lets
136 them refine their visualization and direct changes to the plan and their preparations. Even with this, any
137 plan will change once operations begin. It may need to be scrapped if there is a major shift in the environ-
138 ment. But a plan is still needed, and developing it gives leaders a simple robust idea of what to achieve,
139 even if the methods change. Directing changes to it based on continuous assessment is one aspect of the art
140 of command.

141 A-22. One planning approach is to identify phases of the operation in terms of major objectives to achieve:
142 for example, establish dominance, build local networks, and marginalize the enemy. Make sure the force
143 can easily transition between phases, both forward to exploit successes and backward to recover from set-
144 backs. Insurgents can adapt their activity to friendly tactics. The plan must be simple enough to survive set-
145 backs without collapsing. This plan is the solution that began with the shared analysis and consensus that
146 began preparation. It must be simple and known to everyone.

EXECUTE

148 A-23. *Execute* means to put a plan into action by applying combat power to accomplish the mission and us-
149 ing situational understanding to assess progress and make execution and adjustment decisions (FM 6-0).
150 The execution of counterinsurgency operations demands all of the skills required to execute conventional
151 operations. In addition, it also requires mastery of building alliances and personal relationships, paying at-
152 tention to the local and global media, and a number of additional skills that are not as heavily tasked in
153 conventional operations.

ESTABLISH AND MAINTAIN A PRESENCE

155 A-24. The first rule of COIN is to establish the force's presence in the AO. If troops are not present when
156 an incident happens, there is usually little they can do about it. The force can't be everywhere at once;
157 however, the more time troops spend in the AO the more likely they are to be where the action is. If the
158 force is not large enough to establish a presence throughout AO, then determine the most important places
159 and focus on them. This requires living in the AO close to the population. Raiding from remote, secure
160 bases doesn't work. Movement on foot, sleeping in villages, and night patrolling all seem more dangerous
161 than they are—and they are what ground forces are trained to do. Being on the ground establishes links
162 with the locals. They begin to see troops as real people they can trust and do business with, rather than
163 aliens who descend from armored boxes. Driving around in an armored convoy actually degrades situ-
164 ational awareness. It makes troops targets and is ultimately more dangerous than moving on foot and re-
165 maining close to the population.

ASSESS DURING EXECUTION: AVOID HASTY ACTIONS

167 A-25. Don't act rashly; get the facts first. Continuous assessment, important during all operations, is vital
168 during COIN. Violence can indicate several things. It may be part of the insurgent strategy, interest groups
169 fighting among themselves, or individuals settling vendettas. Or, it may just be daily life. Take the time to
170 learn what normality looks like. Insurgents may try to goad troops into lashing out at the population or
171 making a similar mistake. Unless leaders are on the spot when an incident occurs, they receive only sec-
172 ond-hand reports and may misunderstand the local context or interpretation. This means that first impres-
173 sions are often highly misleading, particularly in urban areas. Of course, leaders cannot avoid making
174 judgments. But when there is time, ask an older hand or a trusted local for their opinion. If possible, keep
175 one or two officers from your predecessor unit for the first part of the tour. Avoid rushing to judgment.

BUILD TRUSTED NETWORKS

177 A-26. Once the unit is settled into the AO, its next task is to build trusted networks. This is the true mean-
178 ing of the phrase "hearts and minds," which comprises two separate components. "Hearts" means persuad-

179 ing people their best interests are served by the COIN's success. "Minds" means convincing them that the
180 force can protect them and that resisting it is pointless. Note that neither concerns whether people like the
181 troops. Calculated self-interest, not emotion, is what counts. Over time, successful trusted networks grow
182 like roots into the population. They displace enemy networks, forcing enemies into the open. That lets the
183 force seize the initiative and destroy them.

184 A-27. Trusted networks are diverse. They include local allies, community leaders, and local security forces.
185 Nongovernmental organizations, other friendly or neutral nonstate actors in the AO, and the media should
186 also be included.

187 A-28. Building trusted networks begins with conducting village and neighborhood surveys to identify com-
188 munity needs. Then follow through to meet them, build common interests, and mobilize popular support.
189 This is the true main effort; everything else is secondary. Actions that help build trusted networks support
190 the COIN effort. Actions that undermine trust or disrupt these networks, even those that provide a short-
191 term military advantage, help the enemy.

GO WITH THE GRAIN AND SEEK EARLY VICTORIES

193 A-29. Don't try to crack the hardest nut first. Don't go straight for the main insurgent stronghold or try to
194 take on villages that support the insurgents. Instead, start from secure areas and work gradually outwards.
195 Extend influence through the locals' own networks. Go with, not against, the grain of local society. First
196 win the confidence of a few villages, and then work with those with whom they trade, intermarry, or do
197 business. This tactic develops local allies, a mobilized population, and trusted networks.

198 A-30. Seek a victory early in the operation to demonstrate dominance of the AO. This may not be a combat
199 victory. Early combat without an accurate situational understanding may create unnecessary collateral
200 damage and ill will. Instead, victories may involve resolving a long-standing issue or co-opting a key local
201 leader. Achieving even a small early victory can set the tone for the tour and help commanders seize the
202 initiative.

PRACTICE DETERRENT PATROLLING

204 A-31. Establish patrolling tactics that deter enemy attacks. An approach using combat patrols to provoke,
205 then defeat, enemy attacks is counterproductive. It leads to a raiding mindset, or worse, a bunker mentality.
206 Deterrent patrolling is a better approach. The aim of deterrent patrolling is to keep the enemy off balance
207 and the population reassured. Constant, unpredictable activity over time deters attacks and creates a more
208 secure environment. Accomplishing this requires one- to two-thirds of the force to be on patrol at any time,
209 day or night.

BE PREPARED FOR SETBACKS

211 A-32. Setbacks are normal in COIN, as in all operations. Leaders make mistakes and lose people. Troops
212 occasionally kill or detain the wrong person. It may not be possible to build or expand trusted networks. If
213 this happens, drop back to the previous phase of the plan, recover, and resume operations. It is normal in
214 company-level COIN operations for some platoons to be doing well while others are doing badly. This
215 situation is not necessarily evidence of failure. Give subordinate leaders the freedom to adjust their posture
216 to local conditions. This creates flexibility that helps survive setbacks.

REMEMBER THE GLOBAL AUDIENCE

218 A-33. The omnipresence and global reach of today's news media affects the conduct of military operations
219 more than ever before. Satellite receivers are common, even in developing countries. Bloggers and print,
220 radio, and television reporters monitor and comment on everything military forces do. Insurgents use ter-
221 rorist tactics to produce graphic images that they hope will influence public opinion—both locally and
222 globally.

223 A-34. Train troops to consider how the global audience might perceive their actions. Troops should assume
224 that everything they say or do will be publicized. Also, treat the media as an ally. Help reporters get their

225 story. That helps them portray military actions favorably. Trade information with media representatives.
226 Good relationships with nonembedded media, especially indigenous media, can dramatically increase situ-
227 ational awareness.

ENGAGE THE WOMEN, BEWARE THE CHILDREN

229 A-35. Most insurgent fighters are men. However, in traditional societies, women are hugely influential in
230 forming the social networks that insurgents use for support. When the women support COIN efforts, family
231 units support COIN efforts. Getting the support of families is a big step toward mobilizing the population
232 against the insurgency. Co-opting neutral or friendly women through targeted social and economic pro-
233 grams builds networks of enlightened self-interest that eventually undermine the insurgents. Female coun-
234 terinsurgents, including interagency people, are required to do this effectively.

235 A-36. Conversely, do not allow troops to fraternize with local children. Homesick troops want to drop their
236 guard with kids. But children are often sharp-eyed, lacking in empathy, and willing to commit atrocities
237 their elders would shrink from. The insurgents are watching. They notice any friendships between troops
238 and children. They may either harm the children as punishment, or use them as agents. Do not allow troops
239 to throw candy or presents to children. It attracts them to military vehicles, creates crowds the enemy can
240 exploit, and can lead to children being run over. It requires discipline to children at arm's length while
241 maintaining the empathy necessary to win local support.

ASSESS CONTINUOUSLY DURING EXECUTION

243 A-37. Develop measures of effectiveness early and refine them as the operation progresses. They should
244 cover a range of social, informational, military, and economic issues. Use them to develop an in-depth op-
245 erational picture and how it is changing, not in a mechanistic, traffic-light, fashion. Typical measures of ef-
246 fectiveness include the following:

247 ● Percentage of engagements initiated by friendly forces versus those initiated by insurgents.

248 ● Longevity of friendly local leaders in positions of authority.

249 ● Number and quality of tips on insurgent activity that originate spontaneously.

250 ● Economic activity at markets and shops.

251 These mean virtually nothing as a snapshot; trends over time are the true indicators of progress.

252 A-38. Avoid using body counts as a measure of effectiveness. They actually measure very little and may
253 provide misleading numbers. Using body counts to measure effectiveness accurately requires the following
254 information:

255 ● How many insurgents there were to start with.

256 ● How many moved into the area.

257 ● How many transferred from supporter to combatant status.

258 ● How many new fighters the conflict has created.

259 Accurate information of this sort is usually not available.

MAINTAIN MISSION FOCUS THROUGHOUT

261 A-39. Once a unit is established in its AO, troops settle into a routine. A routine is good as long as the mis-
262 sion is being accomplished. However, leaders should be alert for the complacency that often accompanies
263 routines.

264 A-40. It will probably take troops at least one-third of the tour to become effective. Toward the tour's end,
265 leaders struggle against the "short-timer" mentality. So the middle part of the tour is often the most produc-
266 tive. However, leaders must work to keep troops focused on the mission and attentive to the environment.

EXPLOIT A SINGLE NARRATIVE

268 A-41. Since COIN is a competition to mobilize popular support, it pays to know how people are mobilized.
269 Most societies include opinion-makers: local leaders, religious figures, media personalities, and others who

270
271
272
273
274
275
set trends and influence public perceptions. This influence often follows a single narrative—a simple, unifying, easily expressed story or explanation that organizes people's experience and provides a framework for understanding events. Nationalist and ethnic historical myths and sectarian creeds are examples of such narratives. Insurgents often try to use the local narrative to support their cause. Undercutting their influence requires exploiting an alternative narrative. An even better approach is tapping into an existing narrative that excludes the insurgents.

276
277
278
279
280
281
282
A-42. Higher headquarters usually establishes the COIN narrative. However, only leaders and troops at the lowest levels have the detailed knowledge needed to tailor it to local conditions and generate leverage from it. For example, a nationalist narrative can be used to marginalize foreign fighters. A narrative of national redemption can undermine former regime elements seeking to regain power. Company level leaders apply the narrative gradually. They get to know local opinion-makers, win their trust, and learn what motivates them. Then they build on this knowledge to find a single narrative that emphasizes the inevitability and rightness the COIN's success. This is art, not science.

283 **LOCAL FORCES SHOULD MIRROR THE ENEMY, NOT U.S. FORCES**

284
285
286
287
288
289
290
A-43. By mid tour, U.S. forces should be working closely with local forces, training or supporting them and building an indigenous security capability. The natural tendency is to create forces in a U.S. image. This is a mistake. Instead, local indigenous forces need to mirror the enemy's capabilities and seek to supplant the insurgent's role. This does not mean they should be irregular in the sense of being brutal or outside proper control. Rather, they should move, equip, and organize like the insurgents but have access to U.S. support and be under the firm control of their parent societies. Combined with a mobilized population and trusted networks, this allows local forces to isolate the enemy from the population.

291
292
293
294
295
296
A-44. U.S. forces should support indigenous forces. At the company level, this means raising, training, and employing local indigenous auxiliary forces (police and military). These tasks require high-level clearance, but if permission is given, companies should each establish a training cell. Platoons should aim to train one local squad and then use that squad as a nucleus for a partner platoon. The company headquarters should train an indigenous leadership team. This process mirrors the development of trusted networks. It tends to emerge naturally with the emergence of local allies willing to take up arms to defend themselves.

297 **CONDUCT CIVIL-MILITARY OPERATIONS**

298
299
300
301
302
303
A-45. COIN can be characterized as armed social work. It includes attempts to redress basic social and political problems while being shot at. This makes civil-military operations a central COIN activity, not an afterthought. Civil-military operations are one means of restructuring the environment to displace the enemy from it. They must focus on meeting basic needs first. A series of village or neighborhood surveys, regularly updated, are invaluable to understanding the population's needs and tracking progress in meeting them.

304
305
306
307
308
309
310
A-46. Effective civil-military operations require close cooperation with national, international, and local interagency partners. These partners are not under military control. Many NGOs, for example, do not want to be too closely associated with military forces because they need to preserve their perceived neutrality. Interagency cooperation may involve a shared analysis of the problem, building a consensus that allows synchronization of military and interagency efforts. The military's role is to provide protection, identify needs, facilitate civil-military operations, and use improvements in social conditions as leverage to build networks and mobilize the population.

311
312
313
314
A-47. There is no such thing as impartial humanitarian assistance or civil-military operations in COIN. Whenever someone is helped, someone else is hurt—not least the insurgents. So civil and humanitarian assistance personnel are often targeted. Protecting them is a matter not only of close-in defense, but also of creating a secure environment by co-opting the local aid beneficiaries and their leaders.

315 **SMALL IS BEAUTIFUL**

316
317
A-48. Another tendency is to attempt large-scale, mass programs. In particular, U.S. forces tend to apply ideas that succeed in one area in another. They also try to take successful small programs and replicate

318
319
320
321
322
323

them on a larger scale. This usually does not work. Often small-scale programs succeed because of local conditions or because their size kept them below the enemy's notice and helped them flourish unharmed. Company-level programs that succeed in one AO often also succeed in another; however, small-scale projects rarely proceed smoothly into large programs. Keep programs small. This makes them cheap, sustainable, low-key, and (importantly) recoverable if they fail. New programs—also small, cheap, and tailored to local conditions—can be added as the situation allows.

324
FIGHT THE ENEMY'S STRATEGY

325
326
327
328
329
330

A-49. When COIN efforts are succeeding, insurgents often transition to the offensive. COIN successes create a situation dangerous to insurgents by threatening to separate them from the population. Insurgents attack military forces and the population to reassert their presence and continue the insurgency. This activity does not necessarily indicate an error in COIN tactics (though it may: it depends on whether the population has been successfully mobilized). It is normal, even in the most successful operations, to have spikes of offensive insurgent activity.

331
332
333
334
335
336
337

A-50. The obvious military response is a counteroffensive to destroy the enemy's forces. This is rarely the best choice at company level. Only attack insurgents when they get in the way. Try not to be distracted or forced into a series of reactive moves by a desire to kill or capture them. Provoking combat usually plays into the enemy's hands by undermining the population's confidence. Instead, attack the enemy's strategy. If insurgents are seeking to recapture a community's allegiance, co-opt that group against him. If they are trying to provoke a sectarian conflict, transition to peace enforcement operations. The possible situations are endless, but the same principle governs the response: fight the enemy's strategy, not enemy forces.

338
ASSESS DURING EXECUTION: RECOGNIZE AND EXPLOIT SUCCESS

339
340
341

A-51. Implement the plan developed early in the campaign and refined through interaction with local partners. Focus on the environment, not the enemy. Aim at dominating the whole district and implementing solutions to its systemic problems. Continuously assess the results and adjust as needed.

342
343
344
345
346
347
348

A-52. Achieving success means that, particularly late in the campaign, it may be necessary to negotiate with the enemy. Members of the population supporting the COIN operation know the enemy's leaders. They may have grown up together. Valid negotiating partners sometimes emerge as the campaign progresses. Again, close interagency relationships are needed to exploit opportunities to co-opt segments of the enemy. This helps you wind down the insurgency without alienating potential local allies who have relatives or friends among the insurgents. As an insurgency ends, a defection is better than a surrender, a surrender better than a capture, and a capture better than a kill.

349
PREPARE DURING EXECUTION: GET READY FOR HANDOVER FROM DAY ONE

350
351
352
353
354
355
356
357
358
359

A-53. It is unlikely the insurgency will end during the tour. There will be a relief in place, and the relieving unit will need as much knowledge as can be passed to them. Start handover folders in every platoon and specialist squad immediately upon arrival, if they are not available from the unit being relieved. The folders should include lessons learned, details about the population, village and patrol reports, updated maps, and photographs—anything that will help newcomers master the environment. Computerized databases are fine, but keep good back-ups and ensure you have hard copy of key artifacts and documents. Developing and keeping this information current is boring, tedious work. But it is essential to both short- and long-term success. The corporate memory this develops gives troops the knowledge they need to stay alive. Passing it on to the relieving unit does the same for them. It also reduces the loss of momentum that occurs during any handover.

360
ENDING THE TOUR

361
362
363

A-54. As the end of the tour approaches, the key leadership challenge becomes keeping the troops focused. They must not drop their guard. They must continue to monitor and execute the many programs, projects, and operations underway.

364
365
366
367
368
369
370
371
372
373
A-55. The tactics discussed above remain applicable as the end-of-tour transition approaches. However, there is an important new one: keep the transition plan secret. The temptation to talk about home becomes almost unbearable toward the end of a tour. The locals know you are leaving and probably have a good sense of the generic transition plan. They have seen units come and go. But the details of the transition plan must be protected; otherwise, the enemy might use the handover to undermine any progress made during the tour. Insurgents may stage a high-profile attack. They may try to recapture the population's allegiance by scare tactics that convince them they will not be protected after the transition. They may try to persuade the locals that the successor unit will be oppressive or incompetent. Keep the transition plan details secret within a tightly controlled compartment in the headquarters. Tell the troops to resist the temptation to say goodbye to local allies. They can always send a postcard from home.

374 THREE "WHAT IFS"

375
376
A-56. The discussion above describes what should happen, but things do go wrong. Here are some "what ifs" to consider.

377 WHAT IF YOU GET MOVED TO A DIFFERENT AREA?

378
379
380
381
382
A-57. Efforts made preparing for operations in one AO are not wasted if a unit is moved to another. In mastering the first area, troops learned techniques they can apply to the new one. For example, they know how to analyze an AO and decide what matters in the local societal structure. The experience provides a mental structure for analyzing the new AO. They can focus on what is different, making the process easier and faster. They need to apply this same skill when they are moved within a battalion or brigade AO.

383 WHAT IF YOU HAVE NO RESOURCES?

384
385
386
387
388
389
390
A-58. Things can be things done in a low-priority AO. However, leaders need to focus on self-reliance, keeping things small and sustainable, and ruthlessly prioritizing efforts. Local community leaders can help. They know what matters to them. Leaders should be honest with them, discuss possible projects and options, and ask them to recommend priorities. Often they can find translators, building supplies, or expertise. They may only expect support and protection in making their projects work. And negotiation and consultation can help mobilize their support and strengthen social cohesion. Setting achievable goals is key to making the situation work.

391 WHAT IF THE THEATER SITUATION SHIFTS?

392
393
394
395
A-59. Sometimes everything goes well at the tactical level, but the theater situation changes and invalidates those efforts. When that happens, drop back a stage, consolidate, regain balance, and prepare to expand again when the situation allows. A flexible, adaptive plan is helpful in such situations. Friendly forces may have to cede the initiative for a time; however, they must regain it as soon as the situation allows.

396 CONCLUSION

397
398
399
400
401
A-60. This appendix has summarized one set of tactics for conducting COIN operations. Like all tactics they need interpretation. Constant study of the AO is needed to apply them to the specific circumstances a unit faces. Observations and experience helps Soldiers and Marines apply them better. Whatever else is done, the focus must remain on gaining and maintaining the support of the population. With their support, victory is assured; without it, COIN efforts cannot succeed.

402

403

Appendix B

Intelligence Preparation of the Battlefield

The purpose of the appendix is to provide commanders and staffs with a framework for performing intelligence preparation of the battlefield during counterinsurgency operations. This appendix does not supersede any Army or Marine Corps doctrine. It is to aid in the conduct of a specific mission type and is meant as to supplement FM 34-130/FMFRP 3-23-2.

SECTION I – INTELLIGENCE PREPARATION OF THE BATTLEFIELD STEPS

B-1. As explained in chapter 3, intelligence preparation of the battlefield (IPB) in (COIN) follows the same process described in FM 34-130/FMFRP 3-23-2. However, it places greater emphasis on civil considerations, especially the people of the area of operations (AO). In addition, the threats evaluated in COIN differ greatly from a conventional military threat and must therefore be evaluated differently. The steps of IPB are—

- Define the operational environment.
- Describe the effects of the operational environment.
- Evaluate the threat.
- Determine threat courses of action.

SECTION II – DEFINE THE OPERATIONAL ENVIRONMENT

B-2. When defining the area of interest for a unit, commanders and staffs must account for the movement of people and information to and from the AO, and how this might affect the AO. Some of the important factors which must be evaluated include the following:

- Cultural geography: family, tribal, ethnic, religious, or other social links that go beyond the AO.
- Communication links to other regions.
- Economic links to other regions.
- Media influence on the local populace, U.S. populace, and multinational partners.
- External financial, moral, and or logistic support of the insurgency.

B-3. Another consideration of defining the operational environment is to identify the other organizations with operating in the AO:

- Higher and adjacent units.
- Host-nation (HN) security forces and government personnel.
- Multinational forces.
- U.S. government agencies.
- Nongovernmental agencies.

SECTION III – DESCRIBE THE EFFECTS OF THE OPERATIONAL ENVIRONMENT

B-4. Describing the effects of the operational environment involves developing an understanding of that environment and is critical to the success of operations. This step includes—

38 39	● Civil considerations (ASCOPE), with emphasis on the people, history, and HN government in the AO.
40 41	● Terrain analysis (physical geography), with emphasis on complex terrain, suburban and urban terrain, and lines of communications.
42 43	● Weather analysis, with attention given to the weather's effects on activities of the population, such as agriculture, smuggling, or insurgent actions.

44 CIVIL CONSIDERATIONS (ASCOPE)

45 B-5. *Civil considerations* concern how the manmade infrastructure, civilian institutions, and attitudes and
46 activities of the civilian leaders, populations, and organizations within an area of operations influence the
47 conduct of military operations (FM 6-0). Because the purpose of COIN is to support a HN government in
48 gaining legitimacy and the support of the population, civil considerations are often the most important fac-
49 tors to consider during mission analysis.

50 B-6. Civil considerations generally focus on the immediate impact of civilians on operations in progress;
51 however, they also include larger, long-term diplomatic, informational, and economic issues at higher lev-
52 els. At the tactical level, they directly relate to key civilian areas, structures, capabilities, organizations,
53 people, and events within the AO.

54 B-7. COIN cannot be conducted effectively without a thorough appreciation of civil considerations con-
55 sistently applied to all operations. Civil considerations comprise six characteristics, expressed in the mem-
56 ory aid ASCOPE:

57 ● Areas.
58 ● Structures.
59 ● Capabilities.
60 ● Organizations.
61 ● People.
62 ● Events.

63 AREAS

64 B-8. Key civilian areas are localities or aspects of the terrain within an AO that have significance to the
65 lives of the people there. This characteristic approaches terrain analysis from a civilian perspective. Com-
66 manders analyze key civilian areas in terms of how they affect the missions of individual units as well as
67 how military operations affect these areas. Examples of key civilian areas are—

68 69 ● Areas defined by political boundaries, such as districts/neighborhoods within a city, municipali-
ties within a region, or provinces within a country.
70 ● Areas of high economic value, such as industrial centers, farming regions, and mines.
71 ● Centers of government and politics.
72 ● Culturally important areas.
73 ● Social, ethnic, tribal, political, religious, criminal, or other important enclaves.
74 ● Trade routes and smuggling routes.
75 ● Possible sites for the temporary settlement of dislocated civilians or other civil functions.

76 STRUCTURES

77 B-9. Analyzing a structure involves determining how its location, functions, and capabilities can support
78 the operation. Commanders also consider the consequences of using it. Using a structure for military pur-
79 poses often competes with civilian requirements for it. Commanders carefully weigh the expected military
80 benefits against costs to the community that will have to be addressed in the future. Some of the important
81 structures in an AO may include—

82 ● Government centers—necessary for the government to function.
83 ● Headquarters and bases for security forces—necessary for the security forces to function.

84 • Police stations, courthouses, and jails—necessary for countering crime and very beneficial for
85   COIN operations.
86 • Communications and media infrastructure, such as radio towers, television stations, cellular tow-
87   ers, newspaper offices, or printing presses—important to information flow and the opinions of
88   the populace.
89 • Roads—allow for movement of populace, goods, insurgents, and COIN forces.
90 • Bridges—allow for movement of populace, goods, insurgents, and COIN forces.
91 • Ports of entry, such as airports and sea ports—allow for movement of populace, goods, insur-
92   gents, and COIN personnel.
93 • Dams—provide electric power, drinking water, and flood control.
94 • Electrical power stations and substations—enable functioning of the economy and often impor-
95   tant for day-to-day life of the populace.
96 • Refineries and other sources of fuel—enable functioning of the economy and often important for
97   day-to-day life of the populace.
98 • Sources of potable water—important for public health.
99 • Sewage systems—important for public health.
100 • Clinics and hospitals—important for the health of the populace; these are protected sites.
101 • Schools and universities—affect the opinions of the populace; these are protected sites.
102 • Places of religious worship—affect opinions of the populace; often of great cultural importance;
103   these are protected sites.

104 **CAPABILITIES**

105 B-10. Capabilities can refer to the ability of local authorities—those of the host nation or some other
106 body—to provide a populace with key functions or services. Commanders and staffs analyze capabilities
107 from different levels, but generally put priority on understanding the capability of the government to sup-
108 port the mission. The most essential capabilities are those required to save, sustain, or enhance life, in that
109 order. Some of the more important capabilities are—
110 • Public administration—effectiveness of bureaucracy, courts, and other parts of the government.
111 • Public safety—provided by security forces, military, police, and intelligence organizations.
112 • Emergency services—fire departments, ambulance services, and so forth.
113 • Public health—clinics, hospitals.
114 • Food.
115 • Water.
116 • Sanitation.

117 B-11. In populated areas, the public services important that need to be assessed may be remembered using
118 the acronym SWEAT-MS. Providing these services is a measure of a government's capabilities. In addi-
119 tion, the relative demand for these services is a measure of some of the interests of the populace:
120 • Sewage.
121 • Water.
122 • Electricity.
123 • Academic (schools and universities).
124 • Trash.
125 • Medical.
126 • Security.

127 **ORGANIZATIONS**

128 B-12. Organizations are nonmilitary groups or institutions in the AO. They influence and interact with the
129 populace, COIN forces, and each other. They generally have a hierarchical structure, defined goals, estab-

130 lished operations, fixed facilities or meeting places, and a means of financial or logistic support. Some or-
131 ganizations may be indigenous to the area. These may include—

132 • Religious organizations.
133 • Political parties.
134 • Patriotic or service organizations.
135 • Labor unions.
136 • Criminal organizations.
137 • Community organizations.

138 B-13. Other organizations may come from outside the AO. Examples of these include—
139 • Multinational corporations.
140 • United Nations agencies.
141 • Nongovernmental organizations (NGOs), such as the International Red Cross.

142 B-14. Operations often require commanders to coordinate with international organizations and NGOs. Re-
143 quired information for evaluation includes these groups' activities, capabilities, and limitations. Necessary
144 situational understanding includes knowing how the activities of different organizations may affect military
145 operations and how military operations may affect these organizations' activities. From this analysis, com-
146 manders can determine how organizations and military forces can work together toward common goals.

147 B-15. In almost every case, military forces have more resources than civilian organizations. However, ci-
148 vilian organizations may possess specialized capabilities that they may be willing to share. Commanders do
149 not command civilian organizations in their AOs. However some operations require achieving unity of ef-
150 fort between these groups and COIN forces. These situations require commanders to influence the leaders
151 of these organizations through persuasion, relying on the force of argument and the example of actions.
152 (See FM 22-100.)

153 B-16. See socio-cultural factors analysis below for a more in-depth means of evaluating organizations.

154 **PEOPLE AND SOCIO-CULTURAL FACTORS ANALYSIS**

155 B-17. "People" refers to nonmilitary personnel encountered by military forces. The term includes all civil-
156 ians within an AO as well as those outside the AO whose actions, opinions, or political influence can affect
157 the mission.

158 B-18. There can be many different kinds of people living and operating in and around an AO. As with or-
159 ganizations, people may be indigenous or introduced from outside the AO. An analysis of people should
160 identify them by their various capabilities, needs, and intentions. It is useful to separate people into distinct
161 categories. When analyzing people, commanders consider historical, cultural, ethnic, political, economic,
162 and humanitarian factors. They also identify the key communicators and the formal and informal processes
163 used to influence people.

164 B-19. An understanding of the people in an AO is developed using socio-cultural factors analysis. This
165 analysis addresses the following factors—

166 • Society.
167 • Social structure.
168 • Culture.
169 • Power.
170 • Interests.

171 **Society**

172 B-20. Populations with a shared political authority and identity. There will usually be one society in an
173 AO; the presence of a more than one society in a country may contribute to the development of an insur-
174 gency.

175 **Social Structure**

176 B-21. Social structure describes the relationships between groups within a society. It can be analyzed in
177 terms of the following:
178 - Groups.
179 - Networks.
180 - Institutions.
181 - Organizations.
182 - Roles and statuses.

183 *Groups*

184 B-22. Racial, ethnic, religious, or tribal identity groups exist within any society. It is important to know the
185 following about them:
186 - The types of groups present in the AO.
187 - The size of the groups.
188 - The locations and distribution of the groups.
189 - Formal connections and relationships between and within groups, such as treaties or alliances.
190 - Informal connections and relationships between and within groups, such as tolerance, vendettas,
191 and cooperation.
192 - Is the insurgent leadership composed of one particular group?
193 - Is the insurgent rank and file composed of one particular group?
194 - Is the government leadership composed of one particular group?
195 - Do key leaders belong to more than one group? If so, do they have conflicting values that may
196 be exploited?
197 - Are the groups organized as networks, institutions, or organizations?

198 *Networks*

199 B-23. A *network* is a series of direct or indirect ties within a social structure that serve a purpose, such as
200 business, emotional support, or criminal activity. (See appendix E.)

201 *Institutions*

202 B-24. *Institutions* are groups engaged in patterned activity to accomplish a common task. It is important to
203 know which institutions are in the AO and their relative importance.

204 *Organizations*

205 B-25. *Organizations* are institutions that have bounded membership, defined goals, established operations,
206 fixed facilities or meeting places, and a means of financial or logistic support. They include communicating
207 organizations, religious organizations, economic organizations, governance organizations, and social or-
208 ganizations. Important information about them includes the following:
209 - What organizations are in the AO?
210 - What activities are they engaged in?
211 - Are the organizations dominated by members of particular groups?
212 - How do they interrelate to the government and the insurgency?

213 *Roles and Status*

214 B-26. *Status* is an achieved or ascribed position, such as that of doctor. Each status has an associated role
215 or activity that is the expected behavior of an individual based on their status, such as the doctor's role of
216 practicing medicine. Roles help understand how and why individuals behave in a certain way. Important
217 information includes the following:

218   •   What are the statuses and roles for each group, organization, and institution within the society?

219   •   What are the social norms associated with different roles?

220   •   Does this particular society have a role for "guests"?

221   •   Are COIN forces considered guests?

222   •   How are guests supposed to behave?

223   *Social Norms*

224   B-27. *Social norms* are unspoken rules associated with statuses and roles. They may either be moral (incest
225   prohibition, homicide prohibition) or customary (prayer before meals, remove shoes before entering
226   house). When behavior does not conform to social norms, it may be sanctioned. Understanding norms al-
227   lows Soldiers and Marines to interact with people positively. Some norms to be understood for each group
228   within the society include the following:

229   •   Requirement for revenge if honor is lost.

230   •   Appropriate treatment of women and children.

231   •   Common courtesies, such as gift giving.

232   •   Standard business practices, such as bribes and haggling.

233   # Culture

234   B-28. *Culture* is a "web of meaning" shared by members of a particular society or group within a society.
235   There are often multiple cultures within a society.

236   *Identity*

237   B-29. Each individual in the AO belongs to multiple groups, through birth, assimilation, or achievement.
238   Each group to which the individual belongs influences his or her beliefs, values, attitudes, and perceptions.
239   Individuals rank their identities consciously or unconsciously into primary identities (national, racial, reli-
240   gious) and secondary identities (hunter, blogger, coffee drinker). Frequently, individuals' identities may be
241   in conflict, and COIN forces can use these conflicts to influence key leaders' decision-making processes.

242   *Cultural Forms*

243   B-30. *Cultural forms* are the material, concrete aspects of culture that express the belief system of a group.
244   The most important of these are narratives, symbols, rituals.

245   B-31. A *narrative* is a story that explains an event in a group's history, and which also expresses values,
246   character, or self-identity of the group. Understanding its narrative is very beneficial to understanding the
247   culture in an AO and how various groups perceive the world. Important information concerning narratives
248   includes the following:

249   •   What narratives are commonly used by each group in the society to explain history and create
250   meaning?

251   •   What narratives are used by the insurgents to mobilize the population?

252   •   What beliefs do these narratives express?

253   •   How can group narratives be used by COIN forces to shift perceptions, gain support or reduce
254   support for insurgents?

255   B-32. Important information concerning symbols includes the following:

256   •   What symbols are commonly used by each group in the society to explain history and create
257   meaning?

258   •   What symbols are used by the insurgents to mobilize the population?

259   •   What beliefs do these symbols express?

260   •   How can group symbols be used by COIN forces to shift perceptions, gain support or reduce
261   support for insurgents?

262   B-33. Important information concerning rituals includes the following:

263   ●   What rituals are commonly used by each group in the society to explain history and create
264   meaning?

265   ●   What rituals are used by the insurgents to mobilize the population?

266   ●   What beliefs do these rituals express?

267   ●   How can group rituals be used by COIN forces to shift perceptions, gain support or reduce sup-
268   port for insurgents?

269   *Beliefs and Belief Systems*

270   B-34. Beliefs are concepts and ideas accepted as true, such as the existence of God.

271   B-35. Core beliefs are those views that are part of a person's deep identity, and are not easily changed, to
272   include—

273   ●   Religious beliefs.

274   ●   Importance of family.

275   ●   The importance of individual and collective honor.

276   B-36. Intermediate beliefs are beliefs that are derived from authoritative figures and texts. These can be
277   changed by formal and informal leaders within an AO.

278   B-37. Peripheral beliefs change relatively quickly over time. They may be influenced by a variety of fac-
279   tors, but flow from central and intermediate beliefs.

280   B-38. A belief system is the sum of beliefs, values, attitudes, and perceptions, as reflected through cultural
281   forms. Belief systems include ideologies, religions, and "-isms." Belief systems form the "lens" through
282   which people perceive the world. Insurgencies are commonly built around a belief system. It is very impor-
283   tant to know the following about belief systems:

284   ●   What is the belief system of each group in the AO? What are their values, attitudes, and percep-
285   tions?

286   ●   If people are members of more than one group, are there contradictions in their belief systems
287   that can be exploited?

288   ●   What is the belief system of the insurgent group?

289   ●   Does the insurgent group use belief systems to gain support and if so, how?

290   ●   Can some of these beliefs be changed or co-opted by COIN forces?

291   *Values*

292   B-39. A value is an enduring belief that a specific mode of conduct or end state is preferable or desirable.
293   Each group to which an individual belongs inculcates that individual with its values and their ranking of
294   importance. It is important to understand the following about values:

295   ●   What are the values of each group? Such values may include the following: toleration, stability,
296   prosperity, social change, and self-determination.

297   ●   Do the values promoted by the insurgency correspond to the values of all the social groups in
298   the AO?

299   ●   Do the values promoted by the insurgency correspond to the values of the government?

300   ●   Can the differences in values be exploited by U.S. forces?

301   *Attitudes*

302   B-40. Attitudes are affinities for and aversions to groups, persons, and objects. It is very important to
303   evaluate public attitudes toward the government, the insurgents, and U.S. forces. It is important to under-
304   stand the following about values:

305   ●   Attitude toward other social groups.

306   ●   Attitude toward insurgent ideology or ideologies.

307   • Attitude toward the government that may contribute to the insurgency.
308   • Attitude toward U.S. forces.

309   *Perceptions*

310   B-41. Perception is the process by which people organize external information. Understanding perceptions
311   allows commanders to shape attitudes.

312   **Power**

313   B-42. Power is t he probability that one actor within a social relationship will be in a position to carry out
314   his own will despite resistance. In any society, many groups have power. Formal power holders include the
315   following: governments, political interest groups, political parties, unions, government agencies and re-
316   gional and international political bodies. Informal power holders include the following: ethno-religious
317   groups, social elites, or tribes. Informal power is often very important in states with a weak, failed, or ille-
318   gitimate government. For each group within the AO, COIN forces should answer the following questions:

319   • What type of power does the group have?
320   • What do they use their power for? To protect their followers? Amass resources?
321   • How is their power acquired and maintained?
322   • Which leaders have power within particular groups?
323   • What type of power do they have?
324   • What do they use their power for?
325   • How is their power acquired and maintained?

326   B-43. Types of power include the following: coercive force, social capital, authority, and economic re-
327   sources.

328   B-44. *Coercive force* is a form of power expressed by using force or the threat of force to change people's
329   behavior. In a well-functioning state, the government has a monopoly or near monopoly on coercive force.
330   In a weak state, insurgents, militias, criminal networks, and other groups may derive extensive power from
331   coercive force. Coercive force can be positive, in the sense that a group may provide security to its mem-
332   bers (such as policing and defense of territory), or it may be negative, in the sense that a group may intimi-
333   date or threaten group members or outsiders.

334   B-45. *Social capital* is the power of individuals and groups to utilize social networks of reciprocity and ex-
335   change to accomplish their goals. In a system based on patron-client relationships, an individual in a pow-
336   erful position provides goods, services, security or other resources to followers in exchange for political
337   support or loyalty, thereby amassing power. This may be seen as corruption, but it is an expression of
338   power through social capital.

339   B-46. *Authority* is power considered legitimate by certain members of the population, which is attached to
340   positions and is justified by the beliefs of the obedient. The three types of authority are—

341   • Rational-legal (rule by law and elected government).
342   • Charismatic (attraction of followers through charismatic appeal).
343   • Traditional (derived from the historic position of leader or leader's family). Traditional power in
344   religious and tribal groups can be very important to COIN.

345   B-47. *Economic resources* comprise the power of groups or individuals to use economic incentives and
346   disincentives to change people's behavior. In weak or failed states, where the formal economy may be
347   functioning in a diminished capacity, insurgent organizations may attract followers through patronage sys-
348   tems, smuggling, and other criminal activities that provide income.

349   **Interests**

350   B-48. Interests comprise the needs, desires, and other core motivations that drive behavior. An interest that
351   has been frustrated may become a grievance, meaning a resentment that motivates action. A group or indi-

352 vidual's interests can be satisfied or frustrated by insurgents or counterinsurgents to gain the support of the
353 people. Interests fall into the following basic categories:

354 *Physical security*

355 B-49. Use the following questions to assess the status of physical security:
356 - Is the civilian population safe from harm?
357 - Is there a functioning police and judiciary system?
358 - Are the police fair and non-discriminatory?
359 - If the police are not providing civilians with physical security, who is?

360 If the government fails to provide security to civilians, they may look for alternative security guarantees
361 from armed groups such as tribes, militias, and insurgents. Failure to provide security may erode the gov-
362 ernment's legitimacy in the eyes of its citizens.

363 *Economic resources*

364 B-50. Use the following questions to assess the status of physical security:
365 - Does the society have a functioning economy?
366 - Are production, distribution, and consumption systems functioning?
367 - Do civilians have fair access to land and property?
368 - Is a multinational corporation or a central government monopolizing the natural resources?
369 - Who provides basic services (sewage, water, electricity, education, trash/sanitation, medical)?

370 *Political participation.*

371 B-51. Use the following questions to assess the status of physical security:
372 - Do all members of the civilian population have a guarantee of political participation?
373 - Is there ethnic, religious, or other discrimination?
374 - Is the government violating human rights?
375 - Is there an occupying force in the country?
376 - Do all civilians have access to basic government services, such as health care, sewage, water,
377 electricity, and so forth?
378 - Are there legal, social, or other policies that contribute to the insurgency?

379 *Grievances*

380 B-52. COIN forces should use these questions to determine whether these interests have become griev-
381 ances that are motivating insurgents and their followers:
382 - What are the insurgents' grievances?
383 - What are the grievances of the population?
384 - Would a reasonable person consider them to be valid? Validity of grievance is not effectively
385 assessed by objective condition.
386 - Are the articulated grievances of the population and those of the insurgency the same?
387 - What does the government believe to be the grievances of the population? Does it consider
388 those grievances to be valid?
389 - Are the articulated grievances of the population the same as those perceived by the government?
390 - Has the government made genuine efforts to address these grievances?
391 - Are these grievances practically addressable or are they beyond the immediate capacity of the
392 government (for example, major social and economic dislocations caused by globalization)?
393 - Can U.S. forces address these interests or grievances to elicit support from the civilian popula-
394 tion?

Additional Civil Considerations

395

396 B-53. The following civil considerations factors should be evaluated:

397 • Languages and dialects spoken by the populace.

398 • Nonverbal communication, like hand signals and gestures.

399 • Education levels, including literacy rates, and availability of education.

400 • Means of communication and its importance to the populace.

401 ▪ Interpersonal via face-to-face conversation, e-mail, or telephone.

402 ▪ Mass media, such as print publications, radio, television, or the Internet.

403 ▪ National history and political history.

404 ▪ Events leading to the insurgency.

405 ▪ Events contributing to the development of the insurgency.

406 • The availability of weapons to the general population.

407 ## EVENTS

408 B-54. Events are routine, cyclical, planned, or spontaneous activities that significantly affect organizations,
409 people, and military operations. They are often symbols, as described in socio-cultural factors analysis. Ex-
410 amples include the following:

411 • National and religious holidays.

412 • Agricultural crop/livestock and market cycles.

413 • Elections.

414 • Civil disturbances.

415 • Celebrations.

416 B-55. Other events include disasters from natural, manmade, or technological sources. These create hard-
417 ships and require emergency responses. Examples of events precipitated by military forces include combat
418 operations, deployments, redeployments, and paydays. Once significant events are determined, it is impor-
419 tant to template the events and to analyze them for their political, economic, psychological, environmental,
420 and legal implications.

421 # TERRAIN ANALYSIS

422 B-56. Terrain analysis in COIN includes the traditional examination of terrain's effects on the movement
423 of military units and enemy personnel. However, because the focus of COIN is on people, terrain analysis
424 usually centers on populated areas and the effects of terrain on the people. Soldiers and Marines will likely
425 spend a great deal of time in suburban and urban areas interacting with the populace. This is a three dimen-
426 sional battlefield. Multistory buildings and underground lines of communication, such as tunnels or sewers,
427 can be extremely important. Insurgents also commonly use complex natural terrain to their advantage as
428 well. Mountains, caves, jungles, forests, swamps, and other complex terrain are potential bases of operation
429 for insurgents.

430 B-57. An important terrain consideration in COIN is urban and suburban land navigation. This can be dif-
431 ficult in areas that do not have an address system and in cities where 10 digit grids may not be accurate
432 enough to find the correct apartment. If at all possible, foreknowledge of how locals find one another's
433 houses and what type of address system is used are beneficial. Recent, accurate maps that use overhead im-
434 agery are extremely helpful for use in operations. In addition, tourist maps and locally produced maps may
435 be helpful for understanding the names used by locals to describe places.

436 # WEATHER ANALYSIS

437 B-58. Weather analysis in COIN evaluates how weather affects friendly and enemy operations, but must
438 also include affects on the people of the host nation. Weather factors of importance include the following:

439 • Visibility.

440   ●   Wind.

441   ●   Precipitation.

442   ●   Cloud cover.

443   ●   Temperature.

444   ●   Humidity.

445   ●   Thermal crossover.

446   ●   Natural disasters—flood, drought, or massive storms.

447 # MILITARY ASPECTS OF TERRAIN (OAKOC) AND CIVIL
448 # CONSIDERATIONS

449   B-59. At the tactical level, Service members can use a modified version of the military aspects of terrain to
450   describe the operational environment's effects. For example:

451   ●   Observation—surveillance, use of spies and infiltrators, areas with good fields of fire or line of
452   sight.

453   ●   Avenues of approach—sewers, rooftops, roads, sidewalks, smuggling routes, infiltration and/or
454   exfiltration routes into a facility or region, ability to blend with the populace, media and other
455   means of influencing popular opinion.

456   ●   Key terrain—the population, key leaders, key structures, economically and politically important
457   areas, access control points, lines of communications.

458   ●   Obstacles—Traffic control points, electronic security systems, guard plan, rules of engagement
459   (enemy using protected place), translators, and ability to communicate with the population, cul-
460   ture, politics.

461   ●   Cover and concealment—from fire, from view, use of disguises or false identification badges,
462   hiding supplies underground or in buildings.

463 ## SECTION IV – EVALUATE THE THREAT

464   B-60. The purpose of evaluating the insurgency and other related threats to the mission is to understand the
465   enemy, enemy capabilities, enemy vulnerabilities, and opportunities commanders may be able to exploit.
466   The following characteristics of an insurgency provide a basis for evaluating the threat:

467   ●   Insurgent objectives.

468   ●   Insurgent motivations.

469   ●   Popular support/tolerance.

470   ●   Support activities, capabilities, and vulnerabilities.

471   ●   Information activities, capabilities, and vulnerabilities.

472   ●   Political activities, capabilities, and vulnerabilities.

473   ●   Violent activities, capabilities, and vulnerabilities.

474   ●   Organization.

475   ●   Key leaders and personalities.

476 # INSURGENCY-RELATED THREATS

477   B-61. The presence of an insurgency in a state usually means the government of the state is weak and los-
478   ing control. In such situations, armed criminal and nongovernmental elements can become powerful.
479   Therefore, it is also important to track noninsurgent threats including—

480   ●   Criminal networks.

481   ●   Nongovernmental militias and armed groups.

OPPORTUNITIES

B-62. As the threat is evaluated and the interests, attitudes, and perceptions of the people are learned, it will be possible to discover divisions between the insurgents and the people and divisions between the host nation and the people. Determining such divisions provides opportunities for crafting friendly operations that expand splits between the insurgents and the people or lessen the divides between the HN government and the people. Some factors that should be tracked include—

- Discrepancies between the insurgent belief systems and the popular belief systems.
- Appeal of insurgent goals.
- Public attitudes toward insurgent and counterinsurgent actions.
- Effectiveness of insurgent strategy and tactics in generating popular support.
- Changes in public interests.

OBJECTIVE AND MOTIVATION IDENTIFICATION

B-63. Insurgents have political objectives and are motivated by ideology or grievances that may be real or perceived. Identification of insurgent objectives and motivations allows COIN forces to address the underlying causes of the conflict. Broadly speaking, insurgencies can be divided into two varieties, national and liberation. Both can be further classified into the eight types explained in chapter 1.

B-64. In a *national insurgency*, the war is between a country's government and a segment or segments of the country's population. In this type of insurgency, the insurgents seek to change the political system, take control of the country, or secede from the country. A national insurgency polarizes the population of a country and is generally a struggle between the government and insurgents for legitimacy and popular support.

B-65. In contrast, a *liberation insurgency* occurs when insurgents seek to expel or overthrow what they consider a foreign or occupation government. The grievance addressed by the insurgents is a foreign government or foreign intervention into their country. Liberation insurgencies tend to have a unifying effect as the foreign occupation serves to unite insurgents with varying objectives and motivations. However, the insurgency can split into competing factions if the occupiers leave, sending the country into civil war.

B-66. Identification of insurgent goals and motivations can be difficult for a number of reasons. There may be multiple insurgent groups in one country, each with differing goals and motivations. In this case, the goals and motivations of each must be monitored independently. Another problem is that leaders within an insurgent organization may change over time and the goals may change with them. In addition to this, insurgent leaders may have different motivations from their followers. For instance, an insurgent leader may simply want to be a new dictator and may motivate followers through a combination of political ideology and money. A fourth issue is that insurgents may hide their true motivations and make false claims. For instance, liberation insurgencies are generally easier to unify and mobilize due to the clear differences between the insurgents and the outsiders they are fighting. Therefore, insurgents may try to portray a national insurgency as a liberation insurgency. Finally, the goals of the insurgency may change over time due to changes in the operational environment. The entry of a foreign military force into a COIN effort can transform the insurgency from national to liberation. The reverse may happen if a foreign force exits the theater. For all of these reasons, analysts continuously track insurgent actions, internal communications, and public rhetoric to determine insurgent goals and motivations.

POPULAR SUPPORT OR TOLERANCE

B-67. Developing passive support (tolerance) is often critical to survival and growth of an insurgent organization in its early stages. Generating active popular support often has the greatest impact on the insurgency's long-term effectiveness. This is the center of gravity of the insurgency. Popular support results in safe havens, freedom of movement, logistic support, financial support, intelligence, and new personnel for the insurgency. Generating such support has a positive feedback effect on an insurgent organization. As the insurgent group gains in support, its capabilities grow. These enable it to gain more support. Insurgents generally view popular support as a zero-sum game in which a gain in support for the insurgents is a loss for the government, and a loss of support for the government is a gain for the insurgents.

B-68. Popular support comes in many forms. In broad terms, it either originates internally to a country or externally, and is either active or passive. There are four forms of popular support:

- Active external.
- Passive external.
- Active internal.
- Passive internal.

The relative importance of each form of support varies depending on the insurgency, but all of them are beneficial to the insurgency.

B-69. *Active external support* includes finance, logistics, training, fighters, and safe havens provided by a foreign government or by NGOs, such as charities.

B-70. *Passive external support* occurs when a foreign government takes no steps to curtail the activities of insurgents living or operating within its borders. This kind of support can also take the form of a foreign government recognizing the legitimacy of an insurgent group or denying the legitimacy of the government fighting the insurgency.

B-71. *Active internal support* includes the following:

- Individuals or groups joining the insurgency.
- Providing logistic or financial support.
- Providing intelligence.
- Providing safe havens.
- Providing medical assistance.
- Providing transportation.
- Carrying out actions on behalf of the insurgents.

This form of support is usually the most important to an insurgent group.

B-72. *Passive internal support* is also beneficial, however. Passive supporters do not provide material support, but allow insurgents to operate and do not provide information to COIN forces.

B-73. The different kinds of support require commanders to take different approaches to address them. Both active and passive external support often requires diplomatic pressure on the governments or groups supporting the insurgency. In addition, actions like sealing borders may effectively stop active external support. Undermining active and passive internal support is critical to the success of counterinsurgents. In order to deal with internal support effectively, commanders must understand how it is generated by the insurgents. Insurgents use numerous methods to generate popular support. These include the following:

- Persuasion.
- Coercion.
- Encouraging overreaction.
- Apolitical fighters.

B-74. *Persuasion* includes charismatic attraction to a leader or group, esoteric appeal of an ideology, exoteric appeal of insurgent promises to address grievances, and demonstrations of potency such as large-scale attacks or social programs for the poor. Persuasion can be used to get either internal or external support. It can be countered with a counterideology, denying insurgents freedom of movement, and provision of basic services.

B-75. *Coercion* includes terrorist tactics, violence, and the threat of violence used to force people to support or tolerate insurgent activities. Coercion may be used to alter the behavior of people, organizations, governments, or counterinsurgents. It is countered by providing security.

574 B-76. *Encouraging overreaction* is also referred to as provocation of a government response. Insurgents
575 will try to force counterinsurgents to use repressive tactics that alienate the people. It is countered by ensur-
576 ing responses are appropriate to the operational environment.

577 B-77. *Apolitical fighters* may be attracted via many nonideological means, such as monetary incentives,
578 promise of revenge, and the "romance" of fighting a revolutionary war.

579 B-78. All of these methods of generating support are potentially effective, and insurgent groups often use a
580 mix of them. Of these methods, terrorist tactics and intimidation often get the most media attention. How-
581 ever, they are also the most likely to backfire on the insurgent. Both methods employ force or the threat of
582 force to make people, organizations, and governments alter their behavior to the insurgents' benefit. These
583 methods are often very effective in the short term, particularly at the community level. However, terrorism
584 against the general populace and popular leaders or attacks that negatively affecting people's way of life
585 can undermine insurgent popularity. Likewise, though intimidation is an easy means for insurgents to use
586 to generate passive support, this support will exist only as long as the insurgents maintain the ability to in-
587 timidate.

588 B-79. Demonstrations of potency can be the most effective of the techniques because they can create the
589 perception that the insurgency has momentum and will succeed. However, this method will publicly ex-
590 pose shadow governments or large-scale military actions of insurgents and thereby provide analysts with a
591 great deal of insight into the insurgent organization's support and capabilities.

592 B-80. Although difficult to quantify, analysts evaluate the popular support an insurgent group receives and
593 its ability to generate more support. This depends largely on open sources and whatever intelligence report-
594 ing is available. Among these sources, polling data can be a valuable, though imprecise, means of gauging
595 support to the government and support to the insurgents. Media and other open-source publications are im-
596 portant at all echelons. Assessing community attitudes, by gauging such things as the reactions of locals to
597 the presence of troops or governmental leaders, is also a means of estimating popular support at the tactical
598 level. At a minimum, the following should be known:

599 • The overall level of popular support to the insurgency relative to the government.
600 • The forms of popular support the insurgents receive.
601 • Sources of popular support by type (active, passive, internal, external).
602 • Segments of the populace supporting the insurgency.
603 • Foreign government support.
604 • Support from NGOs, including charities and transnational terrorist organizations.
605 • Criminal network support.
606 • Methods used to generate popular support and their relative effectiveness.
607 • Grievances (real or perceived) exploited by insurgents.
608 • Capabilities and vulnerabilities in generating popular support.

609 # SUPPORT ACTIVITIES—CAPABILITIES, AND VULNERABILITIES

610 B-81. Violence is the most noticeable activity of an insurgency but may represent only a small portion of
611 overall insurgent activity. The unseen activity includes recruiting, training, and logistic actions used to
612 maintain the insurgency. These are the support activities of an insurgency. They come from an insur-
613 gency's ability to generate popular support. Like a conventional military force, the "logistic tail" of an in-
614 surgency is usually larger than the number of combatants. Insurgent support networks may be substantial
615 even if overall levels of violence are low. For this reason, it is easy to overlook the early development of an
616 insurgency.

617 B-82. Undermining popular support to insurgents is the most effect way to reduce their support capabili-
618 ties. However, identification of support capabilities and vulnerabilities is still important. This process per-
619 mits analysts to evaluate potential threat courses of action. Such analysis also allows commanders to target
620 vulnerable parts of the insurgent support network and thereby reduce the insurgents' ability to carry out
621 their operations. Support activities to be evaluated include the following:
622 • Means of gaining or retaining popular support, to include recruiting. (See discussion above.)

623   ●   Safe havens.

624   ●   Freedom of movement.

625   ●   Logistic support.

626   ●   Financial support.

627   ●   An ability to train recruits in military or terrorist tactics and techniques.

628   ●   Means of collecting intelligence.

629   ●   Means of maintaining operations security.

630   ●   Means of communications.

631
632
INFORMATION AND MEDIA ACTIVITIES—CAPABILITIES, AND VULNERABILITIES

633 B-83. Information and media activities can be the main effort of an insurgency, with violence used in sup-
634 port. Information activities by insurgents undermine the legitimacy of the government, undermine counter-
635 insurgent forces, compel counterinsurgent forces to behave in accordance with national and international
636 laws and norms (while excusing the insurgents' own transgressions), and generate popular support. In or-
637 der to get these effects, insurgents broadcast their successes, the failures of COIN forces, failures of the
638 government, and illegal or immoral actions taken by counterinsurgents or the government. Their broadcasts
639 need not be factual; they need only find traction with the populace. Insurgents may use any form of media
640 to include the following:

641   ●   Word of mouth.

642   ●   Speeches by elites and key leaders.

643   ●   Fliers and hand-outs.

644   ●   Newspapers.

645   ●   Journals or magazines.

646   ●   Books.

647   ●   Audio recordings.

648   ●   Video recordings.

649   ●   Radio broadcasts.

650   ●   Television broadcasts.

651   ●   Web sites.

652   ●   E-mail.

653 B-84. In addition to producing their own media, insurgents take advantage of existing private and public
654 media companies through press releases and interviews. These efforts, in addition to use of the Internet, al-
655 low the insurgent message to reach a global audience. By broadcasting to a global audience, insurgents
656 gain the ability to directly attack public support for foreign intervention.

657 B-85. Insurgents can lie at will, while governments must continuously maintain their integrity to be credi-
658 ble. This often gives the initiative to the insurgents and put the host nation or COIN personnel on the de-
659 fensive in the information fight. Understanding the information capabilities and vulnerabilities of insur-
660 gents is the first step to effectively counter insurgent information activities and take the initiative away
661 from them. Information and media activities to be evaluated include the following:

662   ●   Commitment of assets and personnel to information activities.

663   ●   Types of media employed.

664   ●   Professionalism of products, such as newspaper articles or videos.

665   ●   Effectiveness and reach of information activities.

666
POLITICAL ACTIVITIES

667 B-86. Insurgents use political activities as a means of achieving their goals and to enhance the legitimacy
668 of their cause. Political activities are tightly linked to both information and violent activities. Political par-

669　ties affiliated with an insurgent organization may negotiate or communicate on behalf of insurgents,
670　thereby serving as the public face of the insurgency. Insurgencies may grow out of political parties, or po-
671　litical parties may grow out of insurgencies. It is important to recognize, however, that the links between
672　insurgents and political parties may be weak or easily broken by disputes between insurgents and politi-
673　cians. In such cases, political parties may not be able to deliver on promises to end violent conflict. It is
674　important to understand not only the links between insurgent groups and political organizations, but also
675　the amount of control each exerts over the other.

676　B-87. An understanding of insurgent political activities enables effective political engagement of insur-
677　gents. Without this knowledge, the wrong political party may be engaged, the wrong messages may be
678　used, or counterinsurgents may make deals with political parties that cannot deliver on their promises. Po-
679　litical activities to be evaluated include the following:

680　　● Links, if any, between the insurgency and political parties.
681　　● Influence of political parties over the insurgency and vice versa.
682　　● Political indoctrination and recruiting by insurgent groups.

683　# VIOLENT ACTIVITIES

684　B-88. As explained in previous chapters, insurgents are an asymmetric threat. They make up for a lack of
685　conventional power by exploiting the weaknesses of counterinsurgents. Information, political, and violent
686　activities are all means of doing this. In the case of using violence, insurgents usually tie their limited ca-
687　pability for violence to information or political actions to multiply the effects of the violence. In addition,
688　they usually attack in such a way that they conserve their forces while inflicting maximum damage.

689
690
691
692
693
694
695
696
697
698
699
700

> ## Asymmetric Tactics in Ireland
>
> In 1847, Irish insurgents were advised to engage the British Army in the following way:
>
> You must draw it out of position; break up its mass; break its trained line of march and manoeuvre... You cannot organize, or train, or discipline your own force to any point of efficiency. You must, therefore, disorganize, and untrain, and undiscipline, that of the enemy; and not alone must you unsoldier it–you must unofficer it also; nullify its tactique and strategy, as well as its discipline; decompose the science and systems of war, and resolve them into their first elements. You must make the hostile army a mob, as your own will be; force it to act on the offensive, and oblige it to undertake operations for which it was never constructed.

701　B-89. Violent actions by insurgents use three types of tactics, which may occur simultaneously:
702　　● Terrorist.
703　　● Guerrilla.
704　　● Conventional.

705　B-90. Terrorist tactics employ violence primarily against noncombatants. Terror attacks generally require
706　fewer personnel than guerrilla warfare or conventional warfare. They allow insurgents greater security and
707　relatively low support requirements. Insurgencies often rely on terrorist tactics early in their formation due
708　to these factors. Terrorist tactics do not involve mindless destruction nor are they employed randomly. In-
709　surgents choose targets that have maximum informational and political effects in support of their goals. As
710　explained above, terrorist tactics can be effective for generating popular support and altering the behavior
711　of governments.

712　B-91. Guerrilla tactics, in contrast, feature hit-and-run attacks by lightly armed groups primarily targeting
713　the government, security forces, or other COIN elements. Insurgents using guerrilla tactics usually avoid
714　decisive confrontation unless they know they can win. Instead, they focus on harassing counterinsurgents.
715　As with terrorist tactics, guerrilla tactics are neither mindless nor random. Insurgents try to choose targets

to ensure maximum informational and political effects. The goal is not to militarily defeat COIN forces, but to outlast them while building popular support for the insurgency. It is important to note that terrorist and guerrilla tactics are not mutually exclusive. An insurgent group may employ both forms of violent action simultaneously.

B-92. The use of conventional tactics is relatively rare in insurgencies and is not always necessary for their success. It can arise after an extended period during which the insurgency develops a great deal of popular support and extensive support capabilities. The insurgents are then able to generate a conventional military force that can fight directly against government forces.

B-93. Knowledge of violent capabilities is used to evaluate insurgent courses of action. Commanders use this knowledge to determine appropriate force protection measures, as well as tactics to be used in counter-ing insurgent actions. In addition, knowledge of how the enemy conducts attacks provides a baseline that aids in determining the effectiveness of COIN operations. The following should be evaluated to determine insurgent violent capabilities:

- Forms of violent action used.
- Weapons available and their capabilities.
- Training.
- Known methods of operating.
  - Frequency of attacks.
  - Timing of attacks.
  - Targets of attacks.
  - Tactics and techniques.
- Known linkages between violent, political, and information actions—how do the insurgents use violence to increase their popular support and undermine counterinsurgents?
- Means of command and control during attacks (including communications means used).

INSURGENT ORGANIZATIONAL STRUCTURE AND KEY PERSONALITIES

B-94. In order to conduct the activities described above, an insurgency requires some form of organiza-tional structure and leadership. Insurgents may use any of a wide variety of organizational structures, each with its own strengths and limitations. The structure used balances the following:

- Security.
- Efficiency and speed of action.
- Unity of effort.
- Survivability
- The human and physical geographies of the operational environment.

The organization also varies greatly by region and time. The result often ends up being based on familial, tribal, ethnic, religious, professional, and other existing social networks. For these reasons, social network analysis is required to determine organizational structure. (See appendix E).

B-95. The organizational structure of an insurgency often determines the effectiveness of targeting enemy or eliminating enemy leaders by COIN forces. For instance, if an insurgent organization is hierarchical and has few leaders, removing them may greatly degrade the capabilities of the insurgent organization. How-ever, if the insurgent organization is nonhierarchical and amorphous, focusing on the leadership is unlikely to have much effect. When creating the organization, insurgents decide whether to make the structure:

- Hierarchical or nonhierarchical.
- Highly structured or amorphous.
- Composed of personnel conducting a few specialized tasks or composed of nonspecialized per-sonnel capable of a wide variety of tasks.
- Use rigid command and control or allow autonomous action and initiative.

763   ●   Have a few leaders for rapid decision making or redundant leadership for survivability.

764   ●   Tie directly into other organizations and networks, such as criminal groups, religious organiza-
765   tions, and political organizations.

766   ●   Weight one activity over another, such as a focus on political organization over combat organi-
767   zation.

768 B-96. As explained in Chapter 1, the members of the insurgent organization fall into five broad categories
769 that may overlap:

770   ●   Leaders, who serve as planners and organizers.

771   ●   Cadre, mid-level leaders in the insurgent organization, who conduct information and political
772   activities, recruiting, and ideological indoctrination. They are believers in the insurgent ideol-
773   ogy.

774   ●   Combatants, who serve as foot soldiers and conduct violent activities.

775   ●   Auxiliaries, active members who provide support services (safe havens, supplies, and so forth).

776   ●   Mass base, consisting of active or passive supporters.

777 Understanding the roles and numbers of each category is important.

778 B-97. The mass base forms the foundation upon which most insurgencies are built. Removing its members
779 often reduces the strength of the insurgency to a level where local law enforcement can deal with it. This is
780 usually the most effective means of countering an insurgency and is a part of winning the "hearts and
781 minds" of the people.

782 B-98. The cadre and combatants carry out the visible actions of the insurgency and reinforce the mass base
783 by increasing popular support. Eliminating them may disrupt an insurgency, but rarely brings about the end
784 of the insurgency by itself.

785 B-99. Leaders are important because they choose the organization, strategy, and tactics of the insurgency.
786 The personalities of the leaders and decisions they make often determine the success or failure of the insur-
787 gency. Therefore, the leaders must be identified and their basic beliefs, intentions, capabilities, and vulner-
788 abilities understood. Leader characteristics to be identified include the following:

789   ●   Role of the leader in the organization.

790   ●   Known activities.

791   ●   Known associates.

792   ●   Background and personal history.

793   ●   Beliefs, motivations, and ideology.

794   ●   Education and training.

795   ●   Temperament, meaning whether the person is careful, impulsive, thoughtful, or violent.

796   ●   Importance of the leader in the organization.

797   ●   Popularity outside of the organization.

798 ## SECTION V – DETERMINE THREAT COURSES OF ACTION

799 B-100. The purpose of determining threat courses of action is to understand insurgent strategies and tac-
800 tics so they may be effectively countered. The initial determination of threat courses of action focuses on
801 two levels of analysis. The first involves defining the overall strategy of an insurgent force for achieving its
802 goals. The second focuses on determining tactical courses of action used in support of the strategy.

803 # INSURGENT STRATEGIES

804 B-101. As described in chapters 1 and 3, insurgents may follow five general strategies. These can be com-
805 bined with one another and may change over time.

806 B-102. The conspiratorial strategy has the following characteristics:

807
808
- Secretive political body seeking to grab power rapidly when the government is sufficiently weak.

809
- Usually a small organization that is tightly controlled by leadership.

810
- Cadre large relative to combatants and has little or no mass base.

811
- Difficult to detect.

812
- Countered by capturing or killing leaders and cadre.

813
- An example is the Bolsheviks in Russia.

814 B-103. The protracted popular war strategy has the following characteristics:

815
- Historically the most effective strategy and most complex to counter.

816
817
- Efforts focused on developing a large mass base while undermining the legitimacy and will of the government and supporting elements.

818
- Use of violent, informational, and political actions together to achieve goals.

819
- Popular support is the center of gravity that must be won by counterinsurgents.

820
- Examples are the Viet Minh in Vietnam and the communists in China.

821 B-104. The traditional strategy has the following characteristics:

822
- Relies on a tribal, ethnic, religious, or other traditional identity.

823
- Builds on group identity and exploits differences between the group and the government.

824
825
- Organization tends to have traditional leaders in leadership positions, a large mass base, and many auxiliaries, though roles shift as required.

826
827
- Countering generally requires political or economic deals between the counterinsurgents and insurgents.

828
- An example is the Dinka insurgency in southern Sudan.

829 B-105. The military-focus strategy has the following characteristics:

830
- Relies almost completely on violence to incite a popular revolution.

831
- Organization tends to have leaders and combatants but little, if any, cadre or mass base.

832
833
- Countered by capturing or killing leaders and combatants, though social, political, and development programs also aid COIN efforts.

834
- An example is the Cuban revolution.

835 B-106. The urban-warfare strategy has the following characteristics:

836
- Relies on terrorism in urban areas to incite repression by COIN forces.

837
- May seek to fix COIN forces in cities so insurgents gain freedom of movement in rural areas.

838
- Emphasizes infiltration and subversion of government and security forces.

839
- Tends to be organized in small, compartmentalized cells, due to urban operations.

840
- Cadre and mass base usually small relative to combatants.

841
- COIN forces must win support of people and avoid repressive measures that fuel insurgency.

842
843
- Examples include the Irish Republican Army in Northern Ireland and the Basque Fatherland and Liberty (ETA) movement in Spain.

844
TACTICAL COURSES OF ACTION

845
846
847
848
849
B-107. Insurgents base their tactical courses of action on their capabilities and intentions. The capabilities of an organization are the support, information, political, and violent capabilities evaluated in the third step of IPB. The intentions come from goals, motivations, strategy, culture, perceptions, and leadership personalities. Insurgents can employ a huge variety of tactics. Typical insurgent tactics and operations include, but are not limited to the following:

850
- **Ambushes**. Guerrilla-style attacks to kill or intimidate counterinsurgents.

851
852
853
- **Assassination**. A term generally applied to the killing of prominent persons and symbolic personnel as well as "traitors" who defect from the group, human intelligence sources, and others who work with/for the government or U.S. forces.

854
855
- **Arson**. Less dramatic than most tactics, arson has the advantage of low risk to the perpetrator and requires only a low level of technical knowledge.

856
857
858
859
860
861
862
863
- **Bombing and high explosives**. The improvised explosive device (IED) is the insurgent's or terrorist's weapon of choice. IEDs can be inexpensive to produce and, because of the various detonation techniques available, may be a low risk to the perpetrator. However, suicidal bombing cannot be overlooked as an employment method. Other advantages include their ability to gain publicity, as well as the ability to control casualties through timed detonation and careful placement of the device. It is also easily deniable should the action produce undesirable results. From 1983 through 1996, approximately half of all terrorist incidents worldwide involved the use of explosives.

864
865
- **Demonstrations**. These can be used to incite violent responses by counterinsurgents and also to display the popularity of the insurgent cause.

866
867
868
- **Hostage taking**. This is an overt seizure of one or more individuals with the intent of gaining publicity or other concessions in return for release of the hostage. While dramatic, hostage and hostage barricade situations are risky for the perpetrator.

869
870
871
872
- **Infiltration and subversion**. These tactics gain intelligence and degrade the effectiveness of government organizations by getting them to hire insurgent agents or by convincing members of the government to support the insurgency. Subversion may be achieved through intimidation, indoctrination of sympathetic individuals, or bribes.

873
874
- **Indirect fire**. Insurgents may use indirect fire to harass counterinsurgents or to cause them to commit forces that are attacked by secondary ambushes.

875
876
877
878
879
880
- **Kidnapping**. While similar to hostage taking, kidnapping has significant differences. Kidnapping is usually a covert seizure of one or more specific persons to extract specific demands. It is normally the most difficult task to execute. The perpetrators of the action may or may not be known for a long time. Media attention is initially intense, but decreases over time. Because of the time involved, successful kidnapping requires elaborate planning and logistics. The risk to the perpetrators may be less than in the hostage situation.

881
882
883
884
- **Hijacking or skyjacking**. Sometimes employed as a means of escape, hijacking is normally carried out to produce a spectacular hostage situation. Although trains, buses, and ships have been hijacked, aircraft are the preferred target because of their greater mobility and because they are difficult to penetrate during terrorist operations.

885
886
- **Propaganda**. Insurgents may disseminate propaganda using any form of media, including face-to-face talks.

887
888
- **Seizure**. Seizure usually involves a building or object that has value in the eyes of the audience. There is some risk to the perpetrator because security forces have time to react.

889
- **Raids or attacks on facilities**. Armed attacks on facilities are usually undertaken to—

890
  - Demonstrate the government's inability to secure critical facilities or national symbols.

891
  - Acquire resources (for example, robbery of a bank or armory).

892
  - Kill COIN or government personnel.

893
  - Intimidate the government and the populace.

894
895
896
897
898
899
900
- **Sabotage**. The objective in most sabotage incidents is to demonstrate how vulnerable a particular society or government is to terrorist actions. Industrialized areas are more vulnerable to sabotage than less highly developed societies. Utilities, communications, and transportation systems are so interdependent that a serious disruption of any affects all of them and gains immediate public attention. Sabotage of industrial or commercial facilities is one means of creating significant disruption while making a statement of future intent. Military facilities and installations, information systems, and information infrastructures may become targets of terrorist sabotage.

901
- **Denial and deception**.

902 ■ *Denial* consists of measures taken by the threat to block, prevent, or impair U.S. intelli-
903 gence collection. Examples include killing human intelligence sources.

904 ■ *Deception* involves deliberately manipulating information and perceptions in order to mis-
905 lead. Examples include providing false intelligence.

906 ● **Hoaxes.** Any insurgent or terrorist group that has established credibility can employ a hoax with
907 considerable success. A threat against a person's life causes that person and those associated
908 with that individual to devote time and efforts to security measures. A bomb threat can close a
909 commercial building, empty a theater, or delay an aircraft flight at no cost to the insurgent or ter-
910 rorist. False alarms dull the analytical and operational efficiency of key security personnel, thus
911 degrading readiness.

912 ● **Use of chemical, biological, radiological, or nuclear weapon.** Some insurgent groups may
913 possess chemical and biological weapons, and there is a potential for use of both chemical and
914 biological weapons in the future. These types of weapons, relatively cheap and easy to make,
915 may be used in place of conventional explosives in many situations. The potential for mass de-
916 struction and the deep-seated fear most people have for these weapons can be attractive to a
917 group wishing to attract international attention. Although an explosive nuclear device is ac-
918 knowledged to be beyond the financial and technical reach of most terrorist groups, a chemical
919 or biological weapon, or even a radiological dispersion device using nuclear contaminants, is
920 not. The technology is simple and the payoff is potentially higher than conventional explosives.

921 B-108. When analyzing possible courses of action, it is crucial that informational, political, and violent
922 actions all be evaluated. It is also important to understand the interrelationship between these activities.

923

Appendix C

Linguistic Support

U.S. military forces conducting COIN operations in foreign nations require linguistic support. Military intelligence (MI) units assigned to brigade and higher-level commands have organic interpreters (linguists) that perform human intelligence (HUMINT) and signal intelligence (SIGINT) functions. However, requirements for interpreters usually exceed organic capabilities, and commanders should focus on obtaining external interpreter support early.

C-1. When possible, interpreters should be U.S. military personnel or CAT II/III linguists. Unit intelligence officers should maintain a language roster at home station to determine which assigned personnel have linguistic capabilities prior to deployment. When requirements exceed organic capabilities, unit commanders can hire host nation (HN) personnel to support their operations. Contracted linguists can provide interpreter support and perform intelligence functions. They are categorized into the three following groups:

- CAT I linguists in most cases are locally hired and require vetting. They do not possess a security clearance. They are the most abundant resource pool; however, their skill level is limited. CAT I linguists should be used for basic interpretation requirements (patrols, base entrance coverage, Open Source Intelligence collection, and civil-military operations). A general planning factor of 30 – 40 CAT I linguists for an infantry battalion is recommended. Brigade headquarters should maintain a pool of Approximately 15 CAT I linguists for surge operations.
- CAT II linguists are U.S. citizens with a SECRET clearance. In most cases they possess good oral and written communication skills. They should be managed carefully due to limited availability, and should generally be used as interpreters for battalion level and higher commanders, or Tactical HUMINT Teams (THT). A general planning factor for CAT II linguists is one linguist for the brigade commander, one for each infantry battalion commander, and approximately 10 linguists for the supporting MI company (three for each THT/Operations Management Team and two for each SIGINT collection platform).
- CAT III linguists are U.S. citizens with a TOP SECRET clearance. They are a scarce commodity, and they are often retained at division and higher levels of command. They generally have excellent oral and written communications skills.

C-2. Some linguistic support is obtained through contracts with private companies, such as Titan Corporation and World Wide. The required statement of work and/or contract should define the linguist requirement and the units' responsibilities. Contracted CAT II/III linguists should arrive with their own equipment items (for example, flak vest, Kevlar, and desert combat uniform). A unit linguist manager should be formally designated as early as possible to identify language requirements and manage distribution of assets. In addition, contractor site managers are generally located at the division level to manage personnel issues (leave, vacation, pay, and equipment).

C-3. The guidelines listed below should be considered when hiring HN personnel to perform CAT I linguist requirements as interpreters.

SELECTING INTERPRETERS

C-4. Every effort should be made to vet interpreters prior to hiring them. They should be selected based on the following criteria:

- **Native Speaker**. Interpreters should be native speakers of the socially or geographically determined dialect. Their speech, background, and mannerisms should be completely acceptable to

45 the target audience so that no attention is given to the way they talk, only to what they say. This
46 capability also allows them to better distinguish dialects of different regions and provinces, to
47 identify personnel from other countries or not from the local area.

48 ● **Social Status and Ethno-Religious Identity**. In some situations and cultures, interpreters may
49 be limited in their effectiveness if their social standing is considerably lower than that of the au-
50 dience. Examples include significant differences in military rank or membership in a shunned
51 ethnic or religious group. Soldiers and Marines should remember that their job is to communi-
52 cate with the local population. Exercise a tolerance of local prejudices and choose an interpreter
53 who is least likely to cause suspicion or miscommunication. Interpreters should also have a good
54 reputation in the community and should not be intimidated when dealing with important audi-
55 ences.

56 ● **English Fluency**. An often-overlooked consideration is how well the interpreter speaks English.
57 As a rule, if the interpreter understands the speaker and the speaker understands the interpreter,
58 then the interpreter's command of English is satisfactory. Soldiers and Marines can check that
59 understanding by saying something to the interpreter and requesting it be paraphrased back in
60 English.

61 ● **Audience Understanding**. Interpreting goes both ways. Interpreters should be able to accu-
62 rately convey information expressed by interviewees or the target audience. This is especially
63 important when commanders are speaking with HN civilian leaders and military personnel. Lin-
64 guists involved in military discussions should understand military terms and doctrine.

65 ● **Intellectual Capabilities**. Interpreters should be quick and alert, able to respond to changing
66 conditions and situations. They should be able to grasp complex concepts and discuss them
67 without confusion in a reasonably logical sequence. Although education does not equate to intel-
68 ligence, it does expose students to diverse and complex topics. As a result, generally speaking,
69 the better educated the interpreter, the better he or she performs.

70 ● **Technical Ability**. In certain situations, Soldiers and Marines may need interpreters with tech-
71 nical training or experience in special subject areas. This type of interpreter is able to translate
72 the meaning as well as the words. For instance, if the subject is nuclear physics, background
73 knowledge is useful.

74 ● **Reliability**. Soldiers and Marines should beware of a potential interpreter who arrives late for
75 the interview. Throughout the world, the concept of time varies widely. In many countries, time-
76 liness is relatively unimportant. Soldiers and Marines should make sure that interpreters under-
77 stand the importance of punctuality.

78 ● **Loyalty**. If interpreters are local nationals, it is safe to assume that their first loyalty is to the HN
79 or ethnic group, not to the U.S. military. The security implications are clear. Soldiers should be
80 very cautious in how they explain concepts and what information interpreters can overhear
81 about operations. Additionally, some interpreters, for political or personal reasons, may have ul-
82 terior motives or a hidden agenda when they apply for an interpreting job. Soldiers and Marines
83 who detect or suspects such motives should tell the commander or security manager.

84 ● **Gender, Age, Race and Ethnicity**. Gender, age, and race can seriously affect the mission effec-
85 tiveness of interpreters. In predominantly Muslim countries, cultural prohibitions may render a
86 female interpreter ineffective in communicating with males, while a female interpreter may be
87 required to communicate with females. In regions featuring ethnic strife such as the Balkans,
88 ethnic divisions may limit the effectiveness of an interpreter from outside the target audience.
89 Since traditions, values, and biases vary from country to country, it is important to conduct a
90 thorough cultural analysis to determine the most favorable characteristics for interpreters.

91 ● **Compatibility**. The target audience quickly recognizes personality conflicts between Soldiers
92 and Marines and their interpreters. Such friction can undermine the effectiveness of the commu-
93 nication effort. If possible, when selecting interpreters, Soldiers and Marines should look for
94 compatible traits and strive for a harmonious working relationship.

95 C-5. If several qualified interpreters are available, Soldiers and Marines should select at least two. This
96 practice is of particular importance if the interpreter is used during long conferences or courses of instruc-
97 tion. When two interpreters are available, they should work for one-half hour periods. Due to the mental

98 strain associated with this task, four hours of active interpreting a day is usually the maximum that can be
99 performed before effectiveness declines. In the case of short duration meetings and conversations, when
100 two or more interpreters are available, one can provide quality control and assistance for the active inter-
101 preter. Additionally, this technique is useful when conducting coordination or negotiation meetings, as one
102 interpreter can be used in an active role while the other pays attention to the body language and side con-
103 versations of the audience. Many times, Soldiers and Marines can gain important auxiliary information that
104 assists in negotiations from listening to what others are saying among themselves outside of the main dis-
105 cussion.

106 C-6. Once interpreters are hired, security measures should be emplaced to ensure their safety and the
107 safety of their families. Insurgents know the value of good interpreters, and often try to intimidate or kill
108 interpreters and their family members. Interpreters may also be coerced to gather information on U.S. op-
109 erations; therefore, an aggressive program against subversion and espionage should be emplaced, to in-
110 clude use of a polygraph if available.

111 C-7. Certain tactical situations may require the use of uncleared indigenous personnel as "field expedient"
112 interpreters. Commanders should recognize the increased security risk involved in using such personnel
113 and carefully weigh the risk versus potential gain. If uncleared interpreters are used, discussions of sensi-
114 tive information should be kept to a minimum.

ESTABLISHING RAPPORT

116 C-8. Interpreters are a vital link between Soldiers, Marines and the target audience. Without supportive,
117 cooperative interpreters, the mission could be in serious jeopardy. Mutual respect and understanding is es-
118 sential to effective teamwork. Soldiers and Marines should establish and maintain rapport early and
119 throughout the operation. Problems establishing rapport stem mostly from a lack of personal communica-
120 tion skills and misunderstandings regarding culture.

121 C-9. Soldiers and Marines begin the process of establishing rapport before they meet interpreters by
122 studying to gain an understanding of the area of operations and its inhabitants. This process was discussed
123 in Chapter 3. Many foreigners have some knowledge about the United States. Unfortunately, much of this
124 is gained from commercial movies and television shows. So Soldiers and Marines may need to teach the in-
125 terpreter something about the United States, as well.

126 C-10. Soldiers and Marines working with an interpreter should research and verify the interpreter's back-
127 ground. They should demonstrate a genuine interest in the interpreter's personal information, such as fam-
128 ily, aspirations, career, and educational level. Many cultures emphasize family roles differently than in the
129 United States, so Soldiers and Marines should begin with understanding the interpreter's home life.
130 Though Soldiers and Marines should gain as much cultural information as possible before deploying to a
131 HN, their interpreters can be valuable sources to fill gaps. However, information received from interpreters
132 is likely to represent the views of the group of which they are a member. Members of opposing groups al-
133 most certainly see things differently, and often have a very different view of culture and history.

134 C-11. Soldiers and Marines should gain an interpreter's trust and confidence before discussing sensitive is-
135 sues, such as religion, likes, dislikes, and prejudices. Soldiers and Marines should approach these areas
136 carefully. Although deeply-held personal beliefs may be very revealing and useful in a professional rela-
137 tionship, Soldiers and Marines should gently and tactfully draw these out of their interpreters.

138 C-12. One way to reinforce the bond between military personnel and their interpreter is to make sure the
139 interpreter has every comfort available. This includes personal protection such as boots, helmets and body
140 armor that the interpreter may not necessarily possess when he or she arrives. Providing interpreters with
141 base comforts (for example, shelter, air conditioning, and heat) is a fundamental requirement if the majority
142 of the military personnel already have them. If and when an interpreter is assigned to a specific unit, the in-
143 terpreter ought to live with that organization to develop a bond. If there are a number of assigned interpret-
144 ers, it may be more effective for the interpreters to live together on the unit compound.

145 # ORIENTING INTERPRETERS

146 C-13. Early in the relationship with interpreters, Soldiers and Marines should explain the nature of the in-
147 terpreters' duties, standards of conduct expected, interview techniques to be used, and any other require-
148 ments and expectations. The orientation may include the following:

149 • Current tactical situation.
150 • Background information obtained on the source, interviewee, or target audience.
151 • Specific objectives for the interview, meeting, or interrogation.
152 • Method of interpretation to be used—simultaneous or consecutive:
153 ▪ Simultaneous—when the interpreter listens and translates at the same time (Not rec-
154 ommended).
155 ▪ Consecutive—when the interpreter listens to an entire phrase, sentence, or paragraph,
156 then translates during natural pauses.
157 • Conduct of the interview, lesson, or interrogation.
158 • Need for interpreters to avoid injecting their own personality, ideas, or questions into the inter-
159 view.
160 • Need for the interpreter to inform the interviewer about inconsistencies in language used by in-
161 terviewee. An example would be someone who claims to be a college professor, yet speaks like
162 an uneducated person. During interrogations or interviews, this information is used as part of the
163 assessment of the information obtained from the individual.
164 • Physical arrangements of site, if applicable.
165 • Potential need for the interpreter to assist in after-action reviews or assessments.

166 # PREPARING FOR PRESENTATIONS

167 C-14. Sites for interviews, meetings, or classes should be carefully selected and arranged. The physical ar-
168 rangement can be especially significant with certain groups or cultures.

169 C-15. Other unique cultural practices should also be identified before interviewing, instructing, or confer-
170 ring with foreign nationals. For example, speakers and interpreters should know when it is proper to stand,
171 sit, or cross one's legs. Gestures are a learned behavior and vary from culture to culture. If properly se-
172 lected, interpreters should be very helpful in this regard.

173 C-16. Interpreters should be instructed to mirror the speaker's tone and personality and not to interject
174 their own questions or emotions. Speakers should instruct interpreters to discreetly inform them if they no-
175 tice inconsistencies or peculiarities of speech, dress, and behavior from the audience.

176 C-17. The preceding points have implied the need for a careful analysis of the target audience. This type of
177 analysis goes beyond the scope of this appendix. Mature judgment, thoughtful consideration of the target
178 audience as individuals, and a genuine concern for their receiving accurate information contributes sub-
179 stantively to accomplishing the mission. Soldiers and Marines should remember that a farmer from a small
180 village is going to have markedly different expectations and requirements than a city businessman.

181 C-18. Soldiers and Marines should remember that working through an interpreter may take double or triple
182 the amount of time normally required for an event. Some time may be saved by providing the interpreter
183 beforehand with the briefing slides, questions to be asked, a lesson plan, or copies of any handouts, as well
184 as a glossary of difficult terms.

185 # CONDUCTING PRESENTATIONS

186 C-19. As part of the initial training for interpreters, Soldiers and Marines should emphasize that interpret-
187 ers should follow their speaker's lead, while also serving as a vital communication link between the
188 speaker and the target audience. Soldiers and Marines should appeal to the interpreters' professional pride
189 by clearly describing how the quality and quantity of the information sent and received is directly depend-

190
191
ent on the interpreters' skills. Although interpreters perform some editing as a function of the interpreting process, it is imperative that they transmit the exact meaning without additions or deletions.

192
193
194
195
196
197
198
199
200
C-20. Whether conducting an interview or presenting a lesson, speakers should avoid simultaneous translations - both the speaker and interpreter talking at the same time. Speakers should talk directly to the individual or audience for a minute or less in a neutral, relaxed manner. The interpreter should watch the speaker carefully and, during the translation, mimic the speaker's body language as well as interpret his or her verbal meaning. Speakers should observe interpreters closely to detect any inconsistencies between their manners. Speakers should present one major thought in its entirety and allow the interpreter to reconstruct it in his or her language. One way to ensure that the interpreter is communicating exactly what the speaker means is to have a senior interpreter observe several conversations and provide feedback along with further training.

201
202
203
204
205
C-21. Soldiers and Marines should be aware some interpreters might attempt to save face or to protect themselves by purposely concealing their lack of understanding. They may attempt to translate what they believe the speaker or audience said or meant without asking for a clarification. This situation can result in misinformation and confusion, and impact on credibility. Soldiers and Marines should ensure interpreters know that when in doubt, they should always ask for clarification.

206
207
208
209
C-22. During an interview or lesson, if questions are asked, interpreters should immediately relay them to the speaker for an answer. Interpreters should never attempt to answer questions, even though they may know the correct answer. Additionally, neither speakers nor interpreters should correct each other in front of an interviewee or class; all differences should be settled away from the subject or audience.

210
211
212
213
C-23. Just as establishing rapport with the interpreter is vital, establishing rapport with interview subjects or the target audience is equally important. Speakers and their interpreters should concentrate on this task. To establish rapport, subjects or audiences should be treated as mature, important people who are worthy and capable.

214
215
216
217
218
C-24. There are several methods for ensuring the speaker is communicating to the target audience and the interpreter is only a mechanism for that communication. One technique is to have the interpreter stand to the side of and just behind the speaker, leaving the speaker face-to-face with the target audience. The speaker should always look at the target audience and talk directly to them, rather than to the interpreter. This method allows the speaker and the target audience to establish a personal relationship.

219
SPEAKING TECHNIQUES

220
221
222
223
224
225
C-25. An important first step for Soldiers and Marines communicating in a foreign language is to reinforce and polish their English language skills. These skills are important, even when no attempt has been made to learn the HN language. The clearer Soldiers and Marines speak in English, including the use of correct words, without idiom or slang, the easier it is for interpreters to translate exactly. For instance, speakers may want to add words usually left out in colloquial English, such as the "air" in airplane, to ensure that they are not misinterpreted as referring to the Great Plains or a carpenter's plane.

226
227
228
C-26. Speakers should not use profanity at all and should avoid slang. In many cases, such expressions cannot be translated. Even those that can be translated do not always retain the desired meaning. Terms of surprise or reaction such as "gee whiz" and "golly" are difficult to translate.

229
230
231
232
233
234
235
C-27. Speakers should avoid using acronyms. While these have become part of everyday military language, in most cases interpreters and target audiences are not be familiar with them, and it becomes necessary for the interpreter to interrupt the interview for clarification regarding the expanded form. This can disrupt the rhythm of the interview or lesson. Moreover, if interpreters should constantly interrupt the speaker for clarification, they could lose credibility in the eyes of the target audience, which could jeopardize the goals of the interview or lesson. In addition, if a technical term or expression should be used, speakers should be sure interpreters convey the proper meaning. This preparation is best done in advance.

236
237
238
C-28. When speaking extemporaneously, Soldiers and Marines should consider, in advance, a framework of themes for what they wish to say. They should break their thoughts into logical bits and articulate them one at a time, using short, simple words and sentences, which can be translated quickly and easily. As a

239 rule of thumb, speakers should never say more in one sentence than they can easily repeat word for word
240 immediately after saying it. Each sentence should contain a complete thought without excess verbiage.

241 C-29. Speakers should avoid American "folk" and culture-specific references. Target audiences may have
242 no idea what is being talked about. Even when interpreters understand the reference, they may find it diffi-
243 cult to quickly identify an appropriate equivalent in the target audience's cultural frame of reference.

244 C-30. Transitional phrases and qualifiers may confuse non-native speakers and waste valuable time. Exam-
245 ples are "for example," "in most cases," "maybe," and "perhaps."

246 C-31. Speakers should be cautious about using American humor because humor is culturally specific and
247 doesn't translate well. Cultural and language differences can lead to misinterpretations by foreigners.

248 C-32. To briefly summarize these suggested techniques, speakers should—
249 ● Keep presentations as simple as possible.
250 ● Use short sentences and simple words (in clear context).
251 ● Avoid idiomatic and slang English and colloquial expressions.
252 ● Avoid flowery and nuanced language that would be difficult for a foreign national to interpret
253 exactly.
254 ● Avoid American "folk" and culture-specific references.

255 SUMMARY

256 C-33. The following are some "dos" and "don'ts" for speakers to consider when working with interpreters.

257 C-34. Speakers should—
258 ● Position the interpreter by their side (or a step back). This keeps the subject or target audience
259 from shifting their attention or fixating on the interpreter rather than on the leader.
260 ● Always look at and talk directly to the subject or target audience. Guard against the tendency to
261 talk to the interpreter.
262 ● Speak slowly and clearly. Repeat as often as necessary.
263 ● Speak to the individual or group as if they understand English. Be enthusiastic and employ the
264 gestures, movements, voice intonations, and inflections that would normally be used before an
265 English-speaking group. Considerable non-verbal meaning can be conveyed through voice and
266 body movements. Encourage interpreters to mimic the same delivery.
267 ● Periodically check an interpreter's accuracy, consistency, and clarity. Request a U.S. citizen
268 who possesses a degree of fluency in the language to sit in on a lesson or interview to ensure the
269 translation is not distorted, intentionally or unintentionally. Speakers learning some of the lan-
270 guage also is very useful in this regard.
271 ● Check with the audience whenever misunderstandings are suspected and clarify immediately.
272 Using the interpreter, ask questions to elicit answers that indicate whether the point is clear. If it
273 is not, re-phrase the instruction and illustrate the point differently. Use repetition and examples
274 whenever necessary to facilitate learning. If the target audience asks few questions, it may mean
275 the instruction is not understood or the message is not clear to them.
276 ● Ensure interpreters understand they are valuable members of the team. Recognize them com-
277 mensurate with the importance of their contributions. Protect interpreters as they are invaluable
278 assets that may be targeted by the insurgency and other criminal elements.

279 C-35. Speakers should not—
280 ● Address the subject or audience in the third person through the interpreter. For example, avoid
281 saying, "Tell them I'm glad to be their instructor." Instead, address the subject or audience di-
282 rectly by saying, "I am glad to be your instructor." Make continual eye contact with the audi-
283 ence. Watch them, not the interpreter.
284 ● Make side comments to the interpreter that are not interpreted. This action is rude and discourte-
285 ous and creates the wrong atmosphere for communication.

286
287
288
289
290
291

- Be a distraction while the interpreter is translating and the subject or target audience is listening. Speakers should not pace, write on the blackboard, teeter on the lectern, drink beverages, or carry on any other distracting activity while the interpreter is translating.

1   **Appendix D**

2   # Legal Considerations

3   Law and policy govern the actions of the United States Armed Forces in all military
4   operations, including counterinsurgency (COIN). There must be a legal basis for U.S.
5   forces to conduct operations, and this legal basis has a profound influence on many
6   aspects of the operations themselves, including the rules of engagement (ROE), the
7   military's participation in organizing and training foreign forces, the authority to
8   spend funds for the benefit of the host nation (HN) confronting insurgency, the au-
9   thority of U.S. forces to detain and interrogate, and a range of other matters. Under
10   the Constitution, the President is Commander-in-Chief of the Armed Forces. There-
11   fore, orders issued by the President or the Secretary of Defense to a combatant com-
12   mander provide the starting point in determining the legal basis. This appendix sum-
13   marizes some of the most significant laws and policies that also bear upon U.S.
14   military operations in support of foreign COIN. Laws are legislation passed by Con-
15   gress and signed into law by the President, as well as treaties to which the United
16   States is party. Policies are executive orders, departmental directives and regulations,
17   and other authoritative statements issued by duly elected or appointed government of-
18   ficials. No summary of the sort provided here can substitute for consultation with the
19   unit's supporting Staff Judge Advocate.

20   ## AUTHORITY TO ASSIST A FOREIGN GOVERNMENT

21   ### AUTHORITY FOR FOREIGN INTERNAL DEFENSE

22   D-1.   Without ever receiving a deployment or execute order from the President or Secretary of De-
23   fense, U.S. forces may be authorized to make limited contributions to U.S. Government efforts in sup-
24   port of a foreign nation's COIN. This might occur if participation of military forces is requested by the
25   Secretary of State and approved by the Secretary of Defense through standing statutory authorities in
26   Title 22, United States Code, which contains the Foreign Assistance Act, the Arms Export Control
27   Act, and other laws that authorize security assistance, developmental assistance and other forms of bi-
28   lateral aid. It might also occur under a variety of provisions in Title 10, United States Code, which au-
29   thorizes certain types of military-to-military contacts, exchanges, exercises, and limited forms of hu-
30   manitarian and civic assistance in coordination with the U.S. Ambassador for the country. In such
31   situations, U.S. military personnel are present in the HN as administrative and technical personnel of
32   the U.S. diplomatic mission or pursuant to a status of forces agreement (SOFA) or exchange of letters.
33   This type of cooperation and assistance is limited to liaison, contacts, training, equipping, and provid-
34   ing defense articles and services. It does not include direct involvement in operations.

35   ### DOD USUALLY NOT LEAD—GENERAL PROHIBITION ON ASSISTANCE TO POLICE

36   D-2.   DOD is usually not the lead governmental department for assisting foreign governments, even
37   for the provision of security assistance—that is, military training, equipment, and defense articles and
38   services—to the host country's military forces. DOD contribution may be large, but the legal authority
39   is typically one exercised by the Department of State (DOS). With regard to provision of training to a
40   foreign government's police or other civil interior forces, the U.S. military typically has no authorized
41   role. The Foreign Assistance Act specifically prohibits assistance to foreign police forces except
42   within carefully circumscribed exceptions, and under a Presidential directive, and the lead role in pro-

43 viding police assistance within those exceptions has been normally delegated to the DOS's Bureau of
44 International Narcotics and Law Enforcement Affairs. However, the President did sign a decision di-
45 rective in 2004 granting authority to train and equip Iraqi police to the Commander, United States
46 Central Command (USCENTCOM).

47 # AUTHORIZATION TO USE MILITARY FORCE

48 ## CONGRESSIONAL RESOLUTION

49 D-3. Because Congressional support is necessary to the success of any prolonged involvement of
50 U.S. forces in actual operations overseas, the central legal basis for such involvement within domestic
51 law is often provided in a Congressional resolution. This is especially likely if U.S. forces are antici-
52 pated, at least initially, to be engaged in combat operations against an identified enemy or hostile
53 force.

54 ## STANDING WAR POWERS RESOLUTION

55 D-4. In the absence of a specific Congressional authorization for use of force, the President—without
56 conceding that the 1973 War Powers Resolution binds his own constitutional authority—makes a re-
57 port to Congress within 48 hours of introducing substantial U.S. forces into a foreign country detailing
58 the circumstances necessitating introduction or enlargement of troops, the Constitutional or legislative
59 authority upon which he bases his action, and the estimated scope and duration of the deployment or
60 combat action. The 1973 Resolution states that if Congress does not declare war or specifically author-
61 ize the deployment or combat action within 60 days of the report, the President is required to terminate
62 U.S. military involvement and redeploy U.S. Armed Forces.

63 # RULES OF ENGAGEMENT

64 ## OPERATION-SPECIFIC RULES OF ENGAGEMENT

65 D-5. ROE are directives issued by competent military authority that delineate the circumstances and
66 limitations under which U.S. forces initiate and/or continue combat engagement with other forces en-
67 countered. In a large-scale deployment, the Secretary of Defense may issue ROE that are specific to
68 the operation to a combatant commander. The combatant commander and subordinate commanders
69 then issue ROE that must be consistent with the ROE received from the Secretary of Defense. In addi-
70 tion to stating the circumstances under which Soldiers or Marines may open fire—that is, upon posi-
71 tive identification of a member of a hostile force or upon clear indications of hostile intent—the ROE
72 may include rules concerning when civilians may be detained, specify levels of approval authority for
73 using heavy weapons, or identify facilities that may be protected with deadly force. All ROE comply
74 with the law of war. ROE in COIN are dynamic and must be regularly reviewed for their effectiveness
75 in the complex COIN environment. ROE training for COIN forces should be reinforced regularly.

76 ## CJCS STANDING RULES OF ENGAGEMENT

77 D-6. In the absence of operation-specific ROE, U.S. forces apply CJCSI 3121.01A, Standing Rules
78 of Engagement (SROE) for United States Forces. The SROE establish fundamental policies and pro-
79 cedures governing the actions to be taken by U.S. force commanders in the event of military attack
80 against the United States and during all military operations, contingencies, terrorist attacks, or pro-
81 longed conflicts outside the territorial jurisdiction of the United States. The SROE do not limit a com-
82 mander's inherent authority and obligation to use all necessary means available and to take all appro-
83 priate action in self-defense of the commander's unit and other U.S. forces in the vicinity. The SROE
84 also prescribe the procedures by which supplemental ROE for specific operations are provided as well
85 as the format by which subordinate commanders may request ROE.

86 COALITION RULES OF ENGAGEMENT

87 D-7. When U.S. forces, under U.S. operational or tactical control, operate in conjunction with a mul-
88 tinational force, reasonable efforts are made to affect common ROE. If such ROE cannot be estab-
89 lished, U.S. forces operate under the SROE or operation-specific ROE provided by U.S. authorities.
90 To avoid misunderstanding, any differences in ROE or ROE interpretation among coalition forces
91 must be thoroughly discussed between commanders and disseminated throughout all units involved.

92 # THE LAW OF WAR

93 D-8. COIN and international armed conflicts often overlap. COIN may take place before, after, or
94 simultaneously with a war occurring between nations. In such situations, U.S. forces obey the law of
95 war, a body of international treaties and customs, recognized by the United States as binding, which
96 regulates the conduct of hostilities and protects noncombatants. The main law of war protections of the
97 Hague and Geneva Conventions applicable at the tactical and operational level are summarized in ten
98 rules:

99 • Soldiers and Marines fight only enemy combatants.
100 • Soldiers and Marines do not harm enemies who surrender. They disarm them and turn them over
101   to their superior.
102 • Soldiers and Marines do not kill or torture enemy prisoners of war.
103 • Soldiers and Marines collect and care for the wounded, whether friend or foe.
104 • Soldiers and Marines do not attack medical personnel, facilities, or equipment.
105 • Soldiers and Marines destroy no more than the mission requires.
106 • Soldiers and Marines treat civilians humanely.
107 • Soldiers and Marines do not steal. Soldiers and Marines respect private property and posses-
108   sions.
109 • Soldiers and Marines should do their best to prevent violations of the law of war.
110 • Soldiers and Marines report all violations of the law of war to their superior.

111 D-9. When insurgency occurs during occupation, the law of war includes rules governing situations
112 in which the military forces of one state occupy the territory of another. Occupation is not a transfer of
113 sovereignty, though it does confer upon the occupying power the authority and responsibility to re-
114 store and maintain public order and safety while respecting, as much as possible, the laws in force in
115 the country. One of the four Geneva Conventions of 1949—the Geneva Convention Relative to the
116 Protection of Civilian Persons in Time of War—becomes a prominent source of law during occupa-
117 tion.

118 # INTERNAL ARMED CONFLICT

119 D-10. Geneva Convention, Common Article 3: Although insurgencies can occur simultaneous with a
120 legal state of war between two nations, they are classically conflicts internal to a single nation, be-
121 tween uniformed government forces and armed elements that do not wear uniforms with fixed distinc-
122 tive insignia, carry arms openly, or otherwise obey the laws of war. As such, the main body of the law
123 of war does not strictly apply to these conflicts—a legal fact that can be a source of confusion to
124 commanders and soldiers. It bears emphasis, however, that one article contained in all four of the Ge-
125 neva Conventions—Common Article 3—is specifically intended to apply to internal armed conflicts:

126 *In the case of armed conflict not of an international character occurring in the territory*
127 *of one of the high contracting parties, each Party to the conflict shall be bound to apply,*
128 *as a minimum, the following provisions:*

129 • *Persons taking no active part in the hostilities, including members of armed forces*
130   *who have laid down their arms and those placed hors de combat by sickness,*
131   *wounds, detention, or any other cause, shall in all circumstances be treated hu-*
132   *manely, without any adverse distinction founded on race, color, religion or faith,*
133   *sex, birth or wealth, or any other similar criteria.*

134   ●   *To this end, the following acts are and shall remain prohibited at any time and in*
135   *any place whatsoever with respect to the above-mentioned persons:*
136   ▪   *Violence to life and person, in particular murder of all kinds, mutilation, cruel*
137   *treatment and torture.*
138   ▪   *Taking of hostages.*
139   ▪   *Outrages upon personal dignity, in particular, humiliating and degrading*
140   *treatment.*
141   ▪   *The passing of sentences and the carrying out of executions without previous*
142   *judgment pronounced by a regularly constituted court, affording all the judi-*
143   *cial guarantees which are recognized as indispensable by civilized peoples.*
144   ●   *The wounded and sick shall be collected and cared for.*
145   ●   *An impartial humanitarian body, such as the International Committee of the Red*
146   *Cross, may offer its services to the Parties to the conflict.*
147   ●   *The Parties to the conflict should further endeavor to bring into force, by means of*
148   *special agreements, all or part of the other provisions of the present Convention.*
149   ●   *The application of the preceding provisions shall not affect the legal status of the*
150   *Parties to the conflict.*

APPLICATION OF CRIMINAL LAWS OF THE HOST NATION

151

152   D-11. The final sentence of Common Article 3 makes clear that insurgents have no special status un-
153 der international law. They are not, when captured, prisoners of war and may legally be prosecuted as
154 criminals for bearing arms against the government and for other offenses, so long as they are accorded
155 the minimum protections described in Common Article 3. U.S. elements conducting COIN should re-
156 main mindful that the insurgents they encounter are, as a legal matter, criminal suspects within the le-
157 gal system of the foreign government. Weapons, witness statements, photographs, and other evidence
158 collected at the scene of their offenses must be carefully preserved so as to introduce them into the
159 criminal process and thus hold the insurgents accountable for their crimes while still promoting the
160 rule of law.

161   D-12. SOFAs establish the legal status of military personnel in foreign countries. Criminal and civil
162 jurisdiction, taxation, and claims for damages and injuries are a few of the topics usually covered in a
163 SOFA. In the absence of an agreement or some other arrangement with the HN, DOD personnel in for-
164 eign countries may be subject to the laws of the HN.

DETENTION AND INTERROGATION

165

DETAINEE TREATMENT ACT OF 2005

166

167   D-13. As Chapter 3, Chapter 5, and Chapter 7 (this manual) indicate, the need for human intelligence
168 in COIN operations can create great pressure to obtain time-sensitive information from detained indi-
169 viduals. Law clearly prohibits U.S. forces, including officials from other government agencies, from
170 certain methods in obtaining such information. In response to documented instances of detainee abuse,
171 including maltreatment involving interrogation, Congress passed, and the President signed into law,
172 the Detainee Treatment Act of 2005, which includes the following sections:

Section 2: Uniform Standards for the Interrogation of Persons Under the Detention of the
Department of Defense

173
174

175   D-14. In General—No person in the custody or under the effective control of the DOD or under deten-
176 tion in a DOD facility shall be subject to any treatment or technique of interrogation not authorized by
177 and listed in FM 34-52.

178   D-15. Applicability—Subsection (a) shall not apply with respect to any person in the custody or under
179 the effective control of the DOD pursuant to a criminal law or immigration law of the United States.

180 D-16. Construction—Nothing in this section shall be construed to affect the rights under the U.S. Con-
181 stitution of any person in the custody or under the physical jurisdiction of the United States.

182 **Section 3: Prohibition on Cruel, Inhuman, or Degrading Treatment or Punishment of Persons**
183 **Under Custody or Control of the United States Government**

184 D-17. In General—No individual in the custody or under the physical control of the U.S. Govern-
185 ment, regardless of nationality or physical location, shall be subject to cruel, inhuman, or degrading
186 treatment or punishment.

187 D-18. Construction—Nothing in this section shall be construed to impose any geographical limitation
188 on the applicability of the prohibition against cruel, inhuman, or degrading treatment or punishment
189 under this section.

190 D-19. Limitation on Supersedure—The provisions of this section shall not be superseded, except by a
191 provision of law enacted after the date of the enactment of this Act which specifically repeals, modi-
192 fies, or supersedes the provisions of this section.

193 D-20. Cruel, inhuman, or degrading treatment or punishment defined—in this section, the term "cruel,
194 inhuman, or degrading treatment or punishment" means the cruel, unusual, and inhumane treatment or
195 punishment prohibited by the fifth, eighth, and fourteenth amendments to the Constitution of the
196 United States, as defined in the U.S. reservations, declarations, and understandings to the United Na-
197 tions Convention Against Torture and Other Forms of Cruel, Inhuman, or Degrading Treatment or
198 Punishment established in New York, NY, December 10, 1984.

199 ### INTERROGATION FIELD MANUAL

200 D-21. The Detainee Treatment Act established FM 34-52 as the legal standard. No techniques other
201 than those prescribed by the field manual are authorized by U.S. military forces. Commanders must
202 ensure that interrogators are properly trained and supervised.

203 ### STANDARDS FOR DETENTION AND INTERNMENT

204 D-22. Regardless of the precise legal status of those persons captured, detained, or otherwise held in
205 custody by U.S. forces, they must receive humane treatment until properly released and are provided
206 the minimum protections of the Geneva Conventions. Prolonged detention should be accomplished by
207 specially trained, organized, and equipped military police units in adequately designed and resourced
208 facilities according to the detailed standards contained in AR 190-8. The military police personnel op-
209 erating such facilities shall not be used to assist in or "set the conditions for" interrogation.

210 ### TRANSFER OF DETAINEES TO HOST NATION

211 D-23. U.S. forces may not transfer custody of a detainee to the authorities of the host country or of any
212 other foreign government where there are substantial grounds for believing that he or she would be in
213 danger of being subjected to torture or inhumane treatment.

214 ## ENFORCING DISCIPLINE OF U.S. FORCES

215 ### UNIFORM CODE OF MILITARY JUSTICE

216 D-24. Although the vast majority of well-led and well-trained U.S. military personnel perform their
217 duties honorably and lawfully, history records that some commit crimes amidst the decentralized com-
218 mand and control, the strains of opposing a treacherous and hidden enemy, and the often complex
219 ROE that characterize the COIN environment. Uniformed personnel remain subject at all times to the
220 Uniform Code of Military Justice (UCMJ) and must be investigated and prosecuted, as appropriate, for
221 violations of orders, maltreatment of detainees, assaults, thefts, sexual offenses, destruction of prop-
222 erty, and other crimes, including homicides, that they may commit during COIN.

223 COMMAND RESPONSIBILITY

224 D-25. In some cases, a military commander may be deemed responsible for crimes committed by sub-
225 ordinates or others subject to his control. This situation arises when the criminal acts are committed
226 pursuant to the commander's order. The commander is also responsible if he has actual knowledge, or
227 should have knowledge, through reports received or through other means, that troops or other persons
228 subject to his control are about to commit or have committed a crime, and he fails to take the necessary
229 and reasonable steps to ensure compliance with the law or to punish violators.

230 GENERAL ORDERS

231 D-26. Orders issued by general officers in command during COIN likely include provisions, such as a
232 prohibition against drinking alcohol or against entering places of religious worship, important to main-
233 taining discipline of the force, to safeguarding the image of U.S. forces, and to promoting the legiti-
234 macy of the host government. These orders are readily enforceable under the UCMJ.

235 CIVILIAN PERSONNEL AND CONTRACTORS

236 D-27. Modern COIN operations feature large numbers of civilian employees of the U.S. government
237 as well as civilian personnel employed by government contractors. Although the means of disciplining
238 such persons for violations differ from the means of disciplining uniformed personnel, these civilians
239 may be made subject to general orders. They are also subject to U.S. laws and to the laws of the host
240 government, and may be prosecuted or receive adverse administrative action by the United States or
241 contract employers. DOD directives contain further policy and guidance pertaining to U.S. civilians
242 accompanying our forces in COIN.

243 # HUMANITARIAN RELIEF AND RECONSTRUCTION

244 D-28. In COIN, like all operations, commands require specific authority to expend funds. That au-
245 thority is normally found in the DOD Appropriations Act, specifically, operation and maintenance
246 funds. In recent COIN operations, commands have had at their disposal additional funds that Congress
247 appropriated for the specific purpose of dealing with the COIN. Recent examples include the com-
248 mander's emergency response program (CERP), the Iraq Relief and Reconstruction Fund, Iraq Free-
249 dom Fund, and Commander's Humanitarian Relief and Reconstruction Program funds.

250 DOD FUNDS GENERALLY NOT EXPENDABLE BY COMMANDERS FOR THIS PURPOSE

251 D-29. Congress specifically appropriates funds for foreign assistance. Most such funds are expended
252 by the United States Agency for International Development under legal authorities in Title 22, United
253 States Code. Relatively small amounts of money are authorized under provisions of Title 10, and ap-
254 propriated annually, for expenditure by commanders to provide humanitarian relief, disaster relief, or
255 civic assistance in conjunction with military operations. These standing authorities are narrowly de-
256 fined and generally require significant advance coordination within the DOD and the DOS. As such,
257 they are of limited value to ongoing COIN operations.

258 COMMANDERS EMERGENCY RESPONSE PROGRAM

259 D-30. Beginning in November of 2003, Congress authorized use of a specific amount of operations
260 and maintenance funds for a CERP in Iraq and Afghanistan. The legislation, which was renewed in
261 successive appropriations and authorization acts, specified that the funds could be spent for urgent
262 humanitarian relief and reconstruction projects that would immediately assist the Iraqi and Afghan
263 peoples within a commander's area of responsibility. Congress did not intend the funds to be used as
264 security assistance (that is, weapons, ammunition, supplies for security forces), as salaries for Iraqi or
265 Afghan forces or employees, as rewards for information, or as payments in satisfaction of claims made
266 by Iraqis or Afghanis against the United States, given that specific legislation must authorize such
267 payments. Still, CERP provided tactical commanders a ready source of cash for small-scale projects to

268 repair public buildings, clear debris from roadways, provide supplies to hospitals and schools, and
269 myriad other local needs. Because Congress had provided special authority for the program, normal
270 federal acquisition laws and regulations did not apply, and reporting requirements were minimal. It is
271 important to emphasize that CERP is not a standing program, and that any future similar program
272 should be governed by whatever specific legislative provision Congress chooses to enact. In any pro-
273 gram similar to CERP, it is imperative for commanders and staffs to make sound, well-coordinated de-
274 cisions on the expenditure of the funds to ensure that maximum goodwill is created, that harmful ef-
275 fects are not caused in the local economy due to the infusion of cash (such as unsustainable wages that
276 divert skilled labor from a host government program essential to its legitimacy), that projects can be
277 responsibly administered to achieve the desired objective, and that insurgents are not being inadver-
278 tently financed.

279 # TRAINING AND EQUIPPING FOREIGN FORCES

280 ## NEED FOR SPECIFIC AUTHORITY

281 D-31. All training and equipping of foreign security forces must be specifically authorized. Usually,
282 DOD involvement is limited to a precise level of man-hours and materiel requested from the DOS un-
283 der the Foreign Assistance Act. The President may, on rare occasions, authorize deployed U.S. forces
284 to train or advise HN security forces as part of the operational mission, in which case DOD personnel
285 and operations/maintenance appropriations provide an incidental benefit to those security forces. All
286 other weapons, training, equipment, logistical support, supplies, and services provided to foreign
287 forces must be paid for with funds appropriated by Congress for that purpose. Examples include the
288 Iraq Security Forces Fund and the Afghan Security Forces Fund of fiscal year 2005. Moreover, DOD
289 must be given specific authority by the President for its role in such "train and equip" efforts. For in-
290 stance, in May of 2004, the President signed a decision directive that made the Commander,
291 USCENTCOM, under policy guidance from the Chief of Mission, responsible for coordinating all
292 U.S. Government efforts to organize, train, and equip Iraqi Security Forces, including police. Absent
293 such a directive, DOD lacks authority to take the lead in assisting a HN to train and equip its security
294 forces.

295 ## HUMAN RIGHTS VETTING

296 D-32. Congress typically prohibits expenditure of funds for foreign security force training or equip-
297 ment if the DOS has credible information that the foreign security force unit identified to receive the
298 training or equipment has committed a gross violation of human rights. Such prohibitions impose a re-
299 quirement upon DOS and DOD to vet the proposed recipient units against a database of credible re-
300 ports of human rights violations.

301 # CLAIMS AND SOLATIA

302 ## FOREIGN CLAIMS ACT

303 D-33. Under the Foreign Claims Act, claims by HN civilians for property losses, injury, or death
304 caused by Service members or the civilian component of the U.S. forces may be paid to promote and
305 maintain friendly relations with the HN. Claims that result from noncombat activities or negligent or
306 wrongful acts or omissions are also payable. Claims that are not payable under the Foreign Claims Act
307 include losses from combat, contractual matters, domestic obligations, and claims which are either not
308 in the best interest of the U.S. to pay, or which are contrary to public policy. Because payment of
309 claims is specifically governed by law and because many claims prove, upon investigation, to be not
310 payable, U.S. forces must be careful not to raise expectations by promising payment.

311 SOLATIA

312 D-34. If U.S. forces are conducting COIN in a country where payments in sympathy or recognition of
313 loss are common, solatia payments to accident victims may be legally payable. Solatia payments are
314 not claims payments. They are payments in money or in kind to a victim or to a victim's family as an
315 expression of sympathy or condolence. The payments are customarily immediate, and generally nomi-
316 nal. The individual or unit involved in the damage has no legal obligation to pay; compensation is
317 simply offered as an expression of remorse in accordance with local custom. Solatia payments should
318 not be made without prior coordination with the combatant command.

319 # ESTABLISHING THE RULE OF LAW

320 D-35. Establishment of the "rule of law" is a key goal/end-state in COIN. Defining that end-state re-
321 quires extensive coordination between the instruments of U.S. power, the HN, and coalition partners.
322 Additionally, attaining that end-state is usually the province of HN authorities, international organiza-
323 tions, DOS, and other U.S. government agencies, with support from the military in some cases. Some
324 key aspects of the rule of law, include:

325 • A government that derives its powers from the governed and competently manages, coordinates,
326 and sustains collective security, as well as political, social, and economic development. This in-
327 cludes local, regional, and national government.

328 • Sustainable security institutions. These include a civilian-controlled military, as well as police,
329 court and penal institutions (the latter should be perceived by the local populace as being fair,
330 just, and transparent).

331 • Fundamental human rights. The UN Declaration on Human Rights and the International Con-
332 vention for Civil and Political Rights provide a guide for applicable human rights; the latter pro-
333 vides for derogation from certain rights, however, during a state of emergency. Respect for the
334 full panoply of human rights should be the goal of the host government; derogation and viola-
335 tion of these rights by HN security forces, in particular, often provides an excuse for insurgent
336 activities.

337 D-36. In periods of extreme unrest and insurgency, HN legal structures (for example, courts, prosecu-
338 tors, defense assistance, and prisons) may cease to exist or function at any level. Under these condi-
339 tions, COIN forces may need to undertake a significant role in the reconstruction of the HN judicial
340 system in order to establish legal procedures and systems for dealing with captured insurgents and
341 common criminals. During judicial reconstruction, COIN forces can expect to be substantially in-
342 volved in providing sustainment and security support as well as legal support and advice to the HN ju-
343 dicial entities. Even when judicial functions are restored, COIN forces, may still have to provide logis-
344 tical and security support to judicial activities for a prolonged period of time if insurgents continue to
345 demonstrate keen interest in disrupting all activities supporting the legitimate "rule of law."
346

Appendix E

Social Network Analysis

Social network analysis (SNA) is characterized by a distinct and unique methodology for collecting data, performing statistical analysis, and making visual representations. Such applications can be useful for devising more effective schemes for promoting ideas or exerting influence in organizations. These are certainly important functions, but the relevance of such analysis to counterinsurgency (COIN) primarily deals with explaining how people behave and how that behavior is affected by their relationships. In the past, SNA contributed to the British success in defeating the Malaysian insurgency. More recently in Iraq, it has been used in the calming of the Fallujah region by the U.S. Marine Corps, and in the capture of Saddam Hussein by the 4th Infantry Division.

NETWORKS AND INSURGENTS

E-1. Most governmental, business, and volunteer organizations today are operating increasingly as networks. In a network organization, independent or semi-independent people and groups act as independent nodes, link across boundaries, and work together for a common purpose. The network organization has multiple leaders, lots of voluntary links and many interacting levels. Even the U.S. military is sliding towards a network organization as junior leaders use cell phones and internet connections to solve problems and resolve conflicts without going up the chain of command.

E-2. The differences between a network organization and a traditional organization are that a network—

- Gains authority not from a hierarch but from an individual's recognized knowledge and skill.
- Links people, groups, and teams across conventional boundaries.
- Contains members and structures that adapt to changing circumstances.
- Accomplishes activities because of a sense of mutual responsibility versus following orders.
- Explores ways to work effectively versus following predefined processes.
- Re-adjusts or disbands teams as needed.

E-3. Whether or not a group is intentionally organized as a network, a group operating as a network has specific enabling characteristics. These characteristics are counter-intuitive to traditional military analysts and commanders who have trained within the efficiency of a hierarchy with strict senior-subordinate relationships and established procedures.

E-4. Commanders facing an insurgency are probably dealing with an enemy organized as a network. No longer can analysts just use an organizational chart to describe the enemy configuration. It is increasingly unlikely that we can expect the enemy to exhibit a coherent pattern of activity that would result from a single leader with a clearly hierarchical subordinate structure. It is also much more difficult for a commander to readily differentiate the enemy from members of the population.

E-5. This appendix provides a tool for understanding an insurgency and demonstrates SNA-based metrics that assist a commander in gauging his performance. SNA supports the commander's understanding of the COIN battle-space.

39
40

The Capture of Saddam Hussein

41
42
43
44
45
46
47
48

The capture of Saddam Hussein in December 2003 was the result of hard work along with continuous intelligence gathering and analysis. Each day another piece of the puzzle fell into place, which led to coalition forces identifying and locating more of the key players in the insurgent network – both highly visible ones like Saddam Hussein, and lesser ones who sustained and supported the insurgency on a daily basis. This process produced very detailed diagrams that showed the structure of Saddam Hussein's personal security apparatus and the relationships among the persons identified.

49
50
51
52
53
54
55
56

The intelligence analysts and commanders in the 4th Infantry Division spent the summer of 2003 building "link diagrams;" graphics showing everyone related to Hussein by blood or tribe Those family diagrams led counterinsurgent forces to lower-level, but nonetheless highly trusted, relatives and clan members harboring Hussein and helping him move around the countryside. The circle of bodyguards and mid-level military officers, drivers, and gardeners protecting Hussein was described as a "Mafia organization," where access to Hussein controlled relative power within the network.

57
58
59
60
61

Over the days and months, coalition forces tracked how the enemy operated. Analysts traced trends and patterns, examined the tactics the enemy employed, and related the enemy tendencies to the names and groups on the tracking charts. This process involved making continual adjustments to the network template and constantly evaluating which critical data points were missing.

62
63
64
65
66
67
68

Late in the year, a series of operations produced an abundance of new intelligence about the insurgency and the whereabouts of Hussein. Commanders then designed a series of raids to capture key individuals and leaders of the former regime who could eventually lead counterinsurgent forces to Hussein. Each mission gained additional information which shaped the next raid. This cycle continued as a number of mid-level leaders of the former regime were caught, eventually leading coalition forces into Hussein's most trusted inner circle, and finally to his own capture.

69 # PERFORMING SOCIAL NETWORK ANALYSIS

70
71
72

E-6.   The social network graph is the building block of SNA. A social network graph consists of individuals and connections between them. In a social network graph, individuals are represented as nodes and relationships are represented by links between the nodes.

73
74
75
76
77
78
79
80
81
82
83

E-7.   Figure E-1 (below) shows a simple social network of key individuals and relationships. The nodes in this data set are from a modified, sub-network of the link diagram representing Saddam Hussein and his connections to various family members, former regime members, friends, and associates. The original diagram contained hundreds of names and took shape on a large 36-inch by 36-inch board. Each "box" in the network contained personal information on the particular individual. This included assigning roles and positions to certain people within the network – for example, chief of staff, chief of operations, and personal secretary. These were not necessarily positions the individuals occupied prior to the fall of Hussein, but instead were based on an understanding of the role they were filling in support of the insurgency or Saddam's "underground" operations. These roles were assigned based on an assessment of various personalities and recent intelligence and information reports. Such a process helped coalition forces focus their efforts in determining those who were closest to Hussein and their importance.

84
85
86

E-8.   For an insurgency, a social network is not just a description of who is in the insurgent organization, but a picture of the population, how they are put together, and how they interact with one another. Often, social networks are large, complex, and amorphous. They can be beyond the cognitive limitations of a hu-

87
88
man analyst. Figure E-2 (below) depicts a large, complex network. It attempts to show a graph of every person and relationship found in Baghdad-area newspapers on 4 December 2005.

89
90
91
92
93
94
E-9. SNA enables the commander to simultaneously understand individual and group-level performance. SNA allows for the weighting of relationships to represent the strength of the relationships, information capacity, rates or flow of traffic, distance between nodes, and probabilities of information being passed. It is these values that allow social network analysts to quantify the relationships, thus presenting a means for mathematically testing the network.

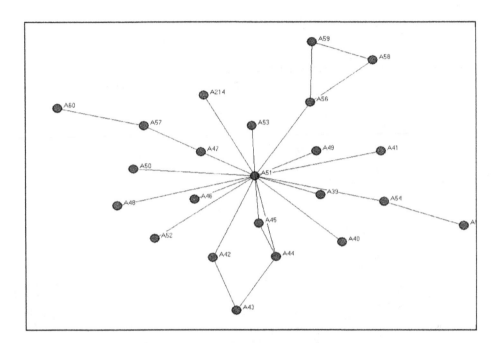

95
Figure E-1. Simple network

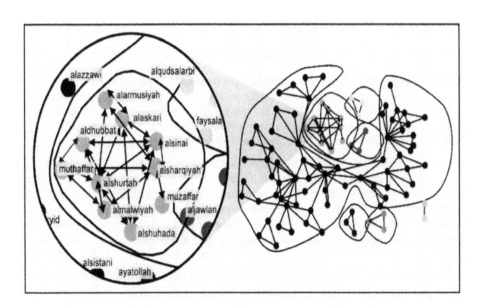

96
Figure E-2. Large complex network

97 E-10. Computers and software support SNA. They assist the commander in finding the target among the
98 noise. Social network algorithms are not complex. However, the software and computer are necessary to
99 crunch the large amount of data in real-time. If an insurgency is so small that the graph can fit on a small
100 map board, full SNA is still helpful. For the diagrams and analysis in this appendix, two readily download-
101 able SNA programs, UCINET and ORA, were used. Both programs are extremely flexible and support the
102 myriad of node-types (people, organizations, places, activities) and relationships that exist in a regional in-
103 surgency. Less flexible, but more specialized software such as Crimelink and Analyst's Notebook are
104 available in computer systems at the brigade and above.

105 # SOCIAL NETWORK DATA COLLECTION

106 E-11.   To draw an accurate picture of an insurgent network, we need to identify ties between the mem-
107 bers. Strong bonds formed over time either by family, friendship, or organizational association characterize
108 these ties. Units gather information on these ties by pouring over historical documents and records, inter-
109 viewing individuals in the network (and those outside of it), and looking over photos and books. It is pains-
110 taking work, but there is really no alternative when we our trying to piece together a network that does not
111 want to be identified.

112 E-12. The basic element of a social network graph is the pair or dyad. A dyad consists of two nodes and a
113 single line. In the simplest form of SNA, the two nodes represent people and the line represents a relation-
114 ship between the two people. A large number of interconnected dyads make the organizational graph. (See
115 Figure E-3, below.)

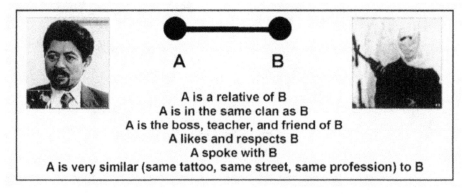

116 **Figure E-3. Dyad examples**

117 E-13. Links in an insurgency are not easy to determine. Insurgents may be productive members of the
118 population one day and carrying out covert attacks the next day. Further, insurgents hide their communica-
119 tions and do not publicize their relationships. Lastly, membership is difficult to determine since there is lit-
120 tle difference between the insurgent, the population that supports the insurgents, the population that is neu-
121 tral to the insurgents, and the population supporting the counter-insurgency.

122 E-14. The typical connection between two nodes is discovered from facts learned by soldiers on patrol or
123 in interactions with the local police. Which nodes are brothers? Which nodes are friends? What street does
124 each node live on? A link can be based on one or more criteria. Examples are kinship (brother of), role-
125 based (boss of), affective (likes), interactive (prays with, demonstrates with, communicates with), and af-
126 filiation (same clan, club). A commander on the ground who best understands his region, the culture, the
127 economy, and the role of regional socio-political organizations is best qualified for selecting the criteria for
128 defining a link. SNA is best done from within the area of operations under the commander's control.

129 E-15. Relationships (links) in large data sets are established by similarities between the nodes (people). In
130 the figure E-4 (below), people are identified by their participation in independent activities. When graphed,
131 pairs who have engaged in the same activity (columns with dots) are designated with a link.

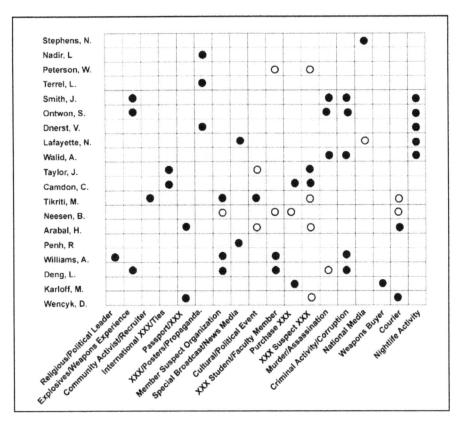

132

Figure E-4. Activities matrix

SOCIAL NETWORK GRAPHS AND INSURGENT ORGANIZATION
133

134 E-16. As you build your situational awareness of the environment you should create easy to understand,
135 adaptable, and accurate diagrams and information sheets. Charts and diagrams contribute to understanding
136 the insurgents' means of operations. These same diagrams are also useful for understanding tribal, family,
137 nongovernmental organization, and trans-national terrorist elements. Each diagram and chart may have
138 linkages to another or several others, but they are not created overnight. It takes time, patience, detailed pa-
139 trolling, and reporting and recording of efforts. These diagrams and charts serve to shape your understand-
140 ing of the enemy and aid you in planning and focusing your efforts.

141 E-17. As the commander dispatches patrols to collect information on the nodes and links, he can begin to
142 build a graph of the population in his area of operations. As the graph grows, he quickly finds that tradi-
143 tional, static organizational line charts may not produce viable explanations of insurgent organizational be-
144 havior. As a network, the individual insurgents are constantly adapting to the environment, their own capa-
145 bilities, and our own counter-insurgency tactics. The commander's understanding of the insurgency is only
146 as good as the patrol's last collection.

147 E-18. A regional insurgency can be fragmented within itself. A fully connected network is an unlikely de-
148 scription of the enemy insurgent order of battle. (See figure E-5, below.) In fact a region may actually con-
149 tain multiple sub-insurgencies that are either unaware of or even competing with other sub-insurgent
150 groups. (See figure E-6, below.)

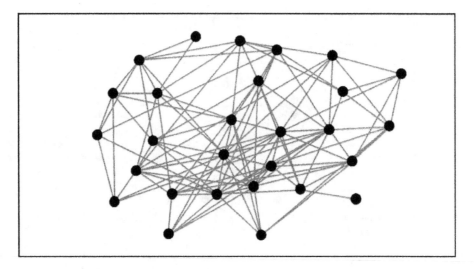

151   **Figure E-5. Network organization with high connections**

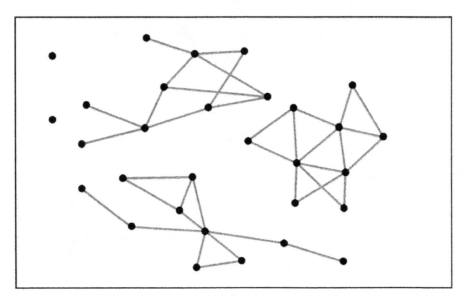

152   **Figure E-6. Fragmented network**

153 E-19. Without some ability to monitor and assess the connectivity and fluidity of a network, the behavior
154 of the insurgent organization becomes unclear. Leaders may make inappropriate assessments of enemy dis-
155 position and intent. COIN analysts can misidentify insurgent leaders and make incorrect assessments of the
156 degree of collusion between regional sub-insurgencies. Figure E-7 (below) depicts two approaches to un-
157 derstanding insurgent organizations. On the left side, the analyst imposes traditional military structure to
158 assess the organizational shape and key leader. The right side represents a regional insurgency as actually
159 substantiated by the relationship data collect by patrols and interviews. This latter view represents a more
160 accurate organization and leadership assessment.

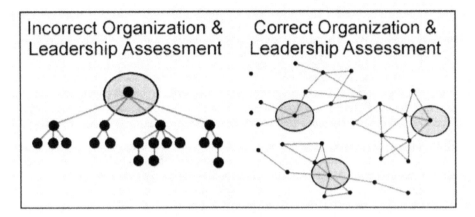

161

Figure E-7. Assessments

162 ## SOCIAL NETWORK MEASURES

163   E-20. In Figure E-8 (below) each individually identified insurgent is identified as a node, and intelligence
164 gathered relationships are identified as a line. One can readily identify three sub-groups (A, B, C) and two
165 individuals without direct ties to the other groups (ellipse). Thus, a social network is not just a description
166 of who is in a group, but also how they interact with one another.

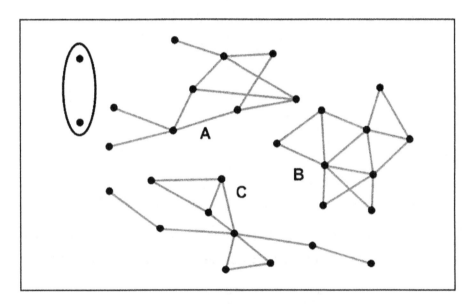

167

Figure E-8. Hypothetical regional insurgency

168   E-21. SNA simultaneously supports a commander's understanding of individual and group behavior. In
169 addition to producing the relationship graphs shown, SNA provides data to assign values to these connec-
170 tions in order to represent the strength of the relationships, information capacity, rates or flow of traffic,
171 distance between nodes, and probabilities of information being passed. It is these values that allow SNA to
172 quantify the relationships, thus presenting a means for mathematically testing the network.

173 **ORGANIZATION LEVEL ANALYSIS**

174 E-22. Organizational-level analysis provides insight about the organization's form, efficiency, and cohe-
175 sion. A regional insurgency may consist of large numbers of disconnected sub-insurgencies. As a result,
176 each group should be analyzed based on its capacities as compared to the other groups.

177 E-23. Organizational-level capacities can be described by the social network measures of network density,
178 cohesion, efficiency, and core-periphery. Each measure explains a characteristic of the network organiza-
179 tion's structure. Different network structures can support or hinder an organization's capabilities. There-
180 fore, each organizational measure supports the analyst's assessment of sub-group capacity.

181 E-24. Analysis is only as good as the data it is based upon. Unquestioningly using someone else's data to
182 come to conclusions that determine military operations is inadvisable. A commander should take owner-
183 ship of the collection process, decide which people are tracked as nodes in the data, and what constitutes a
184 relationship between nodes. Lastly, how a commander collects and organizes his or her data changes which
185 social network measures are most relevant to the situation.

186 E-25. Network density is a general indicator of how connected people are in the network. Network or
187 global-level density is the proportion of ties in a network relative to the total number possible. Comparing
188 network densities between insurgent subgroups provides the commander with an indication of which group
189 is most capable of a coordinated attack and which group is the most difficult to disrupt. (See figure E-9, be-
190 low.)

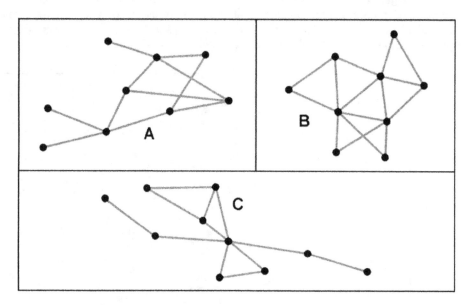

191 **Figure E-9. Comparison of subgroup densities**

192 E-26. Most network measures, to include network density, can be mapped out temporally to evaluate per-
193 formance over time. (See figure E-10, below.) Based on changes in network density over time, a com-
194 mander can—

195 ● Monitor the capabilities of the enemy.
196 ● Monitor the effect of recent operations.
197 ● Develop strategies to further fragment the insurgency.

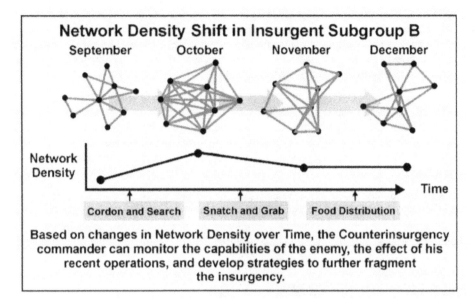

Based on changes in Network Density over Time, the Counterinsurgency commander can monitor the capabilities of the enemy, the effect of his recent operations, and develop strategies to further fragment the insurgency.

198   **Figure E-10. Density shift**

199   E-27. An increase in network density indicates the likelihood that the insurgent group can plan and execute
200   coordinated attacks. A decrease in network density means the group is reduced to fragmented or individ-
201   ual-level attacks. A well executed COIN eventually faces only low network density sub-groups. This is be-
202   cause high network density sub-groups only require the capture of one highly connected insurgent to lead
203   to the rest of the group. So, while a high network density group is the most dangerous, the high network
204   density group is also the easiest to defeat and disrupt.

205   E-28. Network density does not consider how distributed the connections are between the nodes in a net-
206   work. It is possible that a few nodes with a high number of connections push up the group network density
207   even though the majority of the people-nodes are marginally linked to the group. In the case of a highly
208   centralized network dominated by one or a few very connected nodes, these nodes can be removed or dam-
209   aged to fragment the group further into sub-networks. Better metrics of group/organizational performance
210   would be network centrality, core-periphery, and diameter. These individual-level metrics account for link
211   distribution among the nodes and are available in most network analysis software packages.

212   **INDIVIDUAL-LEVEL ANALYSIS**

213   E-29. Individual-level analysis characterizes every member of the organization and identifies key mem-
214   bers. SNA is best known for its ability to find key individual actors in a large mass of data. SNA describes
215   individuals based on their network position as compared to the network position of every other individual
216   in the network.

217   E-30. Individual network centralities provide insight into the individual's location in the network. The rela-
218   tionship between the centralities of all nodes can reveal much about the overall network structure.

219   E-31. One or a very few central nodes dominate a very centralized network. If these nodes are removed or
220   damaged, the network may quickly fragment into unconnected sub-networks. Hubs are nodes with a very
221   high degree of centrality. A network centralized around a well connected hub can fail abruptly if that hub is
222   disabled or removed.

223   E-32. A less centralized network has no single points of failure. It is resilient in the face of many inten-
224   tional attacks or random failures—many nodes or links can fail while allowing the remaining nodes to still
225   reach each other over other, redundant network paths.

226 E-33. Degree centrality describes how active an individual is in the network. We measure network activity
227 for a node by using the concept of degrees—the number of direct connections a node has. In the networks
228 above, the nodes with the most direct connections are the most active in their networks. Common wisdom
229 in organizations is "the more connections, the better." This is not always so. What really matters is where
230 those connections lead to—and how they connect the otherwise unconnected. If a node has many ties, it is
231 often said to be either prominent or influential.

232 E-34. A node with high "betweenness" has great influence over what flows in the network. Betweenness
233 centrality indicates whether an individual lies between other individuals in the network, serving as an in-
234 termediary, a liaison, or a bridge. Depending on position, a person with high betweenness plays a "broker"
235 role in the network. A major opportunity exists for counterinsurgents if, as in Group C (Figure E-9), the
236 high betweenness centrality person is also a single point of failure, which if removed would fragment the
237 organization.

238 E-35. Most people would view the nodes on the periphery of a network as not being very important. Nodes
239 on the periphery receive very low centrality scores. However, peripheral nodes are often connected to net-
240 works that are not currently mapped. The outer nodes may be resource gatherers or actors that have their
241 own network outside of their insurgent group—making them very important resources for fresh informa-
242 tion not available inside their insurgent group.

243 # THE NETWORK PERSPECTIVE

244 E-36. The network organization is a new concept in our military lexicon but an old concept in application.
245 Whether in insurgent groups or business organizations we can expect the network organization to rapidly
246 become the norm. Unfortunately, given the flexibility, speed, and connectedness in a network organization,
247 organizational behaviors can rapidly become an invisible box with hidden connections and transactions.
248 The network organization has a different pulse than an organization tied to strict procedural norms and ex-
249 pectations. Further, the network organization inherently deletes or bypasses stovepipes that slow or inhibit
250 decision making and coordination.

251 E-37. Insurgents are a network organization embedded in a sympathetic population. Differentiating be-
252 tween the insurgent, the insurgent supporting population, the neutral population, and the friendly popula-
253 tion is difficult. With every success of the counter insurgency, the insurgent organization is further frag-
254 mented, but it still can remain dangerous.

255 E-38. SNA assists with formalizing the informality of an insurgent network; with visualizing a structure
256 that we may not readily observe. Network concepts allow us to highlight the structure of a previously un-
257 observed association by focusing on the pre-existing relationships and ties that bind together such a group.
258 By focusing on the roles and organizational positions of those actors who are prominent and/or influential,
259 we get a sense of how the association is structured and thus how the group functions, how members are in-
260 fluenced and power is exerted, and how resources are exchanged.
261
262

1 **Appendix F**

2 # Airpower in Counterinsurgency

3 Airpower is an important asset for the U.S., multinational and host nation (HN)
4 forces fighting an insurgency. This manual defines airpower as all the assets of all the
5 services, multinational partners, and the HN that fly or operate from space. These air-
6 power assets include strike aircraft, reconnaissance aircraft, air transport, helicopters,
7 unmanned aerial systems (UASs), and space satellites.

8 ## OVERVIEW

9 F-1. In counterinsurgency (COIN), airpower will serve predominantly in a supporting role. At times air-
10 craft can strike insurgents but, given the nature of COIN, the most common roles of airpower will be pro-
11 viding reconnaissance and transporting troops, equipment and supplies. Rough terrain and poor transporta-
12 tion nets can be serious obstacles to COIN operations and serve as an aid to insurgents. Airpower helps
13 overcome these obstacles. Insurgencies cannot be defeated from the air—but airpower serves as a tremen-
14 dous force enhancer and enables COIN forces to operate more effectively.

15 F-2. Airpower provides a significant asymmetric advantage to the counterinsurgent. If the insurgents field
16 a force to engage U.S., multinational, or HN forces, air assets can respond quickly with precision fire-
17 power. In a sudden crisis, air mobility can shift troops immediately to threatened points or to surround in-
18 surgent elements. The vital supporting role played by airpower has been demonstrated in numerous COIN
19 operations. In many cases airpower has been a key element in COIN strategy and tactics. In Malaya (1948-
20 1960) and El Salvador (1980-1992), as well as more recently in Colombia and Afghanistan, the ability to
21 airlift U.S. Army and police units to remote locations has proven exceptionally important in tracking down
22 and eliminating insurgent groups. Airpower enables U.S., multinational, and government forces to operate
23 in rough and remote terrain, areas that traditionally were havens for insurgents.

24 F-3. Airpower can be important in many ways that do not involve delivering ordnance. In Colombia ae-
25 rial crop dusters sprayed and eradicated the coca fields that provided the insurgent's drug income. During
26 the El Salvador insurgency the provision of six medical evacuation (MEDEVAC) helicopters to the Salva-
27 doran forces in 1984 played a central role in improving the fighting capabilities of the Salvadorans on the
28 battlefield. Salvadoran force morale improved noticeably when the soldiers knew that, if wounded,
29 MEDEVAC helicopters would get them to a hospital within minutes. With this air support, the Salvadoran
30 army became much more aggressive in tracking down and engaging rebel forces.

31 F-4. Using air transport to airlift or drop food and medical supplies to civilians in isolated regions is a
32 very effective means of delivering humanitarian support quickly and in winning the support of the popula-
33 tion. As was highlighted in chapter 8 air transport is also important for logistics in COIN. In areas of high
34 insurgent activity along the roads, supplies can be airlifted rather than transported in highly vulnerable land
35 convoys. Air transport also enables the U.S., multinational, and HN forces to maintain forces in remote, but
36 strategically important, locations.

37 F-5. This appendix does not aspire to provide a comprehensive doctrine for employing airpower in COIN.
38 It does, however, lay out some general principles for the most effective employment of airpower in support
39 of U.S., multinational, and HN forces.

40 ## THE ADVANTAGES OF AIRPOWER

41 F-6. In COIN the direct attack role of airpower becomes more important as the conflict becomes more
42 conventional. If the insurgents mass forces and try to fight from established positions, airpower can be em-

43
44
45
ployed with devastating and precise effects. However, unless the insurgents assemble large units and try to engage government forces in a conventional battle, employment of airpower in the strike role is likely to be relatively rare.

46
47
48
49
50
51
F-7. Airpower provides the U.S., multinational, and HN forces a flexibility and initiative that is usually the advantage of the insurgent. The presence of airpower strike capability alone provides a deterrent and denies the insurgent many options. If insurgents mass forces for a major operation, they are vulnerable to close air support or gunships. If the insurgents stand and fight, counterinsurgent forces can be quickly air-lifted in to contain the enemy and destroy them. Chapter 8 addresses the significant advantages air transport can provide for COIN logistics.

52 AIRPOWER IN THE STRIKE ROLE

53
54
55
56
57
58
59
60
61
62
63
F-8. The employment of airpower in the strike role should be done with exceptional care. Bombing, even with the most precise weapons, can cause unintended civilian casualties. The benefits of every air strike should be weighed against the risks, the primary danger being collateral damage that turns the population against the government and provides the insurgents with a major propaganda victory. Even when justified under the law of war, bombing a target that results in civilian casualties will bring media coverage that works to the benefit of the insurgents. A standard insurgent and terrorist tactic for decades against Israel has been to fire rockets or artillery from the vicinity of a school or village in the hope that the Israelis would carry out a retaliatory air strike that kills or wounds civilians—who are then displayed to the world media as victims of aggression. Insurgents and terrorists elsewhere have shown few qualms in provoking attacks that ensure civilian casualties if such attacks fuel anti-government and anti-U.S. propaganda. Indeed, insurgents today can be expected to use the civilian population as a cover for their activities.

64
65
66
67
68
69
70
71
72
73
74
75
F-9. Even in a clear case of taking out an insurgent headquarters or command center, care has to be taken to accomplish the mission while minimizing civilian casualties. New, precise munitions with smaller blast effects have been developed and employed to limit collateral damage. There are other means, as well. At the start of the campaign in Afghanistan in 2001, U.S. intelligence identified Taliban armored vehicles parked in built up areas. A miss, or even a direct hit, by a precision weapon would be likely to kill civilians and give the Taliban a propaganda advantage. The United States Air Force (USAF) came up with the idea of employing concrete-filled practice bombs with precision guidance against such Taliban weapons systems. If the bomb hit the target, the kinetic energy of 2,000-pounds of steel and concrete dropped from the air would assure destruction. If the bomb missed the target, it would bury itself deep in the ground with no explosion and little chance of major collateral damage. The destruction of the weapons systems was accomplished without any collateral damage that could have turned the population against the U.S. and multinational forces.

76 AIRPOWER IN INTELLIGENCE COLLECTION

77
78
79
80
81
82
83
F-10. COIN relies on good intelligence—and airpower has a major role in intelligence collection. In a conflict where the insurgents operate in the countryside or remote areas, aerial reconnaissance and surveillance can employ imagery and infrared systems to find hidden base camps and insurgent defensive positions. Aerial surveillance can often identify people, vehicles and buildings even when they are hidden under heavy growth. UASs and other aircraft can patrol roads to locate insurgent ambushes and improvised explosive devices. Air-mounted, signal intelligence collection can detect insurgent communications and locate the point of origin.

84
85
86
87
88
F-11. Air assets have proven important in convoy and route protection and in tactical operations. Helicopters especially have shown great utility in providing overwatch, fire support, alternate communications, and MEDEVAC. At the tactical level, air support requires a decentralized command and control (C2) system in which the supported units have immediate access to information gained by air reconnaissance and support assets, and have combat air assets available to immediately engage insurgent forces.

89
90
91
F-12. However, intelligence obtained through air and space platforms is most useful if it is tied in quickly and efficiently with a joint intelligence center where human intelligence (HUMINT) and other forms of intelligence can be fused. Unlike conventional war where air and space intelligence platforms are the primary

92
93
94
95

means of locating enemy heavy forces and equipment, in COIN aerial intelligence might contribute just one small piece to a larger puzzle. To understand the whole puzzle, air and space intelligence has to be combined with HUMINT. Aerial reconnaissance assets can tell you that the civilians in a village are evacuating the area, but air and space intelligence cannot tell you why.

96
97
98

F-13. In COIN, HUMINT is an essential requirement for properly employing airpower in the strike role. The USAF and other service aviation branches need to be skilled in employing HUMINT in operational planning.

THE ROLE OF HIGH-TECH ASSETS

99

100
101
102
103
104
105
106
107
108

F-14. "High-tech" air and space systems in use today have proven their worth in COIN. UASs, such as the Predator, give U.S. and multinational forces unprecedented capabilities in surveillance and target acquisition. Aerial surveillance platforms with long loiter times can place an entire region under constant surveillance. Predators have even been equipped with precision munitions and employed in the strike role to successfully target senior terrorist leaders. Air and space-based signals intelligence gives the U.S. and multinational partners important intelligence collection capabilities and enabled U.S. forces to monitor insurgent communications in Iraq and Afghanistan. New munitions, such as the JDAM bomb, can bomb accurately through clouds and bad weather, making it possible to destroy insurgent targets even under the most adverse conditions.

THE ROLE OF LOW-TECH ASSETS

109

110
111
112
113
114
115
116
117
118
119

F-15. The "low-tech" aspects of airpower have also proven effective in COIN. In many cases light, slow, and inexpensive civilian aircraft have been successfully employed in patrolling border areas. In the 1980s, Guatemala mobilized its civilian light aircraft, formed them into an air force reserve and used them to patrol the main roads to report any suspected ambushes. This was successful in deterring insurgent attacks along Guatemala's major communications routes. In Africa in the 1980s, South African light aircraft flying low and slow spotted many of the small guerrilla bands trying to infiltrate into Namibia from Angola. In Iraq light aircraft have been used for patrols to spot bands crossing the border. Israel and the United States have employed stationary balloons equipped with video camera and infrared sensors to watch for border incursions. The unmanned, stationary balloons have proven to be a simple, inexpensive and effective means of monitoring activity in remote areas.

120
121
122
123
124
125
126

F-16. The U.S. and many small nations have employed aerial gunships as very effective close air support weapons in COIN. At its most basic, the aerial gunship is essentially a converted transport plane modified to fire heavy machine guns or light artillery pieces from the side. Many different models of a gunship have been created, ranging from the USAF AC-130 to a variety of smaller transports modified to carry weapons ranging from .50-caliber machine guns to 40mm rapid-fire cannon. Gunships are inexpensive and relatively easy to operate. The major limitation to their employment is the insurgent capability to employ antiaircraft weapons and missiles, as gunships require a relatively benign environment for operations.

THE AIRPOWER COMMAND STRUCTURE

127

128
129
130
131
132
133

F-17. The effective use of airpower in COIN requires an effective joint and multinational command structure. It is not only an issue of jointness among U.S. forces, but also coordinating the air assets of multinational partners and the host nation. One of the first steps in COIN planning is the establishment of a joint and multinational airpower command and control system, and agreed policies on the rules and conditions for employing airpower in the theater. All air operations in the theater should be coordinated at one center for safety and efficiency.

134
135

F-18. Expeditionary airfields are likely used by U.S. and multinational air units, along with HN forces. These bases should be properly protected, and defense should be coordinated among all occupants.

BUILDING THE HOST-NATION'S AIRPOWER CAPABILITY

F-19. As the objective of U.S. and multinational operations is to enable the HN to provide for its own internal and external defense, U.S. and multinational planners need to set up a long-term program to develop a HN aviation capability. The HN air force should be appropriate for that nation's requirements. For conducting effective COIN operations a HN air force requires the following basic capabilities:

- Aerial reconnaissance and surveillance.
- Air transport.
- Close air support for land troops.
- Helicopter troop lift.
- MEDEVAC.

F-20. The first step is to develop the right organizational model for a HN air force. The HN air assets should be centrally managed, under command of a joint headquarters, and tied into the joint intelligence center. Planning should identify gaps in the HN's ability to command, control, and employ airpower in COIN operations.

F-21. The next step is to help the HN develop its aviation infrastructure under a long term plan. Most developing nations need considerable assistance to develop an appropriate organization, force structure, and basing plans. As aviation assets represent a large cost for a small nation, an effective airfield security program is also necessary.

F-22. An important training asset is the Air Force Special Operations command, which has teams that are qualified to operate the most common equipment used in developing nations. These teams also have the language and cultural training to effectively support aircrew and personnel training. The USAF also has the capability to train pilots and aircrew through the International Military Education and Training Program.

F-23. Equipment should be selected with the economic and technological resources of the HN in mind. In most cases, aircraft and supporting systems acquired by the HN, or provided by the U.S. and multinational allies, should be effectively operated and maintained over the long term by a small air force with limited resources. Multinational support in training and equipping the HN air force can be very important. U.S. aircraft have tremendous capabilities, but they can be too expensive and too complex for some developing nations to operate and maintain. U.S. allies with capable, but less expensive and sophisticated aircraft, can be very helpful in equipping the HN.

F-24. Training and developing a capable HN air force takes considerable time due to the requirements to qualify aircrew, maintenance personnel and other specialists. Training HN personnel to work effectively in joint operations and to coordinate air-land support also requires a high level of skill. Even when the HN army and police are trained to an acceptable standard, it is likely that U.S. personnel will be required to stay with the HN forces to conduct liaison for supporting U.S. air assets and advise the HN forces in the employment of their own airpower.

F-25. Because developing capable HN aviation forces usually takes longer than developing land forces, U.S. aviation units, advisors, and trainers are likely to be still deployed to the HN after U.S. and multinational land force trainers and advisors have completed their mission. Because effective air/land operations are such a complex task and require so many resources, it is also likely that the HN will rely on U.S. air liaison personnel, land controllers, and aircraft for an extended period. Thus, long-term U.S. air support requirements need to be considered in comprehensive COIN planning.

Appendix G

Learning Counterinsurgency

All insurgencies are different, but there are broad historical trends underlying the causes that motivate insurgents, the successful and unsuccessful counterinsurgency (COIN) practices adopted by the government attempting to defeat them, and the overall course of most insurgencies.

G-1. One of the common features of almost all insurgencies is that the intervening government invariably takes a while to understand that an insurgency is taking place, while the insurgents take advantage of the time lag to gather strength. Thus the intervening government always "comes from behind" when fighting an insurgency. Another common feature is that armies conducting counterinsurgencies tend to begin poorly. Western armies tend to neglect the study of insurgency, falsely believing that if they are able to win large conventional wars, they can also win small unconventional ones. The truth is exactly the opposite; in fact, the very things that make a conventional army successful when fighting conventional adversaries— operational maneuver and the ability to mass firepower—are often counterproductive in COIN warfare. Nonetheless, they are almost invariably the first moves made by armies confronting an insurgency, and they almost invariably fail.

G-2. Armies that successfully defeat an insurgency do so because they are able to overcome their institutional proclivities to wage conventional war and learn how to practice COIN despite themselves. The principles of COIN are well known and form the skeleton of this manual. In essence, the counterinsurgent should—

- Understand the environment in which the war is being fought.
- Isolate the insurgents from their cause and their base.
- Secure the population under the rule of law.
- Generate intelligence from the population to drive actions against the insurgents.
- Apply all elements of national power in unison to support the legitimacy of the host nation's government.
- Be prepared for a long commitment, measured in years if not decades.

G-3. Applying these principles allows the counterinsurgent to gain the support of the local people, who provide either a shield to the enemy or vital intelligence to the counterinsurgent. Their choice of whom to support depends to a large extent on who provides them with the security they need.

G-4. Traditionally, armies have had to unlearn much of their doctrine and (re)learn the principles of COIN while waging COIN campaigns. The intent of this essay is to place into the hands of planners, trainers and field commanders a tool to ensure that this learning process begins faster, builds on a higher knowledge base, and costs fewer lives and less national treasure than it has in the past. For this reason "Learn and Adapt" has been identified as a modern COIN imperative.

G-5. Just as there are historical principles underlying success in COIN, there are organizational traits shared by most successful learning organizations. Armies that learn COIN effectively have generally—

- Developed COIN doctrine locally.
- Established local training centers during counterinsurgencies.
- Regularly challenged their assumptions both formally and informally.
- Requested outside assistance in understanding scenarios beyond their experience.
- Promoted suggestions from the field.
- Fostered truly open communication between senior officers and their subordinates.

- Established rapid avenues to ensure dissemination of lessons learned.
- Coordinated closely with governmental and non-governmental partners at all levels of command.

G-6. These are not easy traits for any organization to adopt, and are particularly challenging objectives for an army during the course of a conflict. However, they are essential when confronting an enemy that does not fight according to a prescribed doctrine and that adapts while waging irregular warfare. The enemies the United States is fighting today—and is likely to fight for many years to come—are continuously and consciously evaluating the strengths and weaknesses of our military, aiming to avoid our power and attack our vulnerabilities. The US government should therefore constantly evaluate its effectiveness in using all elements of national power to learn and adapt more quickly than do its enemies. Learning organizations defeat insurgencies; armies that are not learning organizations do not.

G-7. Effective learning organizations encourage individuals to pay attention to the rapidly changing situations that characterize COIN campaigns—rapid enemy innovation, shifting attitudes of local populations, local civilian leadership turmoil. When current organizational behavior is no longer appropriate in such situations, learning organizations engage in a directed search for more appropriate ways to defeat the enemy, rapidly develop consensus on the new doctrine, publish the doctrine, and carefully observe the impact of the doctrine on organizational behavior. In effective learning organizations, this learning cycle repeats continuously—and faster than that of the insurgent enemy. He who learns faster and adapts more rapidly wins.

G-8. When designing a learning institution, it is hard to improve upon Lieutenant General Jack Cushman's instruction to "Let insight evolve from an atmosphere of open, shared thought...from a willing openness to a variety of stimuli, from intellectual curiosity, from observation and reflection, from continuous evaluation and discussions, from review of assumptions, from listening to the views of outsiders, from study of history, and from the indispensable ingredient of humility." The Army, United States Marine Corps, and Department of Defense exhibited these traits to a remarkable extent during the early years of the insurgencies in Afghanistan and Iraq, making up for lost time and learning COIN from the ground up, but it was a costly way to learn.

G-9. The annotated bibliography can serve as a starting point for study. The books and articles it lists are not the only good ones on insurgency and counterinsurgency; the field is vast and rich. They are, however, some of the more useful ones for Soldiers and Marines who wish to improve their ability to defeat enemies who practice one of the most dangerous and difficult forms of warfare. The insights gained from such study are likely to prove useful and encourage even more learning during the long war in which the Nation is currently engaged.

Glossary

The glossary lists acronyms and terms with Army, multi-Service, or joint definitions, and other selected terms. Where Army and joint definitions are different, *(Army)* follows the term. The proponent manual for other terms is listed in parentheses after the definition. Terms for which the Army and Marine Corps have agreed on a common definition are followed by *(Army-Marine Corps)*.

SECTION I – ACRONYMS AND ABBREVIATIONS

ABL	ammunition basic load
ACR	armored cavalry regiment
AO	area of operations
CAP	Combined Action Program
CMOC	civil-military operations center
COIN	counterinsurgency
CONUS	continental United States
COP	common operational picture
CORDS	Civil Operations and Revolutionary/Rural Development Support (A program used by U.S. forces during the Vietnam war.)
COTS	commercial off-the-shelf
DOCEX	document exploitation
DOD	Department of Defense
DOTMLPF	doctrine, organization, training, materiel, leadership and education, personnel, and facilities
FARC	*Fuerzas Armadas Revolucionarias de Colombia* (Revolutionary Armed Forces of Colombia)—an insurgent movement that began in the 20th century and continues today
FID	foreign internal defense
FM	field manual
FMI	field manual–interim
FMFM	fleet Marine force manual
FMFRP	fleet Marine force reference publication
HMMWV	high-mobility, multipurpose wheeled vehicle
HN	host-nation
HUMINT	human intelligence
IDP	international displaced person
IED	improvised explosive device
IMET	International Military Education and Training
IMINT	imagry intelligence

IO	information operations
IPB	intelligence preparation of the battlefield
ISR	intelligence, surveillance, and reconnaissance
JIACG	joint interagency coordination group
JP	joint publication
LLO	logical lines of operations
LOC	line of communication
LOG	logistics
LOGPAD	logistical pad
MACV	Military Assistance Command, Vietnam
MCDP	Marine Corps doctrinal publication
MCRP	Marine Corps reference publication
MCWP	Marine Corps warfighting publication
NCO	noncommissioned officer
NGO	nongovernmental organization
NVA	North Vietnamese Army
OCS	Officer Candidate School
OIF	Operation Iraqi Freedom
OP	observation post
OSINT	open-source intelligence
PIR	priority intelligence requirements
PRC	purchase request and committal
PSYOP	psychological operations
RFI	rapid fielding initiative
SIGINT	signals intelligence
SOF	special operations forces
TAREX	target exploitation
TECHINT	technical intelligence
UN	United Nations
U.S.	United States
USAID	United States Agency for International Development
VC	Viet Cong

SECTION II – TERMS

assessment	(Army) The continuous monitoring and evaluation of the current situation and progress of an operation. (FMI 5-0.1)
board	A temporary grouping of selected staff representatives delegated decision authority for a particular purpose or function. (FMI 5-0.1)
centers of gravity	(joint) Those characteristics, capabilities, or sources of power from which a military force derives its freedom of action, physical strength, or will to fight. (JP 1-02)

civil considerations	How the manmade infrastructure, civilian institutions, and attitudes and activities of the civilian leaders, populations, and organizations within an area of operations influence the conduct of military operations. (FM 6-0)
clear	A tactical mission task that requires the commander to remove all enemy forces and eliminate organized resistance in an assigned area. (FM 3-90)
commander's intent	(Army) A clear, concise statement of what the force must do and the conditions the force must meet to succeed with respect to the enemy, terrain, and desired end state. (FM 3-0) (Marine Corps) A commander's clear, concise articulation of the purpose(s) behind one or more tasks assigned to a subordinate. It is one of two parts of every mission statement which guides the exercise of initiative in the absence of instructions.
commander's visualization	The mental process of developing situational understanding, determining a desired end state, and envisioning how to move the force from its current state to that end state. (FMI 5-0.1)
command post cell	A grouping of personnel and equipment by warfighting function or purpose to facilitate command and control during operations. (FMI 5-0.1)
common operational picture	(joint) A single identical display of relevant information shared by more than one command. A common operational picture facilitates collaborative planning and assists all echelons to achieve situational awareness. (Army) An operational picture tailored to the user's requirements, based on common data and information shared by more than one command. Also called **COP**. (FM 3-0)
counterinsurgency	(joint) Those military, paramilitary, political, economic, phschological, and civic actions taken by a government to defeat insurgency. (JP 1-02)
counterstate	A competing structure set up by an insurgent to replace the government in power. It includes the administrative and bureaucratic trappings of political power, and performs the normal functions of government.
counterterrorism	(joint) Operations that include the offensive measures taken to prevent, deter, preempt, and respond to terrorism. Also called CT. (JP 1-02)
decisive point	(joint) geographic place, specific key event, critical system or function that allows commanders to gain a marked advantage over an enemy and greatly influence the outcome of an attack. (JP 1-02)
direct action	(joint) Short-duration strikes and other small-scale offensive actions conducted as a special operation in hostile, denied, or politically sensitive environments and which employ specialized military capabilities to seize, destroy, capture, exploit, recover, or damage designated targets. Direct action differs from conventional offensive actions in the level of physical and political risk, operational techniques, and the degree of discriminate and precise use of force to achieve specific objectives. (JP 1-02)
execute	To put a plan into action by applying combat power to accomplish the mission and using situational understanding to assess progress and make execution and adjustment decisions. (FM 6-0)
foco/focoism	A popular theory that says that small cells of armed revolutionaries can create the conditions for revolution through their actions. Demonstrated revolutionary victories, the successes of the foci, are supposed to lead the masses to revolution. If conditions are ripe, according to focoists, a single spark can start the revolutionary fire.
focoist	A practitioner of focoism as espoused by Che Guevara
forward operations base	(joint) In special operations, a base usually located in friendly territory or afloat that is established to extend command and control or communications or

	to provide support for training and tactical operations. Facilities may be established for temporary or longer duration operations and may include an airfield or an unimproved airstrip, an anchorage, or a pier. A forward operations base may be the location of special operations component headquarters or a smaller unit that is controlled and/or supported by a main operations base. [Note: Army SOF term is "forward operational base."] (JP 1-02)
full spectrum operations	The conduct of simultaneous combinations of the four components of Army operations (offense, defense, stability, and civil support) across the spectrum of conflict (peace, crisis, and war). This draft definition is being staffed as part of FM 3-0 development. Upon approval of FM 3-0, it will replace the definition of full spectrum operations in FM 3-0 (2001).
human intelligence	(joint) A category of intelligence derived from information collected and provided by human sources. (JP 1-02) [Note: in Army and Marine Corps usage, human intelligence operations cover a wide range of activities encompassing reconnaissance patrols, aircrew reports and debriefs, debriefing of refugees, interrogations of prisoners of war, and the conduct of counterintelligence force protection source operations.]
information environment	(joint) The aggregate of individuals, organizations or systems that collect, process, or disseminate information; also included is the information itself. (JP 1-02)
information operations	(Army) The employment of the core capabilities of electronic warfare, computer network operations, psychological operations, military deception, and operations security, in concert with specified supporting and related capabilities, to affect and defend information and information systems and to influence decisionmaking. Also called **IO**. (FM 3-13)
insurgency	(joint) An organized movement aimed at the overthrow of a constituted government through the use of subversion and armed conflict. (JP 1-02)
intelligence discipline	(joint) A well-defined area of intelligence collection, processing, exploitation, and reporting using a specific category of technical or human resources. There are seven major disciplines: human intelligence, imagery intelligence, measurement and signature intelligence, signals intelligence (communications intelligence, electronic intelligence, and foreign instrumentation signals intelligence), open-source intelligence, technical intelligence, and counterintelligence. (JP 1-02)
intelligence preparation of the battlefield	The systematic, continuous process of analyzing the threat and environment in a specific geographic area. Intelligence preparation of the battlefield (IPB) is designed to support the staff estimate and military decisionmaking process. Most intelligence requirements are generated as a result of the IPB process and its interrelation with the decisionmaking process. Also called **IPB**. (FM 34-130)
intuitive decision making	(Army-Marine Corps) The act of reaching a conclusion that emphasizes pattern recognition based on knowledge, judgment, experience, education, intelligence, boldness, perception, and character. This approach focuses on assessment of the situation vice comparison of multiple options. (FM 6-0)
lines of operations	(joint) Lines that define the directional orientation of the force in time and space in relation to the enemy. They connect the force with its base of operations and its objectives. (JP 1-02)
measure of effectiveness	(Army) A criterion used to assess changes in system behavior, capability, or operational environment that is tied to measuring the attainment of an end

state, achievement of an objective, or creation of an effect. (FMI 5-0.1) (This is the Army definition until the revised JP 3-0 is approved. The Army will use the joint definition in JP 3-0 when JP 3-0 is approved.)

measure of performance	(Army) A criterion to assess friendly actions that is tied to measuring task accomplishment. (FMI 5-0.1) (This is the Army definition until the revised JP 3-0 is approved. The Army will use the joint definition in JP 3-0 when JP 3-0 approved.)
narrative	An organizational scheme expressed in story form that is central to the representation of a group's identity.
nongovernmental organization	(joint) A private, self-governing, not-for-profit organization dedicated to alleviating human suffering; and/or promoting education, health care, economic development, environmental protection, human rights and conflict resolution; and/or encouraging the establishment of democratic institutions and civil society. Also called **NGO**.(JP 1-02).
operating tempo	The annual operating miles/hours for systems in a particular unit required to execute the commander's training strategy. It is stated in terms of the miles/hours for the major system in a unit; however, all equipment generating significant operating and support cost has an established operating tempo. (FM 7-0)
operational environment	(joint) The air, land, sea, space, and associated adversary, friendly, and neutral systems (political, military, economic, social, informational, infrastructure, legal, and others) that are relevant to a specific joint operation. (JP 1-02) This is a proposed joint definition taken from JP 3-0 (Revision Final Coordination). When published, FM 3-24 will use the approved joint definition.
operational picture	A single display of relevant information within a commander's area of interest. (FM 3-0)
planning	The process by which commanders (and staff if available) translate the commander's visualization into a specific course of action for preparation and execution, focusing on the expected results. (FMI 5-0.1)
preparation	Activities by the unit before execution to improve its ability to conduct the operation including, but not limited to, the following: plan refinement, rehearsals, reconnaissance, coordination, inspections, and movement.(FM 3-0)
reachback	(Army/Marine Corps) The ability to exploit resources, capabilities, and expertise, not physically located in the theater or a joint operations area, when established.
riverine area	(joint) An inland or coastal area comprising both land and water, characterized by limited land lines of communication, with extensive water surface and/or inland waterways that provide natural routes for surface transportation and communications. (JP 1-02)
rules of engagement	(joint) Directives issued by competent military authority that delineate the circumstances and limitations under which United States forces will initiate and/or continue combat engagement with other forces encountered. Also called **ROE**. (JP 1-02)
stability operation	An operation to establish, preserve, and exploit security and control over areas, populations, and resources. (FM 3-0) This draft definition is being staffed as part of FM 3-0 development. Upon approval of FM 3-0, it will replace the definition of stability operations in FM 3-0 (2001).
strike	(joint) An attack which is intended to inflict damage on, seize, or destroy an

objective. (JP 1-02)

subordinates initiative
The assumption of responsibility for deciding and initiating independent actions when the concept of operations or order no longer applies or when an unanticipated opportunity leading to the accomplishment of the commander's intent presents itself. (FM 6-0)

tempo
(Army) The rate of military action. (FM 3-0) (Marine Corps) The relative speed and rhythm of military operations over time with respect to the enemy.

unified action
(joint) A broad generic term that describes the wide scope of actions (including the synchronization of activities with governmental and nongovernmental agencies) taking place within unified commands, subordinate unified commands, or joint task forces under the overall direction of the commanders of those commands. (JP 1-02)

warfighting function
A group of tasks and systems (people, organizations, information, and processes) united by a common purpose that commanders use to accomplish missions and training objectives. (FMI 5-0.1)

working group
A temporary grouping of predetermined staff representatives who meet to coordinate and provide recommendations for a particular purpose or function. (FMI 5-0.1)

Annotated Bibliography

Traditionally, armies have had to relearn the principles of counterinsurgency and apply them to their doctrine while executing counterinsurgency operations. This bibliography is a tool for Army and Marine leaders to help them begin this learning process sooner. Knowledge of what others have written about insurgency and counterinsurgency provides a foundation for a knowledge base that leaders can use to assess counterinsurgency situations and make appropriate decisions.

As learning organizations, the Army and Marine Corps encourage Soldiers and Marines to pay attention to the rapidly changing situations that characterize counterinsurgency operations. When current doctrine (tactics, techniques, and procedures) no longer achieves the desired results, they engage in a directed search for more appropriate ways to defeat the enemy. The Army and Marine Corps rapidly develop institutional consensus on new doctrine, publish it, and carefully observe its impact on mission accomplishment. This learning cycle repeats continuously as Army and Marine counterinsurgency forces seek to learn faster than the insurgent enemy. During insurgencies, the military force that learns faster and adapts more rapidly—the better learning organization—wins. Because of this, "Learn and Adapt" has been identified as a modern counterinsurgency imperative. Learning begins with study before deployment. Adapting occurs as Soldiers and Marines apply what they have learned, assess the results of their actions, and apply the results to accomplish the mission.

The books and articles that follow are not the only good ones on insurgency and counterinsurgency. The field is vast and rich. They are, however, some of the more useful for Soldiers and Marines who wish to improve their ability to defeat enemies who engage them in one of the most dangerous and difficult forms of warfare.

THE CLASSICS

Bulloch, Gavin. "Military Doctrine and Counterinsurgency: A British Perspective." *Parameters* XXVI, 2 (Summer 1996), 4-16.
> A good summary of the British school, which emphasizes hearts and minds and the primacy of the political, by the author of the British Army's current COIN doctrine.

Calwell, Charles E. *Small Wars: Their Principles and Practice.* Lincoln, NE: University of Nebraska Press, 1996. Reprint of *Small Wars: A Tactical Textbook for Imperial Soldiers.* London: Greenhill Books, 1890.
> A British Major General who fought in small wars in Afghanistan and the Boer War provides lessons learned that remain applicable today.

Galula, David. *Counterinsurgency Warfare: Theory and Practice.* London: Praeger, 1964.
> Lessons derived from the author's observation of insurgency and counterinsurgency in Greece, China and Algeria.

Gurr, Theodore. Why Men Rebel. Princeton, NJ: Princeton University Press, 1971.
> Describes the relative deprivation theory, which states that unmet expectations motivate those who join rebel movements.

Hoffer, Eric. The True Believer: Thoughts on the Nature of Mass Movements. New York: Harper Classics, 2002.

This book, originally published in 1951, explains why people become members of cults and similar groups.

Horne, Alistair. A Savage War of Peace. New York: Viking, 1977.
> The best analysis of the approaches and problems on both sides during the war in Algeria. For more on this conflict, see The Battle of Algiers, a troubling and instructive 1966 movie.

Jeapes, Tony. SAS Secret War. London: Greenhill Books, 2005.
> How the British Special Air Service raised and employed irregular tribal forces to counter Communist insurgency in Oman during the 1960s and 1970s.

Kitson, Frank. Low-intensity Operations: Subversion, Insurgency and Peacekeeping. London: Faber and Faber, 1971.
> Explanation of the British school of counterinsurgency from one of its best practitioners.

Komer, R.W., The Malayan Emergency in Retrospect: Organization of a Successful Counterinsurgency Effort. Washington: RAND, 1972.
> http://www.rand.org/pubs/reports/R957/. Analysis of a successful COIN effort by the first chief of CORDS in South Vietnam.

Larteguy, Jean. The Centurions. New York: Dutton, 1962.
> A fact-based novel about the French experience in Vietnam and Algeria that features an excellent depiction of the leadership and ethical dilemmas involved in COIN.

Lawrence, T.E. Seven Pillars of Wisdom: A Triumph. New York: Anchor, 1991. Reprint of 1917 book published in London by George Doran.
> Autobiographical account of Lawrence of Arabia's attempts to organize Arab nationalism during World War I.

Lawrence, T.E. "The 27 Articles of T.E. Lawrence." The Arab Bulletin, 20 August 1917.
> http://www.d-n-i.net/fcs/lawrence_27_articles.htm.
> Much of the best of Seven Pillars of Wisdom, in easily digestible bullet points.

Linn, Brian McAllister. The Philippine War, 1899–1902. Lawrence: University Press of Kansas, 2002.
> The definitive treatment of successful US COIN operations in the Philippines.

McCuen, John J. The Art of Counter-Revolutionary War. St. Petersburg, Florida: Hailer Publishing, 2005. Originally published by Harrisburg, PA: Stackpole Books, 1966.
> Theory, practice, and historical keys to victory.

Mao, Zedong. On Guerilla Warfare. London: Cassell, 1965.
> The world's most successful insurgent describes the principles that allowed him to conquer China.

Peng, Chin. My Side of History. New York: Media Masters, 2003
> Memoirs of a Malayan insurgent leader who lost.

Race, Jeffrey. War Comes to Long An: Revolutionary Conflict in a Vietnamese Province. Berkeley, CA: University of California Press, 1972.
> Counterinsurgency is scalable. Depicts the evolution of insurgency in one province in Vietnam.

Thompson, Robert. Defeating Communist Insurgency. St. Petersburg, FL: Hailer, 2005.
> Written in 1966, provides lessons from the author's counterinsurgency experience in Malaya and Vietnam.

Tomes, Robert. "Relearning Counterinsurgency Warfare." Parameters XXXIV, 1 (Spring 2004), 16-28.
> A good comparison of Galula, Kitson, and Trinquier.

Trinquier, Roger. Modern Warfare: A French View of Counterinsurgency. New York: Praeger, 1964.
> The French school of counterinsurgency, with a focus on "whatever means necessary."

United States Marine Corps. Small Wars Manual. Washington, DC: Government Printing Office, 1940. http://www.au.af.mil/au/awc/awcgate/swm/index.htm.
> A classic with lessons learned from the Corps' experience in the interwar years.

OVERVIEWS AND SPECIAL SUBJECTS IN COUNTERINSURGENCY

Asprey, Robert. War in the Shadows: The Guerrilla in History. New York: William Morrow, 1994.
First published in 1975. Presents the history of guerrilla war from ancient Persia to modern Afghanistan.

Baker, Ralph O. "The Decisive Weapon." Military Review LXXXVI, 3 (May/June 2006), 13-32.
A Brigade Combat Team Commander in Iraq in 2003-2004 gives his perspective on Information Operations.

Corum, James and Wray Johnson. Airpower and Small Wars. Lawrence, KS: University Press of Kansas, 2003.
The uses and limits of airpower and technology in counterinsurgency.

Ellis, John. From the Barrel of a Gun: A History of Guerrilla, Revolutionary, and Counterinsurgency Warfare from the Romans to the Present. London: Greenhill, 1995.
A phenomenal short overview of counterinsurgency.

Hammes, T.X. The Sling and the Stone. Osceola, WI: Zenith Press, 2004.
The future of warfare for the west is insurgency and terror, by a Marine with OIF experience.

Merom, Gil. How Democracies Lose Small Wars. New York: Cambridge University Press, 2003.
Examines the cases of Algeria, Lebanon, and Vietnam and determines that great powers lose small wars when they lose public support at home.

Nagl, John A. Learning to Eat Soup With a Knife: Counterinsurgency Lessons From Malaya and Vietnam. Chicago: University of Chicago Press, 2005.
How to learn to defeat an insurgency. Foreword by the CSA.

O'Neill, Bard E. Insurgency and Terrorism: From Revolution to Apocalypse. Dulles, VA: Potomac Books, 2005.
A framework for analyzing insurgency operations and a good first book in insurgency studies.

Sepp, Kalev I. "Best Practices in Counterinsurgency." Military Review LXXXV, 3 (May-June 2005), 8-12.
Historical best practices for success in counterinsurgency.

Shy, John and Thomas W. Collier. "Revolutionary War." in Peter Paret, ed., Makers of Modern Strategy: From Machiavelli to the Nuclear Age Princeton, NJ: Princeton University Press, 1986.
The best overview of the various COIN schools, discussing both the writings and the contexts in which they were developed.

Taber, Robert. War of the Flea. Dulles, VA: Potomac Books, 2002.
The flea can't kill the dog, but with enough friends, he can make the dog's life miserable. Explains the advantages of the insurgent and how to overcome them.

VIETNAM

Cassidy, Robert M. "Back to the Street Without Joy: Counterinsurgency Lessons from Vietnam and other Small Wars." Parameters XXXIV, 2 (Summer 2004), 73-83.
A good summary of COIN lessons learned by a serving Army officer.

Davidson, Phillip. Secrets of the Vietnam War. Novato, CA: Presidio Press, 1990.

MACV Commander General Westmoreland's intelligence officer provides an insightful analysis of the intricacies of the North Vietnamese strategy of dau tranh ("The Struggle").

DeForest, Orrin. Slow Burn, The Rise and Bitter Fall of American Intelligence in Vietnam. New York: Simon and Schuster, 1990.
Analysis of the American intelligence effort in Vietnam.

Komer, Robert. Bureaucracy Does Its Thing: Institutional Constraints on U.S.-GVN Performance in Vietnam. Washington: RAND, 1972. http://www.rand.org/pubs/reports/R967/
Bureaucracies do what they do—even if they lose the war.

Krepinevich, Andrew. The Army and Vietnam. Baltimore: Johns Hopkins, 1986.
The author argues that the Army never adapted to the insurgency in Vietnam, preferring to fight the war as a conventional conflict with an emphasis on firepower.

Sorley, Lewis. A Better War. New York: Harvest/HBJ, 2000.
Describes the impact of GEN Creighton Abrams on the conduct of the Vietnam War and his efforts to achieve unity of effort in COIN.

West, Bing. The Village. New York: Pocket Books, 1972.
A stunning first-person account of military advisors embedded with Vietnamese units.

CONTEMPORARY EXPERIENCES AND THE WAR ON TERROR

Alwyn-Foster, Nigel "Changing the Army for Counterinsurgency Operations," Military Review LXXXV, 6 (November-December 2005), 2-15.
A provocative look at US COIN in Iraq in 2003-2004 from a British practitioner.

Chiarelli, Peter W. and Patrick R. Michaelis, "Winning the Peace: The Requirement for Full-Spectrum Operations," Military Review LXXXV, 4 (July-August 2005), 4-17.
The commander of Task Force Baghdad in 2004 describes his lessons learned.

Crane, Conrad and W. Andrew Terrill, Reconstructing Iraq. Carlisle Barracks, PA: Army War College, 2003. http://www.strategicstudiesinstitute.army.mil/pubs/display.cfm?pubID=182
Prescient look at the demands of rebuilding a state after changing a regime.

Filkins, Dexter. "What the War Did to Colonel Sassaman." The New York Times Sunday Magazine, 23 October 2005.
Case study of a Fourth Infantry Division battalion commander in Iraq in 2003-4.

Gunaratna, Rohan. Inside al Qaeda: Global Network of Terror. Berkeley, CA: University of Berkeley Press, 2003.
The story behind the rise of the global Islamofascist insurgency.

Hoffman, Bruce. Insurgency and Counterinsurgency in Iraq. Washington, DC: RAND, 2004.
http://www.rand.org/pubs/occasional_papers/OP127/
Analysis of America's efforts in Iraq in 2003 informed by good history and theory.

Kepel, Gilles. The War for Muslim Minds: Islam and the West. Boston: Harvard University Press, 2004.
A French explanation of Islamofascism with suggestions for defeating it.

Kilcullen, David. "Countering Global Insurgency." Journal of Strategic Studies 28, 4 (August 2005), 597-617.
Describes the War on Terror as a counterinsurgency campaign.

Kilcullen, David. "Twenty-Eight Articles: Fundamentals of Company-Level Counterinsurgency." Military Review LXXXVI, 3 (May-June 2006), 103-108.
Australian counterinsurgent prescribes actions for captains in COIN campaigns.

Krepinevich, Andrew F. "How to Win in Iraq." Foreign Affairs 84, 5 (September / October 2005).
Assessment of US policy in Iraq by a scholar of Vietnam COIN.

Lewis, Bernard. The Crisis of Islam: Holy War and Unholy Terror. New York: Modern Library, 2003.
A controversial but important analysis of the philosophical origins of the global Islamofascist insurgency.

Maass, Peter. "The Counterinsurgent." The New York Times Sunday Magazine, 11 January 2004.
A tank battalion operations officer and counterinsurgency scholar tries out his theories in Al Anbar in 2003-4.

McFate, Montgomery. "Iraq: The Social Context of IEDs." Military Review LXXXV, 3 (May-June 2005).

The insurgents' best weapon doesn't grow next to roads—it's constructed and planted there. Understanding who does that, and why, helps defeat IEDs.

Metz, Steven and Raymond Millen, Insurgency and Counterinsurgency in the 21st Century: Reconceptualizing Threat and Response. Carlisle Barracks, PA: US Army War College, 2004.
Longtime scholars of counterinsurgency put the War on Terror in historical context.

Packer, George. The Assassin's Gate: America in Iraq. New York: Farrar, Straus and Giroux, 2005.
A journalist for The New Yorker talks to Iraqis and Americans about OIF. Runner-up for the Pulitzer Prize in non-fiction in 2005.

Packer, George. "The Lesson of Tal Afar." The New Yorker (10 April 2006), 48-65.
The 2005 success of the 3rd ACR with the Clear-Hold-Build strategy in Tal Afar.

Petraeus, David. "Learning Counterinsurgency: Observations from Soldiering in Iraq." Military Review LXXXVI, 2 (March-April 2006).
Commander of the 101st and MNSTC-I passes on his personal lessons learned from 2 ½ years in Iraq.

Sageman, Marc. Understanding Terror Networks. Philadelphia: University of Pennsylvania Press, 2004.
A former Foreign Service officer with Afghanistan experience explains the motivation of terrorists—not deprivation, but the need to belong.

Military References

REQUIRED PUBLICATIONS

These documents must be available to intended users of this publication.

FM 1-02/MCRP 5-12A. *Operational Terms and Graphics*. 21 Sep 2004.

JP 1-02. *Department of Defense Dictionary of Military and Associated Terms*. 4 Dec 2001.
Available< http://www.dtic.mil/doctrine/jel/doddict/>

RELATED PUBLICATIONS

These sources contain relevant supplemental information.

FM 2-0 (34-1). *Intelligence*. 17 May 2004.

FM 2-22.3. *Human Intelligence*. TBP.

FM 3-0. *Operations*. 14 Jun 2001.
Under revision. Projected for republication during fiscal year 2007.

FM 3-05.301/MCRP 3-40.6A. *Psychological Operations Tactics, Techniques, and Procedures*. 31 Dec 2003.
Distribution limited to government agencies only

FM 3-09.31 (6-71)/MCRP 3-16C. *Tactics, Techniques, and Procedures for Fire Support for the Combined Arms Commander*. 10 Jan 2002.

FM 3-13 (100-6). *Information Operations: Doctrine, Tactics, Techniques, and Procedures*. 28 Nov 2003.
Appendix E addresses information operations targeting.

FM 3-61.1. *Public Affairs Tactics, Techniques, and Procedures*. 1 Oct 2000.

FM 3-90. *Tactics*. 4 Jul 2001.

FM 4-0 (100-10). *Combat Service Support*. 29 Aug 2003. .

FM 5-0 (101-5). *Army Planning and Orders Production*. 20 Jan 2005.
Addresses Army problem solving, the military decision-making process, and troop leading procedures.

FM 6-0. *Mission Command: Command and Control of Army Forces*. 11 Aug 2003.

FM 22-100. *Army Leadership*. 31 Aug 1999.
FM 22-100 will be republished as FM 6-22.

FM 27-10. *The Law of Land Warfare*. 18 Jul 1956.

FM 31-20-3. *Foreign Internal Defense: Tactics, Techniques, and Procedures for Special Forces*. 20 Sep 1994.

FM 34-52. *Intelligence Interrogation*. 28 Sep 1992.

FM 34-130/FMFRP 3-23-2. *Intelligence Preparation of the Battlefield*. 8 Jul 1994.

FM 41-10. *Civil Affairs Operations*. 14 Feb 2000.
Contains information on civil-military operations centers.

FM 100-25. *Doctrine for Army Special Operations Forces*. 1 Auf 1999.

FMI 2-91.4. *Intelligence Support to Operations in the Urban Environment*. 30 Jun 2005.
Expires 30 Jun 2007. Distribution limited to government agencies only. Available in electronic media only. <www.adtdl.army.mil>.

FMI 5-0.1. *The Operations Process.* 31 Mar 2006.
> Expires 31 Mar 2008. When FM 3-0 is republished, it will address the material in FMI 5-0.1 that is relevant to this publication.

MCDP 4. *Logistics.* 21 Feb 1997.

MCWP 4-12. *Operational-Level Logistics.* 30 Jan 2002.

JP 1. *Joint Warfare of the Armed Forces of the United States.* 14 Nov 2000.

JP 3-07.1. *Joint Tactics, Techniques, and Procedures for Foreign Internal Defense.*

JP 3-08. *Interagency Coordination During Joint Operations.* 2 vols. 9 Oct 1996.

JP 3-61. *Public Affairs.* 9 May 2005.

PRESCRIBED FORMS

None

REFERENCED FORMS

None

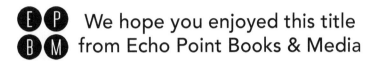 We hope you enjoyed this title
from Echo Point Books & Media

Before Closing this Book, Two Good Things to Know

1. Buy Direct & Save

Go to www.echopointbooks.com (click "Our Titles" at top or click "For Echo Point Publishing" in the middle) to see our complete list of titles. We publish books on a wide variety of topics—from spirituality to auto repair.

Buy direct and save 10% at www.echopointbooks.com

DISCOUNT CODE: EPBUYER

2. Make Literary History and Earn $100 Plus Other Goodies Simply for Your Book Recommendation!

At Echo Point Books & Media we specialize in republishing out-of-print books that are united by one essential ingredient: high quality. Do you know of any great books that are no longer actively published? If so, please let us know. If we end up publishing your recommendation, you'll be adding a wee bit to literary culture and a bunch to our publishing efforts.

Here is how we will thank you:

- A free copy of the new version of your beloved book that includes acknowledgement of your skill as a sharp book scout.

- A free copy of another Echo Point title you like from echopointbooks.com.

- And, oh yes, we'll also send you a check for $100.

Since we publish an eclectic list of titles, we're interested in a wide range of books. So please don't be shy if you have obscure tastes or like books with a practical focus. To get a sense of what kind of books we publish, visit us at www.echopointbooks.com.

If you have a book that you think will work for us,
send us an email at editorial@echopointbooks.com

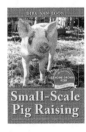    

CPSIA information can be obtained
at www.ICGtesting.com
Printed in the USA
BVHW010143021218
534539BV00009B/454/P